Applied Probability and Statistics (Continued)

CHAKRAVARTI, LAHA and ROY · Handbook of Methods of Applied Statistics, Vol. I
CHAKRAVARTI, LAHA and ROY · Handbook of Methods of Applied Statistics, Vol. II
CHERNOFF and MOSES · Elementary Decision Theory
CHIANG · Introduction to Stochastic Processes in Biostatistics
CLELLAND, deCANI, BROWN, BURSK, and MURRAY · Basic Statistics with Business Applications
COCHRAN · Sampling Techniques, *Second Edition*
COCHRAN and COX · Experimental Designs, *Second Edition*
COX · Planning of Experiments
COX and MILLER · The Theory of Stochastic Processes
DAVID · Order Statistics
DEMING · Sample Design in Business Research
DODGE and ROMIG · Sampling Inspection Tables, *Second Edition*
DRAPER and SMITH · Applied Regression Analysis
GOLDBERGER · Econometric Theory
GUTTMAN and WILKS · Introductory Engineering Statistics
HALD · Statistical Tables and Formulas
HALD · Statistical Theory with Engineering Applications
HOEL · Elementary Statistics, *Second Edition*
HUANG · Regression and Econometric Methods
JOHNSON and LEONE · Statistics and Experimental Design: In Engineering and the Physical Sciences, Volumes I and II
LANCASTER · The Chi Squared Distribution
MILTON · Rank Order Probabilities: Two-Sample Normal Shift Alternatives
PRABHU · Queues and Inventories: A Study of Their Basic Stochastic Processes
SARHAN and GREENBERG · Contributions to Order Statistics
SEAL · Stochastic Theory of a Risk Business
WILLIAMS · Regression Analysis
WOLD and JURÉEN · Demand Analysis
WONNACOTT and WONNACOTT · Econometrics
YOUDEN · Statistical Methods for Chemists

Tracts on Probability and Statistics

BILLINGSLEY · Ergodic Theory and Information
BILLINGSLEY · Convergence of Probability Measures
CRAMÉR and LEADBETTER · Stationary and Related Stochastic Processes
RIORDAN · Combinatorial Identities
TAKÁCS · Combinatorial Methods in the Theory of Stochastic Processes

Order Statistics

A WILEY PUBLICATION IN APPLIED STATISTICS

ORDER
STATISTICS

H. A. DAVID

University of North Carolina

John Wiley & Sons, Inc.

New York · London · Sydney · Toronto

Library of Congress Catalogue Card Number: 78-114915

SBN 471 19675 4

Printed in the United States of America

10 9 8 7 6 5 4 3 2 1

To Vera

Preface

Order statistics make their appearance in many areas of statistical theory and practice. Recent years have seen a particularly rapid growth, as attested by the references at the end of this book. There is a growing recognition that the large body of theory, techniques, and applications involving order statistics deserves study on its own, rather than as a mere appendage to other fields, such as nonparametric methods. Some may decry this increased specialization, and indeed it is entirely appropriate that the most basic aspects of the subject be incorporated in general textbooks and courses, both theoretical and applied. On the other hand, there has been a clear trend in many universities toward the establishment of courses of lectures dealing more extensively with order statistics. I first gave a short course in 1955 at the University of Melbourne and have since then periodically offered longer courses at the Virginia Polytechnic Institute and especially at the University of North Carolina, where much of the present material has been tried out.

In this book an attempt is made to present the subject of order statistics in a manner combining features of a textbook and of a guide through the research literature. The writing is at an intermediate level, presupposing on the reader's part the usual basic background in statistical theory and applications. Some portions of the book, are, however, quite elementary, whereas others, particularly in Chapters 4 and 9, are rather more advanced. Exercises supplement the text and, in the manner of M. G. Kendall's books, usually lead the reader to the original sources.

A special word is needed to explain the relation of this book to the only other existing general account, also prepared in the Department of Biostatistics, University of North Carolina, namely, the multiauthored *Contributions to Order Statistics*, edited by A. E. Sarhan and B. G. Greenberg, which appeared in this Wiley Series in 1962. The present monograph is not meant to replace that earlier one, which is almost twice as long. In particular, the extensive sets of tables in *Contributions* will long retain their usefulness. The present work contains only a few tables needed to clarify the text but provides, as an appendix, an annotated guide to the massive output of tables scattered over numerous journals and books; such tables are essential for the ready use of many of the methods described. *Contributions* was not designed as a textbook and is, of course, no longer quite up to date. However, on a number of topics well developed by 1962 more extensive coverage will be found there than here. Duplication of all but the most fundamental material has been kept to a minimum.

In other respects also the size of this book has been kept down by deferring wherever feasible to available specialized monographs. Thus plans for the treatment of the role of order statistics in simultaneous statistical inference have largely been abandoned in view of R. G. Miller's very readable account in 1966.

The large number of references may strike some readers as too much of a good thing. Nevertheless the list is far from complete and is confined to direct, if often brief, citations. For articles dealing with such central topics as distribution theory and estimation I have aimed at reasonable completeness, after elimination of superseded work. Elsewhere the coverage is less comprehensive, especially where reference to more specialized bibliographies is possible. In adopting this procedure I have been aided by knowledge of H. L. Harter's plans for the publication of an extensive annotated bibliography of articles on order statistics.

It is a pleasure to acknowledge my long-standing indebtedness to H. O. Hartley, who first introduced me to the subject of order statistics with his characteristic enthusiasm and insight. I am also grateful to E. S. Pearson for his encouragement over the years. In writing this book I have had the warm support of B. G. Greenberg. My special thanks go to P. C. Joshi, who carefully read the entire manuscript and made many suggestions. Helpful comments were also provided by R. A. Bradley. J. L. Gastwirth and P. K. Sen. Expert typing assistance and secretarial help were rendered by Mrs. Delores Gold and Mrs. Jean Scovil. The writing was supported throughout by the Army Research Office, Durham, North Carolina.

H. A. DAVID

Chapel Hill, North Carolina
December, 1969

Contents

Order Statistics

CHAPTER 1

Introduction

1.1. THE SUBJECT OF ORDER STATISTICS

If the random variables $X_1, X_2, \ldots, X_n$ are rearranged in ascending order of magnitude and then written as

$$X_{(1)} \leq X_{(2)} \leq \cdots \leq X_{(n)},$$

we call $X_{(i)}$ the ith order statistic $(i = 1, 2, \ldots, n)$. The (unordered) X_i are commonly statistically independent and identically distributed, but need be neither; the $X_{(i)}$, because of the inequality relations among them, are necessarily dependent.

The subject of order statistics deals with the properties and applications of these ordered random variables and of functions involving them. Examples are the *extremes* $X_{(n)}$ and $X_{(1)}$, the *range* $W = X_{(n)} - X_{(1)}$, the *extreme deviate* (from the sample mean) $X_{(n)} - \bar{X}$, and, for a random sample from a normal $N(\mu, \sigma^2)$ distribution, the *studentized range* W/S_v, where S_v is a root-mean-square estimator of σ based on v degrees of freedom. All these statistics have important applications. The extremes arise in the statistical study of floods and droughts, as well as in problems of breaking strength and fatigue failure. The range is well known to provide a quick estimator of σ and has found particularly wide acceptance in the field of quality control. The extreme deviate is a basic tool in the detection of outliers, large values of $(X_{(n)} - \bar{X})/\sigma$ indicating the presence of one or a few excessively large observations. In the same context, the studentized range is useful when outliers are not confined to one direction. However, it also supplies the basis of many quick tests in small samples and is of key importance for ranking "treatment" means in analysis of variance situations.

With the help of the Gauss-Markov theorem of least squares it is possible to use linear functions of order statistics quite systematically for the estimation of parameters of location and/or scale. This application is particularly

1

useful when some of the observations in the sample have been "censored," since in that case standard methods of estimation tend to become laborious or otherwise unsatisfactory. Life tests provide an ideal illustration of the advantages of order statistics in censored data. Since such experiments may take a long time to complete, it is often desirable to stop after failure of the first r out of n (similar) items under test. The observations are the r times to failure, which here, unlike in most situations, arrive already ordered for us by the method of experimentation; from them we can estimate the necessary parameters, such as the true mean life.

In recent years the study of order statistics has received fresh impetus from a number of directions. Computers have made it feasible to look at the same data in many different ways, thus calling for a body of versatile, often rather informal techniques commonly referred to as *data analysis* (cf. Tukey, 1962). Are the data really in accord with (*a*) the assumed distribution and (*b*) the assumed model? Clues to (*a*) may be obtained from a plot of the ordered observations against some simple function of their ranks, preferably on probability paper appropriate for the distribution assumed. A straight-line fit in such a *probability plot* indicates that all is more or less well, whereas serious departures from a straight line may reveal the presence of outliers or other failures in the distributional assumption. Similarly, in answer to (*b*), one can plot the ordered *residuals* from the fitted model. Somewhat in the same spirit is the search for statistics and tests which, although not optimal under ideal (say normal-theory) conditions, perform well under a variety of circumstances likely to occur in practice. An example of these *robust methods* is the use, in samples from symmetrical populations, of the *trimmed mean*, which is the average of the observations remaining after the most extreme k ($k/n < \frac{1}{2}$) at each end have been removed. Loss of efficiency in the normal case may, for suitable choice of k, be compensated by lack of sensitivity to outliers or to other departures from an assumed distribution.

Finally, we may point to a rather narrower but truly space-age application. In large samples (e.g., of particle counts taken on a space craft) there are interesting possibilities for data compression (Eisenberger and Posner, 1965), since the sample may be replaced (on the space craft's computer) by enough order statistics to allow (on the ground) both satisfactory estimation of parameters and a test of the assumed underlying distributional form.

1.2. THE SCOPE AND LIMITS OF THIS BOOK

Although we shall be concerned with all of the topics sketched in the preceding section, and with many others as well, the field of order statistics impinges on so many different areas of statistics that some limitations in coverage have to be imposed. To start with, unlike Wilks (1948), we use "order statistics" in the narrower sense now widely accepted: we shall *not* deal with *rank-order statistics*, as exemplified by the Wilcoxon two-sample statistic, although these also require an ordering of the observations. The point of difference is that rank-order statistics involve the ranks of the ordered observations only, not their actual values, and consequently lead to nonparametric or distribution-free methods—at any rate for continuous random variables. On the other hand, the great majority of procedures based on order statistics depend on the form of the underlying population. The theory of order statistics is, however, useful in many nonparametric problems and also in an assessment of the non-null properties of rank tests, for example, by the power function.

Other restrictions in this book have more of an *ad hoc* character. Order statistics play an important supporting role in multiple comparisons and multiple decision procedures such as the ranking of treatment means. In view of the useful book by Miller (1966) and, to a lesser extent, the recent monograph by Bechhofer *et al.* (1968) there seems little point in developing here the inference aspects of the subject, although the needed order-statistics theory is either given explicitly or obtainable by only minor extensions. However, some multiple decision procedures are considered in Chapter 8 on the treatment of outliers.

Much more could be said about asymptotic methods than we do in Chapter 9. The practical side, however, has been largely preempted by Gumbel's (1958) book. On the other hand, the theory, which has made rapid strides in recent years, tends to be highly mathematical and would justify an advanced monograph on its own. We have thought it best to confine ourselves to a detailed treatment of some of the most important results and to a summary of other developments.

The effective application of order-statistics techniques requires a great many tables. Inclusion even of only the most useful would greatly increase the bulk of this book. We therefore limit ourselves to a few tables needed for illustration; for the rest, we refer the reader to general books of tables, such as Pearson and Hartley (1966), Beyer (1968), and especially the extensive collection of tables in Sarhan and Greenberg (1962). Many references to tables in original papers are given throughout the book, and a guided commentary to these is provided in the Appendix.

1.3. NOTATION

Although this section may serve for reference, the reader is urged to look through it before proceeding further.

As far as is feasible, random variables, or *variates* for short, will be designated by uppercase letters, and their realizations (observations) by the corresponding lowercase letters. By order statistics we will mean either ordered variates or ordered observations. Thus:

$X_1, X_2, \ldots, X_n$	unordered variates
$x_1, x_2, \ldots, x_n$	unordered observations
$X_{(1)} \le X_{(2)} \le \cdots \le X_{(n)}$	ordered variates $\Big\}$ order statistics
$x_{(1)} \le x_{(2)} \le \cdots \le x_{(n)}$	ordered observations
$X_{1:n} \le X_{2:n} \le \cdots \le X_{n:n}$	ordered variates—extensive form

When the sample size n needs to be emphasized, we use the more extensive form of notation, switching rather freely from the extensive to the brief form.

$P(x) = \Pr\{X \le x\}$	cumulative distribution function (cdf) of X
$p(x)$	$\Big($probability density function (pdf) for a continuous variate $\Big)$ probability function (pf) for a discrete variate
$F_r(x),\ F_{r:n}(x)$	cdf of $X_{(r)},\ X_{r:n}$ $r = 1, 2, \ldots, n$
$f_r(x), f_{r:n}(x)$	pdf or pf of $X_{(r)},\ X_{r:n}$
$F_{rs}(x, y) = \Pr\{X_{(r)} < x,$	joint cdf of $X_{(r)}$ and $X_{(s)}$
$\qquad X_{(s)} < y\}$	$1 \le r < s \le n$
$f_{rs}(x, y)$	joint pdf or pf of $X_{(r)}$ and $X_{(s)}$
ξ_p	population quantile of order p, given by $P(\xi_p) = p$ or equivalently by $\xi_p = P^{-1}(p) = Q(p),\ 0 < p < 1$
$\xi_{1/2}$	population median
$X_{([np]+1)}$	sample quantile of order p, where $[np]$ denotes the largest integer $\le$ np
$X_{([n\lambda_i]+1)}$	sample quantile of order $\lambda_i,\ 0 < \lambda_1 < \lambda_2 < \cdots < \lambda_k < 1$

But the sample median is

$X_{(\frac{1}{2}\overline{n+1})}$	n odd
$\frac{1}{2}(X_{(\frac{1}{2}n)} + X_{(\frac{1}{2}n+1)})$	n even
$W,\ W_n = X_{(n)} - X_{(1)}$	(sample) range
$W_{(i)} = X_{(n+1-i)} - X_{(i)}$	ith quasi-range $(W_{(1)} = W)$
$W_{rs} = X_{(s)} - X_{(r)}$	
$\overline{W},\ \overline{W}_{n,k}$	mean of k ranges of n
$_jW$	W for jth sample
$\mu = \mathcal{E}X,\ \sigma^2 = \operatorname{var} X$	mean, variance of X
$\mu_x = \mathcal{E}X,\ \mu_y = \mathcal{E}Y$	means of $X,\ Y$ (bivariate case)

$\sigma_x^2 = \text{var } X,\ \sigma_y^2 = \text{var } Y$	variances of $X,\ Y$
$\sigma_{xy} = \text{cov } (X,\ Y),\ \rho = \sigma_{xy}/\sigma_x\sigma_y$	covariance, correlation coefficient of $X,\ Y$
$\mu_{r:n} = \mathscr{E}X_{r:n}$	mean of $X_{r:n}$
$\mu_{r:n}^{(k)} = \mathscr{E}(X_{r:n}^k)$	kth raw moment of $X_{r:n}$
$\mu_{rs:n} = \mathscr{E}(X_{r:n}X_{s:n})$	
$\sigma_{r:n}^2 = \text{var } X_{r:n}$	
$\sigma_{rs:n} = \text{cov } (X_{r:n},\ X_{s:n})$	
$Q(x) = P^{-1}(x)$	inverse cdf
$p_r = r/(n + 1),\ q_r = 1 - p_r$	
$Q_r = Q(p_r),\ f_r = p(Q_r)$	
$Q_r' = dQ(p_r)/dp_r = 1/f_r$	
S_ν	estimator of σ based on ν DF; for a $N(\mu, \sigma^2)$ distribution $\nu S_\nu^2/\sigma^2 \frown \chi_\nu^2$.
$Q_{n,\nu} = W_n/S_\nu$	studentized range (W_n, S_ν independent)
$S = [\Sigma(X_i - \bar{X})^2/(n - 1)]^{\frac{1}{2}}$	(internal) estimator of σ
$S^{(P)} = \{[(n - 1)S^2 + \nu S_\nu^2]/ (n - 1 + \nu)\}^{\frac{1}{2}}$	pooled estimator of σ
$_jS$	S for jth sample
$\text{B}(a, b) = \int_0^1 t^{a-1}(1 - t)^{b-1}\, dt$ $a > 0, b > 0$	beta function
$I_p(a, b) = \int_0^p t^{a-1}(1 - t)^{b-1}\, dt/ \text{B}(a, b)$	incomplete beta function (1.3.1)
$\beta(a, b)$	beta variate X having cdf $\Pr\{X \leq x\} = I_x(a, b)$ (1.3.2)
χ_ν^2	chi-square variate with ν DF
$\phi(x) = (2\pi)^{-\frac{1}{2}}e^{-\frac{1}{2}x^2}$ $-\infty < x < \infty$	unit normal pdf
$\Phi(x) = \int_{-\infty}^x \phi(t)\, dt$	unit normal cdf
$N(\mu, \sigma^2)$	normal variate, mean μ, variance σ^2
$N(\boldsymbol{\mu}, \boldsymbol{\Sigma})$	multinormal variate, mean vector $\boldsymbol{\mu}$, covariance matrix $\boldsymbol{\Sigma}$
$n^{(k)} = n(n - 1) \cdots (n - k + 1)$ $k = 1, 2, \ldots, n$	
$[x]$	integral part of x, but $\mu_{[k]} = \mathscr{E}X^{(k)}$
pdf	probability density function
cdf	cumulative distribution function
iid	independent, identically distributed
DF	degrees of freedom

ML	maximum likelihood
LS	least squares
BLUE	best linear unbiased estimator
UMVU	uniformly minimum variance unbiased
UMP	uniformly most powerful
PH	Pearson and Hartley (1966)— *Biometrika Tables* 1
SG	Sarhan and Greenberg (1962)— *Contributions to Order Statistics*
Ex.	exercise ("example" is written in full)
D	decimal (e.g., to 3D = to 3 decimal places)
S	significant (e.g., to 4S = to 4 significant figures)
A5.3	appendix listing of tables relating to Section 5.3

CHAPTER 2

Basic Distribution Theory

2.1. DISTRIBUTION OF A SINGLE ORDER STATISTIC

We suppose that $X_1, X_2, \ldots, X_n$ are n independent variates, each with cumulative distribution function (cdf) $P(x)$. Let $F_r(x)$ $(r = 1, 2, \ldots, n)$ denote the cdf of the rth order statistic $X_{(r)}$. Then the cdf of the largest order statistic $X_{(n)}$ is given by

$$F_n(x) = \Pr\{X_{(n)} \le x\}$$
$$= \Pr\{\text{all } X_i \le x\} = P^n(x). \tag{2.1.1}$$

Likewise we have

$$F_1(x) = \Pr\{X_{(1)} \le x\} = 1 - \Pr\{X_{(1)} > x\}$$
$$= 1 - \Pr\{\text{all } X_i > x\} = 1 - [1 - P(x)]^n. \tag{2.1.2}$$

These are important special cases of the general result for $F_r(x)$:

$$F_r(x) = \Pr\{X_{(r)} \le x\}$$
$$= \Pr\{\text{at least } r \text{ of the } X_i \text{ are less than or equal to } x\}$$
$$= \sum_{i=r}^{n} \binom{n}{i} P^i(x)[1 - P(x)]^{n-i} \tag{2.1.3}$$

since the term in the summand is the binomial probability that *exactly* i of X_1, $X_2, \ldots, X_n$ are less than or equal to x. We write (2.1.3) as

$$F_r(x) = E_{P(x)}(n, r) \tag{2.1.4}$$

and note that the E function has been tabled extensively (e.g., Harvard Computation Laboratory, 1955, where the notation $E(n, r, P(x))$ is used). Alternatively, from the well-known relation between binomial sums and the

7

incomplete beta function we have

$$F_r(x) = I_{P(x)}(r, n - r + 1),\qquad\qquad (2.1.5)$$

where $I_p(a, b)$ is defined by (1.3.1). Thus $F_r(x)$ can also be evaluated from tables of $I_p(a, b)$ (K. Pearson, 1934). Percentage points of $X_{(r)}$ may be obtained by inverse interpolation in the above tables or more directly from Table 16 of *Biometrika Tables* (Pearson and Hartley, 1966), which gives percentage points of the incomplete beta function.

Example 2.1. Find the upper 5% point of $X_{(4)}$ in samples of 5 from a unit normal parent.

We require x such that

$$I_{P(x)}(4, 2) = 0.95$$

or

$$I_{1-P(x)}(2, 4) = 0.05.$$

This gives $1 - P(x) = 0.07644$ and hence $x = 1.429$.

It should be noted that results (2.1.1)–(2.1.5) hold equally for continuous and discrete variates. We shall now assume that X_i is continuous with probability density function (pdf) $p(x) = P'(x)$, but will return to the discrete case in Section 2.4. If $f_r(x)$ denotes the pdf of $X_{(r)}$ we have from (2.1.5)

$$f_r(x) = \frac{1}{B(r, n - r + 1)} \frac{d}{dx} \int_0^{P(x)} t^{r-1}(1 - t)^{n-r}\, dt$$

$$= \frac{1}{B(r, n - r + 1)} P^{r-1}(x)[1 - P(x)]^{n-r}p(x). \qquad (2.1.6)$$

In view of the importance of this result we will also derive it otherwise. The event $x < X_{(r)} \le x + \delta x$ may be realized as follows:

$$\begin{array}{c}
1 \\
\overline{r - 1 \quad \| \quad n - r} \\
x\| \ x + \delta x
\end{array}$$

$X_i \le x$ for $r - 1$ of the X_i, $x < X_i \le x + \delta x$ for one X_i, and $X_i > x + \delta x$ for the remaining $n - r$ of the X_i. The number of ways in which the n observations can be so divided into three parcels is

$$\frac{n!}{(r - 1)!1!(n - r)!} = \frac{1}{B(r, n - r + 1)},$$

and each such way has probability

$$P^{r-1}(x)[P(x + \delta x) - P(x)][1 - P(x + \delta x)]^{n-r}.$$

Regarding δx as small, we have therefore

$$\Pr\{x < X_{(r)} \le x + \delta x\} = \frac{1}{B(r, \, n - r + 1)}$$

$$\times P^{r-1}(x)p(x)\,\delta x[1 - P(x + \delta x)]^{n-r} + O(\delta x^2),$$

where $O(\delta x^2)$ means terms of order $(\delta x)^2$ and includes the probability of realizations of $x < X_{(r)} \le x + \delta x$ in which more than one X_i is in $(x, x + \delta x)$. Dividing both sides by δx and letting $\delta x \to 0$, we again obtain (2.1.6).

2.2. JOINT DISTRIBUTION OF TWO OR MORE ORDER STATISTICS

The joint density function of $X_{(r)}$ and $X_{(s)}$ $(1 \le r < s \le n)$ is conveniently denoted by $f_{rs}(x, y)$. An expression corresponding to (2.1.5) may be derived by noting that the compound event $x < X_{(r)} \le x + \delta x,\, y < X_{(s)} \le y + \delta y$ is realized (apart from terms having a lower order of probability) by the configuration

$$\underset{x}{\underset{\|}{\vert\vert}} \overset{1}{\underset{x + \delta x}{\underset{\|}{\vert\vert}}} \quad s - r - 1 \quad \overset{1}{\underset{y}{\vert\vert}} \quad \underset{y + \delta y}{\vert\vert} \quad n - s \,,$$

meaning that $r - 1$ of the observations are less than x, one is in $(x, x + \delta x)$, etc. It follows that for $x \le y$

$$f_{rs}(x, y) = \frac{n!}{(r - 1)!(s - r - 1)!(n - s)!} P^{r-1}(x)p(x)[P(y) - P(x)]^{s-r-1}$$

$$p(y)[1 - P(y)]^{n-s}. \quad (2.2.1)$$

Generalizations are now clear. Thus the joint pdf of $X_{(r_1)}, X_{(r_2)}, \ldots, X_{(r_k)}$ $(1 \le r_1 < r_2 < \cdots < r_k \le n; \, 1 \le k \le n)$ is, for $x_1 \le x_2 \le \cdots \le x_k$,

$$f_{r_1 r_2 \ldots r_k}(x_1, x_2, \ldots, x_k) = \frac{n!}{(r_1 - 1)!(r_2 - r_1 - 1)! \cdots (n - r_k)!}$$

$$\cdot P^{r_1-1}(x_1)p(x_1)[P(x_2) - P(x_1)]^{r_2-r_1-1}p(x_2) \cdots [1 - P(x_k)]^{n-r_k}p(x_k). \quad (2.2.2)$$

If we define $x_0 = -\infty, x_{k+1} = +\infty, r_0 = 0, r_{k+1} = n + 1$, the RHS may be written as

$$n! \left[\prod_{i=1}^{k} p(x_i) \right] \cdot \prod_{i=0}^{k} \left\{ \frac{[P(x_{i+1}) - P(x_i)]^{r_{i+1}-r_i-1}}{(r_{i+1} - r_i - 1)!} \right\}. \quad (2.2.3)$$

In particular, the joint pdf of all n order statistics becomes simply

$$n!p(x_1)p(x_2) \cdots p(x_n).$$

This result is indeed directly obvious since there are $n!$ equally likely orderings of the x_i, and may be used as the starting point for the derivation of the joint distribution of k order statistics ($k < n$) in the continuous case.

The joint cdf $F_{rs}(x, y)$ of $X_{(r)}$ and $X_{(s)}$ may be obtained by integration of (2.2.1) as well as by a direct argument valid also in the discrete case. We have for $x < y$

$$F_{rs}(x, y) = \Pr \{\text{at least } r\ X_i \leq x, \text{at least } s\ X_i \leq y\}$$

$$= \sum_{i=r}^{n} \sum_{j=s-i}^{n-i} \Pr \{\text{exactly } i\ X_i \leq x, \text{exactly } j\ X_i \text{ with } x < X_i \leq y\}$$

unless $i > s$, when j starts at 0. Thus for $x < y$

$$F_{rs}(x, y) = \sum_{i=r}^{n} \sum_{j=\max(0,s-i)}^{n-i} \frac{n!}{i!j!(n - i - j)!}$$
$$\cdot P^i(x)[P(y) - P(x)]^j[1 - P(y)]^{n-i-j}. \quad (2.2.4)$$

Also for $x \geq y$ the inequality $X_{(s)} \leq y$ implies $X_{(r)} \leq x$, so that

$$F_{rs}(x, y) = F_s(y). \quad (2.2.5)$$

2.3. DISTRIBUTION OF THE RANGE AND OF OTHER SYSTEMATIC STATISTICS

From the joint pdf of k order statistics we can by standard transformation methods derive the pdf of any well-behaved function of the order statistics. For example, to find the pdf of $W_{rs} = X_{(s)} - X_{(r)}$ we put $w_{rs} = y - x$ in (2.2.1) and note that the transformation from x, y to x, w_{rs} has Jacobian unity in modulus. Thus, writing C_{rs} for the constant in (2.2.1), we have on integrating out over x

$$f(w_{rs}) = C_{rs} \int_{-\infty}^{\infty} P^{r-1}(x)p(x)[P(x + w_{rs}) - P(x)]^{s-r-1}p(x + w_{rs})$$
$$\cdot [1 - P(x + w_{rs})]^{n-s}\, dx. \quad (2.3.1)$$

Of special interest is the case $r = 1$, $s = n$, when w_{rs} becomes the range w and (2.3.1) reduces to

$$f(w) = n(n - 1) \int_{-\infty}^{\infty} p(x)[P(x + w) - P(x)]^{n-2}p(x + w)\, dx. \quad (2.3.2)$$

The cdf of W is somewhat simpler. On interchanging the order of integration we have

$$F(w) = n \int_{-\infty}^{\infty} p(x) \int_0^w (n-1)p(x+w')[P(x+w') - P(x)]^{n-2} \, dw' \, dx$$

$$= n \int_{-\infty}^{\infty} p(x)[P(x+w') - P(x)]^{n-1} \Big|_{w'=0}^{w'=w} dx$$

$$= n \int_{-\infty}^{\infty} p(x)[P(x+w) - P(x)]^{n-1} \, dx. \qquad (2.3.3)$$

This important result may also be obtained by noting that

$$np(x) \, dx [P(x+w) - P(x)]^{n-1}$$

is the probability given x that one of the X_i falls into the interval $(x, x + dx)$ and all of the $n-1$ remaining X_i fall into $(x, x + w)$.

In applying formulae (2.3.1)–(2.3.3) a little care has to be taken if the range of x is finite.

Example 2.3. Find the distribution of the order statistics and of W_{rs} when $p(x)$ is uniform: $p(x) = 1 \ (0 \le x \le 1)$; $p(x) = 0$ elsewhere.

From (2.1.6) we have at once

$$f_r(x) = \frac{1}{B(r, n-r+1)} x^{r-1}(1-x)^{n-r} \qquad 0 \le x \le 1$$

$$= 0 \quad \text{elsewhere.}$$

Thus $X_{(r)}$ is a beta $\beta(r, n-r+1)$ variate as defined in (1.3.2). By (2.2.1)

$$f_{rs}(x, y) = C_{rs} x^{r-1}(y-x)^{s-r-1}(1-y)^{n-s} \qquad 0 \le x \le y \le 1$$

$$= 0 \quad \text{elsewhere.}$$

Since $p(x + w_{rs}) = 0$ for $x \ge 1 - w_{rs}$, (2.3.1) gives

$$f(w_{rs}) = C_{rs} \int_0^{1-w_{rs}} x^{r-1} w_{rs}^{s-r-1}(1-x-w_{rs})^{n-s} \, dx.$$

Putting $x = y(1 - w_{rs})$, we obtain

$$f(w_{rs}) = \frac{1}{B(s-r, n-s+r+1)} w_{rs}^{s-r-1}(1-w_{rs})^{n-s+r} \qquad 0 \le w_{rs} \le 1. \quad (2.3.4)$$

This simple result shows that W_{rs} has a beta distribution which depends only on $s - r$, and not on s and r individually. See also Section 5.4.

In addition to the range, simple systematic statistics of interest include the quasi-ranges $w_{r,\,n-r+1}$ $(r = 2, 3, \ldots, [\tfrac{1}{2}n])$, the sample median when n is even $\tfrac{1}{2}(x_{(1/2n)} + x_{(1/2n+1)})$, and the extremal quotient $x_{(n)}/x_{(1)}$, the last for $x \ge 0$.

2.4. ORDER STATISTICS FOR A DISCRETE PARENT

When $p(x)$ is discrete over $x = 0, 1, 2, \ldots$, let $f_r(x) = \Pr\{X_{(r)} = x\}$ be the probability function (pf) of $X_{(r)}$. From (2.1.4) and (2.1.5) we have the alternative expressions

$$
\begin{aligned}
f_r(x) &= F_r(x) - F_r(x-1) \\
&= E_{P(x)}(n, r) - E_{P(x-1)}(n, r) \\
&= I_{P(x)}(r, n-r+1) - I_{P(x-1)}(r, n-r+1).
\end{aligned}
\tag{2.4.1}
$$

The bivariate pf $f_{rs}(x, y) = \Pr\{X_{(r)} = x, X_{(s)} = y\}$ follows from (2.2.4) and (2.2.5) since

$$
f_{rs}(x, y) = F_{rs}(x, y) - F_{rs}(x-1, y) - F_{rs}(x, y-1) + F_{rs}(x-1, y-1)
$$
$$
x \le y.
$$

Although computationally this appears to be the most convenient expression available, another form due to Khatri (1962) is more useful for theoretical work. By an argument similar to that used in the derivation of (2.2.1) we have, corresponding to the typical configuration shown, that for $x < y$

$$
\underbrace{r-1-i}_{} \overset{1+i+t}{\underset{x}{\diagdown\diagup}} \underbrace{s-r-1-u}_{} - t^{1+j+u} \overset{}{\underset{y}{\diagdown\diagup}} \underbrace{n-s-j}_{}
$$

$$
f_{rs}(x, y) = \sum_{i=0}^{r-1} \sum_{j=0}^{n-s} \sum_{u,t}
$$

$$
\cdot \frac{n!}{(r-1-i)!(1+i+t)!(s-r-1-u-t)!(1+j+u)!(n-s-j)!}
$$
$$
\cdot [P(x-1)]^{r-1-i}[p(x)]^{1+i+t}[P(y-1) - P(x)]^{s-r-1-u-t}
$$
$$
\cdot [p(y)]^{1+j+u}[1 - P(y)]^{n-s-j},
$$

where $\sum_{u,t}$ denotes summation over non-negative integral values of u, t subject to $u + t \le s - r - 1$. With

$$
C_{rs} = n!/[(r-1)!(s-r-1)!(n-s)!]
$$

we may write

$$
f_{rs}(x, y) = C_{rs} \sum_{i=0}^{r-1} \sum_{j=0}^{n-s} \sum_{u,t} \binom{r-1}{i}\binom{n-s}{j} \frac{(s-r-1)!}{(s-r-1-u-t)!u!t!}
$$
$$
\cdot [P(x-1)]^{r-1-i}[1 - P(y)]^{n-s-j}[P(y-1) - P(x)]^{s-r-1-u-t}
$$
$$
\cdot [p(x)]^{1+i+t}[p(y)]^{1+j+u} \int_0^1 \int_0^1 z^i(1-z)^t z'^j(1-z')^u \, dz \, dz'.
$$

Interchanging the summation and integral signs and putting $v = P(y) - z'p(y)$, $w = P(x - 1) + zp(x)$, we find

$$f_{rs}(x, y) = C_{rs} \int_{P(x-1)}^{P(x)} \int_{P(y-1)}^{P(y)} w^{r-1}(v - w)^{s-r-1}(1 - v)^{n-s} \, dv \, dw. \quad (2.4.2)$$

When $x = y$ we obtain similarly

$$f_{rs}(x, x) = C_{rs} \iint w^{r-1}(v - w)^{s-r-1}(1 - v)^{n-s} \, dv \, dw, \quad (2.4.3)$$

where the integration is now over $P(x - 1) \leq w \leq v \leq P(x)$. Since in (2.4.2) $w \leq v$ is automatically satisfied, we have the general result

$$f_{rs}(x, y) = C_{rs} \iint w^{r-1}(v - w)^{s-r-1}(1 - v)^{n-s} \, dv \, dw, \quad (2.4.4)$$

the integration being over $w \leq v$, $P(x - 1) \leq w \leq P(x)$, $P(y - 1) \leq v \leq P(y)$.

2.5. DISTRIBUTION-FREE CONFIDENCE INTERVALS FOR QUANTILES

Suppose first that X is a continuous variate with strictly increasing cdf $P(x)$. Then the equation

$$P(x) = p \qquad 0 < p < 1, \quad (2.5.1)$$

has a unique solution, say $x = \xi_p$, which we call the (population) *quantile of order p*. Thus $\xi_{1/2}$, is the *median* of the distribution. If $P(x)$ is not strictly increasing, $P(x) = p$ may hold in some interval, in which case any point in the interval would serve as a quantile of order p.

When X is discrete, ξ_p can still be defined by a generalization of (2.5.1), namely,

$$\Pr\{X < \xi_p\} \leq p \leq \Pr\{X \leq \xi_p\}. \quad (2.5.2)$$

This gives ξ_p uniquely unless $P(\xi_p)$ equals p, in which case ξ_p again lies in an interval.

We shall now show that if X is continuous the random interval $(X_{(r)}, X_{(s)})$ covers ξ_p with a probability which depends on r, s, n, and p, but not on $P(x)$, thus allowing the construction of distribution-free confidence intervals for ξ_p. To this end, note that the event $X_{(r)} \leq \xi_p$ is the union of the disjoint compound events $X_{(r)} \leq \xi_p$, $X_{(s)} \geq \xi_p$ and $X_{(r)} \leq \xi_p$, $X_{(s)} < \xi_p$. Thus, since $X_{(s)} < \xi_p$ implies $X_{(r)} \leq \xi_p$, we have

$$\Pr\{X_{(r)} \leq \xi_p\} = \Pr\{X_{(r)} \leq \xi_p \leq X_{(s)}\} + \Pr\{X_{(s)} < \xi_p\}$$

or

$$\Pr\{X_{(r)} \leq \xi_p \leq X_{(s)}\} = \Pr\{X_{(r)} \leq \xi_p\} - \Pr\{X_{(s)} < \xi_p\}. \quad (2.5.3)$$

It follows from (2.5.1), (2.1.5), and (2.1.3) that in the continuous case $(X_{(r)}, X_{(s)})$ covers ξ_p with probability $\pi(r, s, n, p)$ given by

$$\pi(r, s, n, p) = I_p(r, n - r + 1) - I_p(s, n - s + 1)$$

$$= \sum_{i=r}^{s-1} \binom{n}{i} p^i (1 - p)^{n-i}. \tag{2.5.4}$$

This is the required result, essentially due to Thompson (1936); for an alternative proof see Ex. 2.5.1.

In the discrete case the inequalities $\Pr\{X \leq \xi_p\} \geq p$ and $\Pr\{X < \xi_p\} \leq p$ imply that

$$\Pr\{X_{(r)} \leq \xi_p\} \geq I_p(r, n - r + 1), \quad \Pr\{X_{(s)} < \xi_p\} \leq I_p(s, n - s + 1)$$

$$\tag{2.5.5}$$

so that from (2.5.3)

$$\Pr\{X_{(r)} \leq \xi_p \leq X_{(s)}\} \geq \pi(r, s, n, p). \tag{2.5.6}$$

By a similar argument it follows also that

$$\Pr\{X_{(r)} < \xi_p < X_{(s)}\} \leq \pi(r, s, n, p). \tag{2.5.7}$$

The LHS's of (2.5.6) and (2.5.7) are no longer independent of $P(x)$, but we see that they have lower and upper bounds, respectively, which are distribution-free. These results were first obtained (by a different method) by Scheffé and Tukey (1945).

Confidence intervals with confidence coefficient $\geq 1 - \alpha$ for given n and p result from any choice of r and s making $\pi \geq 1 - \alpha$. Proper choice is somewhat arbitrary, but it is reasonable to try to make $s - r$ as small as possible subject to $\pi \geq 1 - \alpha$. If $p = \frac{1}{2}$, this procedure evidently results in taking $s = n - r + 1$, in which case π reduces to

$$\pi(r, n - r + 1, n, \tfrac{1}{2}) = 2I_{1/2}(r, n - r + 1) - 1 = 2^{-n} \sum_{i=r}^{n-r} \binom{n}{i}.$$

Confidence intervals for the median are closely related to the sign test, and the same table serves for both purposes. Extensive tables together with a review of related tables are given by MacKinnon (1964). From the normal approximation to the binomial we also have the very simple rule of thumb: For $n > 10$ an approximate $1 - \alpha$ confidence interval for the median is obtained by counting off $\frac{1}{2}n^{1/2}u_\alpha$ observations to the left and the right of the sample median and rounding *out* to the next integer, where u_α is the upper $\alpha\frac{1}{2}$ significance point of a unit normal variate.

Example 2.5. For $n = 100$, $\alpha = 0.05$, the rule gives $\frac{1}{2}n^{1/2}u_\alpha = 5(1.96) = 9.8$. Rounding out 50.5 ∓ 9.8, we obtain the interval $(x_{(40)}, x_{(61)})$, agreeing with MacKinnon.

When $p(x)$ is known to be symmetric and continuous, it is possible to construct confidence intervals for $\xi_{1/2}$ that are generally shorter and have a wider choice of confidence coefficients. Instead of being based on single order statistics these intervals have as end points two of the $\frac{1}{2}n(n+1)$ averages $\frac{1}{2}(x_{(i)} + x_{(j)})$ with $i \geq j$. It is interesting to note that the intervals are closely related to the signed-rank test. See Walsh (1949a,b) and Tukey (1949).

Differences of order statistics, $x_{(s)} - x_{(r)}$, may be used in a similar manner to set confidence intervals for *quantile differences* $\xi_q - \xi_p$ $(q > p)$. Such quantile differences may be of interest in their own right, especially the *interquartile distance* $\xi_{3/4} - \xi_{1/4}$; perhaps more important, confidence intervals for $\xi_q - \xi_p$ may readily be converted into confidence intervals for the standard deviation if $p(x)$ is a pdf depending on location and scale parameters only. In the latter case the confidence intervals are no longer distribution-free (see Chapter 6). We now show (Chu, 1957; cf. Ex. 2.5.5) that

$$\Pr\{X_{(s)} - X_{(r)} \geq \xi_q - \xi_p\} \geq E_p(n, r) - E_q(n, s) = L, \qquad (2.5.8)$$

$$\Pr\{X_{(v)} - X_{(u)} \leq \xi_q - \xi_p\} \geq E_q(n, v) - E_p(n, u) = L'. \qquad (2.5.9)$$

Proof. $\Pr\{X_{(s)} - X_{(r)} \geq \xi_q - \xi_p\} \geq \Pr\{X_{(s)} \geq \xi_q, X_{(r)} \leq \xi_p\}$

$$\geq \Pr\{X_{(s)} \geq \xi_q\} + \Pr\{X_{(r)} \leq \xi_p\} - 1$$

$$= \Pr\{X_{(r)} \leq \xi_p\} - \Pr\{X_{(s)} < \xi_q\}$$

$$\geq E_p(n, r) - E_q(n, s) \qquad \qquad \text{by (2.5.5).} \quad \blacktriangleright$$

Inequality (2.5.9) follows likewise.

For n sufficiently large it is easily shown that for any α $(0 < \alpha < 1)$ there exists at least one set of integers r, s, u, and v for which

$$L \geq 1 - \alpha \quad \text{and} \quad L' \geq 1 - \alpha. \qquad (2.5.10)$$

The corresponding $X_{(s)} - X_{(r)}$ and $X_{(v)} - X_{(u)}$ are then, respectively, upper and lower confidence limits for $\xi_q - \xi_p$ with confidence coefficient $\geq 1 - \alpha$. In the symmetric case $q = 1 - p$ it seems natural to use quasi-ranges.[1] With $s = n - r + 1$ and $v = n - u + 1$ condition (2.5.10) reduces to

$$E_p(n, r) \geq 1 - \tfrac{1}{2}\alpha, \quad E_p(n, u) \leq \tfrac{1}{2}\alpha.$$

Sequential procedures giving confidence intervals for ξ_p have been studied by Farrell (1966). Some confidence sets for multivariate medians are put

---

[1] Note that Chu uses "quasi-range" to denote $x_{(s)} - x_{(r)}$ for any $s > r$, whereas we confine the term to the case $s = n - r + 1$.

forward by Hoel and Scheuer (1961). Procedures for stratified samples are considered by McCarthy (1965) and Loynes (1966).

2.6. DISTRIBUTION-FREE TOLERANCE INTERVALS

Like a confidence interval, a tolerance interval has random terminals, say L and V. However, whereas a confidence interval is designed to cover, with prescribed probability, a population parameter such as mean, variance, or a quantile, the requirement of a tolerance interval (L, V) is that it contain at least a proportion γ of the population with probability β, both β and γ being preassigned constants $(0 \leq \beta, \gamma \leq 1)$. Thus if $p(x)$ is continuous we seek L, V such that

$$\Pr \left\{ \int_L^V p(x)\, dx \geq \gamma \right\} = \beta. \tag{2.6.1}$$

It turns out that the LHS of (2.6.1) has a value not depending on $p(x)$ if (Wilks, 1942) and only if (Robbins, 1944) L and V are chosen to be order statistics [including possibly $X_{(0)} = -\infty$ and $X_{(n+1)} = +\infty$]. To see the first part, note that with $L = X_{(r)}$, $V = X_{(s)}$ $(s > r)$ the LHS of (2.6.1) can be written as

$$\Pr \{P(X_{(s)}) - P(X_{(r)}) \geq \gamma\}. \tag{2.6.2}$$

But the probability integral transformation $u = P(x)$ is order-preserving and transforms $X_{(1)}, X_{(2)}, \ldots, X_{(n)}$ to $U_{(1)}, U_{(2)}, \ldots, U_{(n)}$ with $U_{(i)} = P(X_{(i)})$ $(i = 1, 2, \ldots, n)$, where the $U_{(i)}$ are now order statistics from a distribution uniform in $(0, 1)$. From (2.3.4) probability (2.6.2) is therefore simply

$$\Pr \{W_{rs} \geq \gamma\} = 1 - I_\gamma(s - r, n - s + r + 1).$$

Clearly (2.6.1) cannot in general be satisfied exactly, but r and s can be chosen to give $\Pr \{W_{rs} \geq \gamma\} \geq \beta$. For a one-sided tolerance interval we take either $r = 0$ or $s = n + 1$; for a two-sided interval it is usual to have $s = n - r + 1$. Then unique values of r and s will make $\Pr \{W_{rs} \geq \gamma\}$ as little in excess of β as possible. The problem may also be turned round: For given r, s (as well as β, γ) how large must n be?

Example 2.6. For $r = 1$, $s = n$ (2.6.1) reduces to

$$1 - I_\gamma(n - 1, 2) = \beta$$

or

$$\frac{n!}{(n - 2)!} \int_0^\gamma z^{n-2}(1 - z)\, dz = 1 - \beta,$$

that is,

$$n\gamma^{n-1} - (n - 1)\gamma^n = 1 - \beta.$$

This may be solved numerically for n and the result rounded to the next highest integer. For $\gamma = 0.95$, $\beta = 0.90$ we find $n = 77$.

Tables helpful in the general situation are given by Murphy (1948) and Somerville (1958).

As in Section 2.5 it can be shown (Scheffé and Tukey, 1945) that for a discrete parent distribution

$$\Pr\left\{\sum_{x=X_{(r)}}^{X_{(s)}} p(x) \geq \gamma\right\} \geq 1 - I_\gamma(s - r, n - s + r + 1) \geq \Pr\left\{\sum_{x=X_{(r+1)}}^{X_{(s-1)}} p(x) \geq \gamma\right\}$$

For interesting generalizations to distribution-free tolerance regions when the parent is multidimensional, see, for example, Fraser (1957) and Wilks (1962). Sequential procedures are studied by Saunders (1963). See also Ex. 2.6.2.

2.7. INDEPENDENCE RESULTS—ORDER STATISTICS AS A MARKOV CHAIN

Let $z_{(1)} \leq z_{(2)} \leq \cdots \leq z_{(n)}$ denote the order statistics in a sample of n from the exponential distribution

$$p(z) = e^{-z} \qquad 0 \leq z < \infty. \tag{2.7.1}$$

Then the joint pdf of the $Z_{(r)}$ is

$$n!\, \exp\left(-\sum_{r=1}^{n} z_r\right) \qquad 0 \leq z_1 \leq \cdots \leq z_n < \infty,$$

which may be written (Sukhatme, 1937) as

$$n!\, \exp\left[-\sum_{r=1}^{n} (n - r + 1)(z_r - z_{r-1})\right],$$

where $z_0 = 0$. Making the transformation

$$y_r = (n - r + 1)(z_{(r)} - z_{(r-1)}) \qquad r = 1, 2, \ldots, n \tag{2.7.2}$$

and noting that the range of each y_r is $(0, \infty)$, we see that the Y_r are statistically independent variates, each with pdf (2.7.1).

This simple result has important applications in life testing since, apart from a scale factor, the $Z_{(r)}$ may be interpreted as the successive lifetimes of n items subjected to simultaneous testing when the individual lifetime $X = \lambda Z$ ($\lambda > 0$) follows an exponential law with mean λ. The intervals of length $X_{(r)} - X_{(r-1)}$ between successive failures are then independently distributed as $\lambda Z/(n - r + 1)$. We return to this application in Chapter 6, Section 6.4.

Relations (2.7.2) allow $z_{(r)}$ to be expressed as

$$z_{(r)} = \sum_{i=1}^{r}(z_{(i)} - z_{(i-1)}) = \frac{\sum_{i=1}^{r} y_i}{n - i + 1}, \qquad (2.7.3)$$

that is, as a linear function of *independent* exponential variates. It follows at once that the distribution of $Z_{(r)}$, given $Z_{(j)} = z_{(j)}$ for all $j < r$, is the same as the distribution of $Z_{(r)}$, given only $Z_{(r-1)} = z_{(r-1)}$; in other words, $Z_{(1)}$, $Z_{(2)}, \ldots, Z_{(n)}$ form an (additive) Markov chain (Rényi, 1953).

Consider now the order statistics $x_{(1)} \leq x_{(2)} \leq \cdots \leq x_{(n)}$ in a sample drawn from any continuous parent whose cdf $P(x)$ is a strictly increasing function of x. Then, as noted in Section 2.6, the transformation $u = P(x)$ converts the $x_{(r)}$ into $u_{(r)}$ $(r = 1, 2, \ldots, n)$, the order statistics from a uniform $R(0, 1)$ distribution. Since $z = -\log u$ is a monotonic decreasing function of u and $-\log U$ has the exponential distribution (2.7.1), it follows that the $Z_{(r)}$, defined by

$$Z_{(r)} = -\log U_{(n-r+1)},$$

are the order statistics introduced at the outset. By (2.7.3) $x_{(n-r+1)}$ can therefore be expressed as

$$x_{(n-r+1)} = P^{-1}(u_{(n-r+1)}) = P^{-1}(e^{-z_{(r)}})$$

$$= P^{-1}\left[\exp - \left(\frac{y_1}{n} + \frac{y_2}{n-1} + \cdots + \frac{y_r}{n-r+1}\right)\right]. \quad (2.7.4)$$

We may now write

$$x_{(n-r)} = P^{-1}\left\{\exp\left[\log P(x_{(n-r+1)}) - \frac{y_{r+1}}{n-r}\right]\right\},$$

which shows, in view of the independence of $X_{(n-r+1)}$ and Y_{r+1} by (2.7.4), that the variates $X_{(n)}, X_{(n-1)}, \ldots, X_{(1)}$, form a Markov chain. So do the variates $X_{(1)}, X_{(2)}, \ldots, X_{(n)}$, as is apparent on replacing X by $-X$. This has the following important implication:

Theorem 2.7. *For a random sample of n from a continuous parent, the conditional distribution of $X_{(s)}$, given $X_{(r)} = x_{(r)}$ $(s > r)$, is just the distribution of the $(s - r)$th order statistic in a sample of $n - r$ drawn from the parent distribution truncated on the left at $x = x_{(r)}$.*[2]

From (2.7.4) we see also that the ratios

$$\frac{U_{(r)}}{U_{(r+1)}} = \exp\left(\frac{-Y_{n-r+1}}{r}\right) \qquad (2.7.5)$$

---

[2] This theorem follows also from what Tukey (1947) calls "Wald's principle."

are mutually independent variates $(r = 1, 2, \ldots, n; U_{(n+1)} = 1)$. It follows easily that the quantities

$$\left(\frac{U_{(r)}}{U_{(r+1)}}\right)^r = \exp\left(-Y_{n-r+1}\right)$$

are mutually independent uniform $R(0, 1)$ variates, a result due to Malmquist (1950).

A rather different set of independence results holds when the parent distribution is normal. From the well-known property that the joint distribution of the differences $X_i - \bar{X}$ $(i = 1, 2, \ldots, n - 1)$ is independent of the distribution of the mean $\bar{X}$, it follows that $\bar{X}$ is independent of any statistic expressible purely in terms of the $X_i - \bar{X}$, that is, of any location-free statistic, such as the range which may be written as $W = \max (X_j - \bar{X}) - \min (X_j - \bar{X})$ $(j = 1, 2, \ldots, n)$ (cf. Daly, 1946).

Characterizations

The independence of the Y_r in (2.7.2) may be used to characterize the exponential distribution. Simplest among theoretical results of this sort is the following: If X_1 and X_2 are iid with absolutely continuous distribution, and if $X_{(1)}$ and $X_{(2)} - X_{(1)}$ are independent, then X_1 and X_2 follow a (general) exponential distribution.

A good discussion of characterizations of various distributions by properties of order statistics is given by Ferguson (1967), where further references may be found. Also in the same spirit are Rossberg (1965a) and Govindarajulu (1967). An interesting but different kind of characterization is due to Chan (1967c): Let X and Y have distributions such that $\mathscr{E}X$ and $\mathscr{E}Y$ exist; then a necessary and sufficient condition that the two distributions are identical is that, in random samples of n, $\mathscr{E}X_{(n)} = \mathscr{E}Y_{(n)}$ for all $n \geq 1$.

EXERCISES

2.1.1. Let $X_1, X_2, \ldots, X_n$ be independent variates, X_i having a geometric distribution with parameter p_i, namely,

$$p(x_i) = q_i^{x_i-1} p_i \qquad q_i = 1 - p_i, x_i = 1, 2, \ldots .$$

Show that $X_{(1)}$ is distributed geometrically with parameter $1 - q_1 q_2 \cdots q_n$.

(Margolin and Winokur, 1967)

2.1.2. For a random sample of n from a continuous population whose pdf $p(x)$ is symmetrical about $x = \mu$ show that $f_r(x)$ and $f_{n-r+1}(x)$ are mirror images of each other in $x = \mu$ as mirror, that is,

$$f_r(\mu + x) = f_{n-r+1}(\mu - x).$$

Generalize this result to joint distributions of order statistics.

2.1.3. For the exponential distribution

$$p(x) = e^{-x} \quad x \geq 0,$$
$$= 0 \quad x < 0,$$

show that the cdf of $X_{(n)}$ in a random sample of n is

$$F_n(x) = (1 - e^{-x})^n.$$

Hence prove that, as $n \to \infty$, the cdf of $X_{(n)} - \log n$ tends to the limiting form $e^{-e^{-x}}$ ($-\infty \leq x \leq \infty$).

2.1.4. Let $x_1' < x_2' \cdots < x_N'$ be the elements of a finite population from which a sample $x_{(1)} < x_{(2)} < \cdots < x_{(n)}$ ($n \leq N$) is taken without replacement. Show that

$$\Pr\{X_{(i)} = x_t'\} = \frac{\binom{t-1}{i-1}\binom{N-t}{n-i}}{\binom{N}{n}} \quad t = i, i+1, \ldots, N - n + i.$$

(Wilks, 1962, p. 243)

2.1.5. Show that, in odd-sized random samples from a continuous population, the median of the distribution of sample medians is equal to the population median (i.e., the sample median is a median-unbiased estimator of the population median).

(van der Vaart, 1961b)

2.1.6. Suppose that particles are distributed over an area in such a way that (a) the number per unit area follows a Poisson law with mean λ, (b) the particles vary in magnitude so that the cdf of their size x is $P(x)$ ($a \leq x \leq b$). Show that the nth smallest particle in a unit area has size $\leq x$ with probability

$$F_n(x) = 1 - \sum_{i=0}^{n-1} e^{-\lambda P(x)} \frac{[\lambda P(x)]^i}{i!} \quad x < b,$$
$$= 1 \quad x \geq b.$$

(Epstein, 1949a)

2.2.1. Let $x_{r:n}$ denote the rth order statistic in a random sample of n, and $x_{r+1:n+1}$ the $(r+1)$th order statistic in the sample of $n+1$ obtained by taking an additional observation from the same population. If $P(x)$ is the parent cdf, show that for $x \leq y$

$$\Pr\{X_{r:n} \leq x, X_{r+1:n+1} > y\} = \binom{n}{r} P^r(x) [1 - P(y)]^{n-r+1}.$$

2.3.1. (a) Find the pdf of $X_{(r)}$ in a random sample of n from the exponential parent

$$p(x) = \theta^{-1} e^{-x/\theta} \quad \theta > 0, x \geq 0,$$
$$= 0 \quad x < 0.$$

(b) Show that $X_{(r)}$ and $X_{(s)} - X_{(r)}$ ($s > r$) are independently distributed.
(c) What is the distribution of $X_{(r+1)} - X_{(r)}$?
(d) Interpret (b) and (c) in the context of a life test on n items with exponential lifetimes.

2.3.2. Let $X_1, X_2, \ldots, X_n$ be independent variates, and let X_i have pdf $p_i(x)$ and cdf $P_i(x)$. Prove that:
(a) the pdf of $X_{(n)}$ is

$$f_n(x) = \left[\prod_{i=1}^{n} P_i(x) \right] \sum_{i=1}^{n} \left(\frac{p_i(x)}{P_i(x)} \right);$$

(b) the cdf of $W = X_{(n)} - X_{(1)}$ is given by

$$F(w) = \sum_{i=1}^{n} \int_{-\infty}^{\infty} p_i(x) \prod_{\substack{j=1 \\ j \neq i}}^{n} [P_j(x + w) - P_j(x)] \, dx.$$

2.3.3. Let X_{ij} $(i = 1, 2, \ldots, k; j = 1, 2, \ldots, n)$ be k independent random samples of n, with X_{ij} having cdf $P_i(x)$, $j = 1, 2, \ldots, n$. Show that the k sample maxima are the k largest of the kn variates with probability

$$n^k \int_{-\infty}^{\infty} \left[\prod_{m=1}^{k} P_m^{n-1}(x) \right] \sum_{i=1}^{k} \left[\prod_{\substack{j \neq i}}^{k} (1 - P_j(x)) \right] dP_j(x).$$

(Cohn *et al.*, 1960)

2.3.4. Show that the cdf of the midpoint (or midrange) $M = \frac{1}{2}(X_{(1)} + X_{(n)})$ in random samples of n from a continuous parent with cdf $P(x)$ is

$$F(m) = n \int_{-\infty}^{m} [P(2m - x) - P(x)]^{n-1} p(x) \, dx.$$

(Gumbel, 1958, p. 108)

2.3.5. (a) Show that the joint pdf of range W and midpoint M in random samples of n from a population rectangular on the interval $(-\frac{1}{2}, \frac{1}{2})$ is

$$f(w, m) = n(n - 1)w^{n-2} \qquad 0 \leq w \leq 1 - 2\,|m| \leq 1.$$

(b) Hence show that the pdf of M is

$$f(m) = n(1 - 2\,|m|)^{n-1} \qquad\qquad |m| \leq \frac{1}{2}$$

and that

$$\text{var } (M) = \frac{1}{2(n + 1)(n + 2)}.$$

(Neyman and Pearson, 1928; Carlton, 1946)

2.3.6. Show that the pdf of the range W in samples of 3 from a normal parent with unit variance is

$$f(w) = \frac{6e^{-\frac{1}{4}w^2}}{\pi\sqrt{2}} \int_{0}^{w/\sqrt{6}} e^{-\frac{1}{2}t^2} \, dt.$$

(McKay and Pearson, 1933)

2.3.7. Show that for a continuous parent symmetric about zero the cdf of the range in samples of n may be written as

$$F(w) = [P(\tfrac{1}{2}w) - P(-\tfrac{1}{2}w)]^n + 2n \int_{\frac{1}{2}w}^{\infty} p(x)[P(x) - P(x - w)]^{n-1} \, dx.$$

(Hartley, 1942)

2.3.8. If the parent distribution is unlimited, differentiable, symmetrical, and unimodal, show that the distribution of the midrange is also unlimited, differentiable, symmetrical, and unimodal.

(Gumbel *et al.*, 1965)

2.3.9. Let $V = X_{(n)}^{(1)} X_{(n)}^{(2)} \cdots X_{(n)}^{(k)}$ be the product of k maxima in independent random samples of n drawn from a uniform $R(0, 1)$ population. Show that the pdf of V is

$$f(v) = \frac{n^k}{\Gamma(k)} v^{n-1}(-\log v)^{k-1} \qquad 0 \leq v \leq 1.$$

(Rider, 1955; Rahman, 1964)

2.3.10. Let W_1, W_2 be the ranges in independent random samples of n_1, n_2 $(n_1 + n_2 = N)$ drawn from a uniform $R(0, c)$ parent. Prove that the pdf of $U = W_1/W_2$ is given by

$$f(u) = \frac{n_1(n_1 - 1)n_2(n_2 - 1)}{N(N - 1)(N - 2)} [Nu^{n_1-2} - (N - 1)u^{n_1-1}] \qquad 0 \leq u \leq 1,$$

$$= \frac{n_1(n_1 - 1)n_2(n_2 - 1)}{N(N - 1)(N - 2)} [Nu^{-n_2} - (N - 2)u^{-n_2-1}] \qquad 1 \leq u < \infty.$$

(Rider, 1951)

2.3.11. W_1, W_2 are the ranges in independent random samples of 3 and 2 taken from a normal parent. Prove that $R = W_1/W_2$ has cdf

$$F(r) = \frac{6}{\pi}\left[\tan^{-1} (3 + 4r^2)^{1/2} - \frac{\pi}{3} \right]$$

(Link, 1950)

2.3.12. Let $X_{(n_1)}$, $Y_{(n_2)}$ be the maxima in independent random samples of n_1, n_2 $(n_1 + n_2 = N)$ variates drawn from a uniform $R(0, c)$ parent. Prove that the pdf of $V = X_{(n_1)}/Y_{(n_2)}$ is given by

$$f(v) = \frac{n_1 n_2 v^{n_1-1}}{N} \qquad 0 \leq v \leq 1,$$

$$= \frac{n_1 n_2 v^{-n_2-1}}{N} \qquad 1 \leq v \leq \infty.$$

(Murty, 1955)

2.3.13. Prove that in samples of size $2m + 1$ (m integral) from a continuous cdf $P(x)$ $(0 \leq a \leq x \leq b)$ the pdf of the "peak to median ratio" $Z = X_{(2m+1)}/X_{(m+1)}$ is

$$f(z) = \frac{(2m + 1)!}{m!(m - 1)!} \int_a^{b/z} x P^m(x)[P(zx) - P(x)]^{m-1} p(x) p(zx)\, dx.$$

(Morrison and Tobias, 1965)

2.3.14. Suppose that points $X_1, X_2, \ldots, X_n$ are randomly and independently selected on the interval $0 \leq x \leq L$. Let

$$D = \min_j |X_i - X_j| \qquad \text{for some fixed } i.$$

Show that the cdf of D is

$$F(d) = 1 - \left[1 - \left(\frac{2d}{L} \right) \right]^n + \frac{2\{[1 - (d/L)]^n - [1 - (2d/L)]^n}{n} \qquad 0 \leq d \leq \tfrac{1}{2}L,$$

$$= 1 - (2/n)[1 - (d/L)]^n. \qquad \tfrac{1}{2}L \leq d \leq L.$$

(Halperin, 1960)

2.3.15. In a random sample of 3 from the continuous parent $p(x)$ let x', x'' ($x' \leq x''$) be the two *closest* observations. Show that the joint pdf of X' and X'' is

$$f(x', x'') = 6p(x')p(x'')[1 - P(2x'' - x') + P(2x' - x'')].$$

Hence show that, when $p(x)$ is the unit normal, $U = X'' - X'$, $V = U/(X_{(3)} - X_{(1)})$ have respective pdf's

$$f(u) = \frac{3\sqrt{3}}{\pi} \int_u^\infty e^{-\frac{1}{4}(3t^2 + u^2)} \, dt \qquad 0 \leq u < \infty,$$

$$f(v) = \frac{3\sqrt{3}}{\pi(1 - v + v^2)} \qquad 0 \leq v \leq \tfrac{1}{2}.$$

(Seth, 1950; Lieblein, 1952)

2.4.1. Prove equation (2.4.3).

2.4.2. Let X be a discrete variate taking the values $x = 0, 1, 2, \ldots, c$, where c is a positive integer or $+\infty$. Show that the pf of the range in samples of n is given by

$$f(w) = \sum_{x=0}^{c-w} \{ [P(x + w) - P(x - 1)]^n - [P(x + w) - P(x)]^n$$

$$- [P(x + w - 1) - P(x - 1)]^n + [P(x + w - 1) - P(x)]^n \} \qquad w > 0,$$

$$= \sum_{x=0}^c [p(x)]^n \qquad w = 0.$$

(Abdel-Aty, 1954; Burr, 1955; Siotani, 1957

2.5.1. Obtain (2.5.4) by noting that $X_{(r)} > \xi_p$ implies that at most $r - 1$ of the X_i are less than ξ_p.

2.5.2. Find the smallest value of n such that (a) $(X_{(1)}, X_{(n)})$, (b) $(X_{(2)}, X_{(n-1)})$ contains $\xi_{1/2}$ with probability ≥ 0.99.

2.5.3. Prove inequality (2.5.7).

2.5.4. Let (x_i, y_i) $(i = 1, 2, \ldots, n)$ be a random sample of n pairs of observations from a continuous bivariate distribution with bivariate median $(\xi_{1/2}, \eta_{1/2})$, where $\xi_{1/2} > 0$. Let $z_i(\theta) = y_i - \theta x_i$, and let $z_{(i)}(\theta)$ denote the ordered $z_i(\theta)$. Show that confidence intervals $(\underline{\theta}, \hat{\theta})$, with confidence coefficient $2^{-n} \sum_{i=1}^{n-r} \binom{n}{i}$, for $\eta_{1/2}/\xi_{1/2}$ may be found by solving for $\underline{\theta}$ and $\hat{\theta}$ from

$$z_{(r)}(\underline{\theta}) = \inf_\theta \{ z_{(r)}(\theta) \} = 0 \quad \text{and} \quad z_{(n-r-1)}(\hat{\theta}) = \sup_\theta \{ z_{(n-r+1)}(\theta) \} = 0.$$

(Bennett, 1966)

2.5.5. Show that for a random sample of n from a continuous distribution

$$\Pr \{ X_{(r)} < \xi_p < \xi_q < X_{(s)} \} = \int_0^p \int_q^1 f_{rs}(x, y) \, dx \, dy,$$

where $p < q$ and $f_{rs}(x, y)$ is as in Example 2.3. Hence prove that the interval (ξ_p, ξ_q) is contained in $(X_{(r)}, X_{(s)})$ with probability

$$\frac{n!}{(s - r)!} \sum_{i=0}^{r-1} \frac{(-1)^i p^{r+i}}{i!(n - s + r - i)!} I_p(n - s + 1, r - i).$$

(Wilks, 1962, p. 332; cf. Chu, 1968)

2.6.1. Find the value of n so that the proportion of the continuous population included between $X_{(r)}$ and $X_{(n-r+1)}$ has (a) the average value 0.99 and (b) probability approximately 0.9 of lying between 0.985 and 0.995. (*Ans.* $n = 999$.)

(Wilks, 1941)

2.6.2. Let $P(x)$ be the cdf of a continuous variate X, symmetric about $\xi_{1/2}$. In random samples of n write

$$V = \max (X_{(n)}, 2\xi_{1/2} - X_{(1)}).$$

Show that for $\gamma \geq \frac{1}{2}$

$$\Pr\{P(V) > \gamma\} = 1 - (2\gamma - 1)^n. \tag{A}$$

[Walsh (1962) uses this and further results to obtain distribution-free tolerance intervals for continuous symmetric populations. There are misprints in Walsh's proof of (A).]

2.7.1. Two independent random samples, $x_1, x_2, \ldots, x_{n_1}$ and $y_1, y_2, \ldots, y_{n_2}$, are taken from a common parent uniform over $(0, c)$. The samples are labeled so that $x_{(1)} \leq y_{(1)}$. By considering the joint pdf of $X_{(n_1)}$ and $Y_{(1)}$, conditional on $X_{(1)} = x_{(1)}$, show that the pdf of $T = (Y_{(1)} - X_{(1)})/(X_{(n_1)} - X_{(1)})$ is

$$f(t) = (n_1 - 1)n_2 \sum_{i=0}^{n_2-1} (-1)^i \binom{n_2 - 1}{i} \frac{t^i}{n_1 + i}, \qquad 0 \leq t \leq 1,$$

$$= \frac{(n_1 - 1)(n_1 - 1)!n_2!}{t^{n_1}(n_1 + n_2 - 1)!} \qquad 1 \leq t < \infty.$$

(Hyrenius, 1953)

2.7.2. k mutually independent random samples of size n, each drawn from a continuous distribution with cdf $P(x)$, are ordered on the basis of the largest member in each sample. Define Y_{ij} $(i = 1, 2, \ldots, n; j = 1, 2, \ldots, k)$ to be the ith variate in order of magnitude in the sample whose largest member Y_{1j} has rank j among the k maxima $Y_{11}, Y_{12}, \ldots, Y_{1k}$. Show that

$$\Pr (Y_{ij} < x) = \sum_{\alpha=0}^{j-1} \binom{k}{\alpha} [1 - P^n(x)]^\alpha [P^n(x)]^{k-\alpha}$$

$$+ \sum_{\alpha=0}^{j-1} \sum_{m=1}^{i-1} \sum_{\beta=0}^{m-1} j \binom{k}{j} m \binom{n}{m} \binom{j-1}{\alpha} \binom{m-1}{\beta} (-1)^{j+m-\beta-\alpha}$$

$$\cdot \frac{[P(x)]^{n-1-\beta} - [P(x)]^{nk-n\alpha}}{nk - n\alpha + 1 - n + \beta}$$

where the triple summation is zero for $i = 1$.

(Conover, 1965; David, 1966)

CHAPTER 3

Expected Values and Moments

3.1. BASIC FORMULAE

In this chapter we shall be concerned with the moments of order statistics, particularly with their means, variances, and covariances. As we shall see repeatedly in the sequel, linear functions of the order statistics, of which the range is a simple example, are extremely useful in the estimation of parameters. Knowledge of the means, variances, and covariances of the order statistics involved allows us to find the expected value and variance of the linear function, and hence permits us to obtain estimators and their efficiencies. The means are also of interest in selection problems (e.g., Ex. 3.2.2) and in so-called scoring procedures, where due to uncertainty about the underlying distribution, ordered observations $x_{(i)}$ $(i = 1, 2, \ldots, n)$ are replaced by their "scores" $\mathscr{E}Z_{(i)}$, the $Z_{(i)}$ being ordered variates from some standardized distribution such as the unit normal. When the correct parent distribution has been assumed, then, of all functions of the ranks i, the scores (up to a linear transformation) have the highest correlation coefficient squared with the $X_{(i)}$ (Brillinger, 1966).

It will be convenient at times to emphasize the sample size, and for the rest of this chapter we write $X_{r:n}$ for $X_{(r)}$. The mean or expected value of $X_{r:n}$ we denote by $\mu_{r:n}$. When $p(x)$ is continuous (the discrete case is deferred until Section 3.3), we have therefore

$$\mu_{r:n} = \int_{-\infty}^{\infty} x f_r(x) \, dx$$
$$= n\binom{n-1}{r-1} \int_{-\infty}^{\infty} x[P(x)]^{r-1}[1 - P(x)]^{n-r} \, dP(x). \qquad (3.1.1)$$

Since $0 \leq P(x) \leq 1$, it follows that

$$|\mu_{r:n}| \leq n\binom{n-1}{r-1} \int_{-\infty}^{\infty} |x| \, dP(x),$$

25

showing that $\mu_{r:n}$ exists provided $\mathscr{E}X$ exists.[1] The converse is not necessarily true. To see this, note that by the probability integral transformation $u = P(x)$ we may also write $\mu_{r:n}$ as

$$\mu_{r:n} = n\binom{n-1}{r-1}\int_0^1 P^{-1}(u)u^{r-1}(1-u)^{n-r}\,du,$$

where $P^{-1}(u)$ denotes x expressed as a function of u. Thus, if the mean

$$\mathscr{E}X = \int_0^1 P^{-1}(u)\,du$$

does not exist because of singularities at $u = 0$ or 1, $\mu_{r:n}$ may nevertheless exist for certain (but not all) values of r. For example, in the case of the Cauchy distribution $\mu_{r:n}$ exists unless $r = 1$ or n. See also Exs. 3.1.7 and 3.1.11.

In like manner, if $\mathscr{E}[g(X)]$ exists, where $g(x)$ is some function of x, so will $\mathscr{E}[g(X_{r:n})]$. The special cases $g(x) = x^k$, $(x - \mu_{r:n})^k$, and e^{tx} give, respectively, the raw moments, the central moments, and the moment-generating function (mgf) of $X_{r:n}$. We write the kth raw moment as

$$\mu_{r:n}^{(k)} = \mathscr{E}(X_{r:n}^k).[2] \tag{3.1.2}$$

Product moments may be defined similarly, namely,

$$\mu_{rs:n} = \mathscr{E}(X_{r:n}X_{s:n}). \tag{3.1.3}$$

For the covariance of $X_{r:n}$, $X_{s:n}$ we put correspondingly

$$\sigma_{rs:n} = \mathscr{E}(X_{r:n} - \mu_{r:n})(X_{s:n} - \mu_{s:n}). \tag{3.1.4}$$

As usual $\sigma_{rs:n} = \sigma_{sr:n}$, and $\sigma_{rr:n}$ or $\sigma_{r:n}^2$ is just the variance of $X_{r:n}$. Explicitly we have

$$\sigma_{r:n}^2 = \int_{-\infty}^{\infty} (x - \mu_{r:n})^2 f_r(x)\,dx$$

and, for $r < s$,

$$\sigma_{rs:n} = \int_{-\infty}^{\infty}\int_{-\infty}^{y} (x - \mu_{r:n})(y - \mu_{s:n})f_{rs}(x,y)\,dx\,dy, \tag{3.1.5}$$

where the joint pdf $f_{rs}(x,y)$ is defined by (2.2.1).

---

[1] Existence of $\mathscr{E}X$ implies the separate convergence of $\int^{\infty} x\,dP(x)$ and $\int_{-\infty} x\,dP(x)$, and hence also of $\int_{-\infty}^{\infty} |x|\,dP(x)$.

[2] Sen (1959) has shown that, if $\mathscr{E}|x|^{\delta}$ exists for some $\delta > 0$, then $\mu_{r:n}^{(k)}$ exists for all r satisfying $r_0 \le r \le n - r_0 + 1$, where $r_0\delta = k$.

Example 3.1.1. For $p(x)$ uniform in $(0, 1)$ equation (3.1.1) gives

$$\mu_{r:n} = n \binom{n-1}{r-1} \int_0^1 x \, x^{r-1} (1-x)^{n-r} \, dx$$

$$= \frac{n \binom{n-1}{r-1}}{(n+1) \binom{n}{r}} = \frac{r}{n+1}.$$

In view of the probability integral transformation this result implies that the order statistics divide the area under the curve $y = p(x)$ into $n+1$ parts, each with expected value $1/(n+1)$.

The general approach for the evaluation of product moments may be illustrated for 4 variates. In

$$f_{rstu}(x_1, x_2, x_3, x_4) = \frac{n!}{(r-1)!(s-r-1)!(t-s-1)!(u-t-1)!(n-u)!}$$

$$\times x_1^{r-1}(x_2-x_1)^{s-r-1}(x_3-x_2)^{t-s-1}(x_4-x_3)^{u-t-1}(1-x_4)^{n-u}$$

with $r < s < t < u$, put

$$x_4 = y_4, \quad x_3 = y_3 y_4, \quad x_2 = y_2 y_3 y_4, \quad x_1 = y_1 y_2 y_3 y_4.$$

Writing C for the constant and noting that the Jacobian is $y_2 y_3^2 y_4^3$, we obtain

$$f(y_1, y_2, y_3, y_4) = C y_1^{r-1}(1-y_1)^{s-r-1} \cdot y_2^{s-1}(1-y_2)^{t-s-1} \cdot y_3^{t-1}(1-y_3)^{u-t-1}$$

$$\times y_4^{u-1}(1-y_4)^{n-u} \quad 0 \le y_i \le 1; \, i = 1, 2, 3, 4.$$

This shows incidentally that the Y_i and hence the quantities $X_{r:n}/X_{s:n}$, $X_{s:n}/X_{t:n}$, $X_{t:n}/X_{u:n}$, and $X_{u:n}$ are statistically independent, a result which may be compared with (2.7.5). We therefore have

$$\mathscr{E}(X_{r:n}^a X_{s:n}^b X_{t:n}^c X_{u:n}^d)$$

$$= \frac{1}{B(r, s-r)} \int_0^1 y_1^{r-1+a}(1-y_1)^{s-r-1} \, dy_1 \cdots$$

$$\times \frac{1}{B(u, n-u+1)} \int_0^1 y_4^{u-1+a+b+c+d}(1-y_4)^{n-u} \, dy_4$$

$$= \frac{(r-1+a)!(s-1+a+b)!(t-1+a+b+c)!}{(r-1)!(s-1+a)!(t-1+a+b)!(u-1+a+b+c)!} \cdot \frac{\times (u-1+a+b+c+d)!n!}{\times (n+a+b+c+d)!}.$$

In general, for the order statistics $X_{r_i:n}$ $(i = 1, 2, \ldots, k)$ the result is (F. N. David and Johnson, 1954)

$$\mathscr{E}\left(\prod_{i=1}^{k} X_{r_i:n}^{a_i}\right) = \frac{n!}{\left(n + \sum\limits_{i=1}^{k} a_i\right)!} \prod_{i=1}^{k} \frac{\left(r_i - 1 + \sum\limits_{j=1}^{i} a_j\right)!}{\left(r_i - 1 + \sum\limits_{j=1}^{i-1} a_j\right)!}. \tag{3.1.6}$$

Hence, setting $p_r = r/(n + 1)$, $q_r = 1 - p_r$, we can deduce in particular for $r \le s \le t$

$$\mu_{r:n} = p_r, \qquad \sigma_{rs:n} = \frac{p_r q_s}{n + 2},$$

$$\mathscr{E}[(X_{r:n} - \mu_{r:n})(X_{s:n} - \mu_{s:n})(X_{t:n} - \mu_{t:n})] = \frac{2p_r(q_s - p_s)q_t}{(n + 2)(n + 3)}, \tag{3.1.7}$$

and

$$\mathscr{E}(X_{r:n} - \mu_{r:n})^4 = \frac{3p_r^2 q_r^2}{(n + 2)^2} + \frac{6p_r q_r}{(n + 2)(n + 3)(n + 4)}\left[(q_r - p_r)^2 - \frac{n + 3}{n + 2}p_r q_r\right].$$

When $p(x)$ is exponential, the corresponding explicit formulae are easily obtained (Ex. 3.1.1). However, numerical integration is generally needed for the evaluation of the means, variances, and covariances. Both computation and tabulation are reduced when $p(x)$ is symmetric, say about $x = 0$, by the following relations (cf. Ex. 2.1.2):

$$\mu_{r:n} = -\mu_{n-r+1:n}, \tag{3.1.8}$$

$$\sigma_{rs:n} = \sigma_{n-s+1, n-r+1:n}. \tag{3.1.9}$$

In the normal $N(0, 1)$ case the means have been tabulated extensively (Harter, 1961a), as well as the variances and covariances for $n \le 20$ (Teichroew, 1956; Sarhan and Greenberg, 1956). Other parents for which tables are available include the gamma, the logistic, the extreme-value, and the chi distribution with 1 DF (see A3.1).

Among linear functions of the order statistics the range is of special interest. We have, of course,

$$\mathscr{E}W_n = \mu_{n:n} - \mu_{1:n},$$

$$\text{var } W_n = \sigma_{n:n}^2 - 2\sigma_{n1:n} + \sigma_{1:n}^2,$$

which in the case of symmetry about $x = 0$ reduce further to

$$\mathscr{E}W_n = 2\mu_{n:n},$$

$$\text{var } W_n = 2(\sigma_{n:n}^2 - \sigma_{n1:n}).$$

Example 3.1.2. For $p(x)$ uniform in $(0, 1)$ we have from (3.1.7)

$$\operatorname{var} W_n = \frac{2}{n+2} \frac{(n \cdot 1 - 1 \cdot 1)}{(n+1)^2} = \frac{2(n-1)}{(n+2)(n+1)^2}.$$

As a check, note that from (2.3.4) with $r = 1$, $s = n$,

$$f(w_n) = \frac{1}{B(n-1, 2)} w_n^{n-2}(1 - w_n) \qquad 0 \leq w_n \leq 1,$$

giving

$$\operatorname{var} W_n = \frac{(n-1)n}{(n+1)(n+2)} - \left(\frac{n-1}{n+1}\right)^2 = \frac{2(n-1)}{(n+2)(n+1)^2}.$$

Alternative formulae for $\mu_{r:n}$ may be obtained by integration by parts in

$$\mu_{r:n} = \int_{-\infty}^{\infty} x \, dF_r(x).$$

To this end note that for any cdf $P(x)$ the existence of $\mathscr{E}X$ implies

$$\lim_{x \to -\infty} xP(x) = 0, \qquad \lim_{x \to \infty} x[1 - P(x)] = 0,$$

so that we have

$$\mathscr{E}X = \int_{-\infty}^{0} x \, dP(x) - \int_{0}^{\infty} x \, d[1 - P(x)]$$

$$= \int_{0}^{\infty} [1 - P(x)] \, dx - \int_{-\infty}^{0} P(x) \, dx. \qquad (3.1.10)$$

This general formula gives $\mu_{r:n}$ if $P(x)$ is replaced by $F_r(x)$. Setting $r = n$ and $r = 1$, we obtain on subtraction the well-known formula (Tippett, 1925; Cox, 1954)

$$\mathscr{E}W_n = \int_{-\infty}^{\infty} \{1 - P^n(x) - [1 - P(x)]^n\} \, dx. \qquad (3.1.11)$$

We also may write

$$\mu_{r:n} = \int_{0}^{\infty} [1 - F_r(x) - F_r(-x)] \, dx$$

and when $p(x)$ is symmetric about $x = 0$

$$\mu_{r:n} = \int_{0}^{\infty} [F_{n-r+1}(x) - F_r(x)] \, dx.$$

Useful general checks on computations of the raw moments are provided by noting that

$$\left(\sum_{r=1}^{n} X_{r:n}^k\right)^m = \left(\sum_{r=1}^{n} X_r^k\right)^m \qquad (3.1.12)$$

since the LHS is only a rearrangement of the RHS. Let μ and σ^2 be the population mean and variance. Taking expectations, we obtain from (3.1.12), with (k, m) in turn equal to $(1, 1)$, $(2, 1)$, $(1, 2)$,

$$\sum_{r=1}^{n} \mu_{r:n} = n\mu, \tag{3.1.13}$$

$$\sum_{r=1}^{n} \mathscr{E} X_{r:n}^2 = n\mathscr{E} X^2, \tag{3.1.14}$$

$$\sum_{r=1}^{n} \sum_{s=1}^{n} \mathscr{E}(X_{r:n} X_{s:n}) = n\mathscr{E} X^2 + n(n - 1)\mu^2, \tag{3.1.15}$$

and, subtracting (3.1.14) from (3.1.15),

$$\sum_{r=1}^{n-1} \sum_{s=r+1}^{n} \mathscr{E}(X_{r:n} X_{s:n}) = \tfrac{1}{2} n(n - 1)\mu^2. \tag{3.1.16}$$

Again, squaring the relation

$$\Sigma(X_{r:n} - \mu_{r:n}) = \Sigma(X_r - \mu)$$

gives

$$\sum_{r=1}^{n} \sum_{s=1}^{n} \sigma_{rs:n} = n\sigma^2 \tag{3.1.17}$$

It is clear from the method of derivation of (3.1.12)–(3.1.17) that these results apply equally whether the parent distribution is continuous or discrete.

3.2. THE NORMAL CASE

From a practical point of view Teichroew's (1956) tables, referred to in Section 3.1, go a long way toward meeting all needs for $n \leq 20$. There is nevertheless considerable theoretical interest in considering the normal case in more detail. We take in this section

$$p(x) = \phi(x) = (2\pi)^{-\frac{1}{2}} e^{-\frac{1}{2}x^2} \qquad -\infty < x < \infty,$$

$$P(x) = \Phi(x) = \int_{-\infty}^{x} \phi(t) \, dt.$$

In addition to the general relations (3.1.13)–(3.1.17), which now hold with

$\mu = 0$, $\sigma = 1$, and the symmetry results (3.1.8) and (3.1.9), we have

$$\sum_{s=1}^{n} \mathscr{E}(X_{r:n}X_{s:n}) = 1 \qquad r = 1, 2, \ldots, n \qquad (3.2.1)$$

or equivalently

$$\sum_{s=1}^{n} \sigma_{rs:n} = 1. \qquad (3.2.1')$$

Proof. The independence of $X_{r:n} - \bar{X}$ and $\bar{X}$ (Section 2.7) gives

$$\mathscr{E}(X_{r:n} - \bar{X})\bar{X} = 0$$

or

$$\mathscr{E}(X_{r:n}\bar{X}) = \mathscr{E}\,\bar{X}^2 = \frac{1}{n}.$$

Substituting $n\bar{X} = \sum_{s=1}^{n} X_{s:n}$, we obtain (3.2.1). ▶

For $n \leq 5$ the moments and product moments of order statistics can be expressed in terms of elementary functions (Jones, 1948; Godwin, 1949; Bose and Gupta, 1959). Following the last authors, let

$$I_n(a) = \int_{-\infty}^{\infty} [\Phi(ax)]^n e^{-x^2} \, dx \qquad (3.2.2)$$

so that

$$I(a) = \pi^{1/2}.$$

Now

$$\int_{-\infty}^{\infty} [\Phi(ax) - \tfrac{1}{2}]^{2m+1} e^{-x^2} \, dx = 0 \qquad m = 0, 1, 2, \ldots$$

since the integrand is an odd function of x. Hence

$$I_{2m+1}(a) = \frac{\displaystyle\sum_{i=1}^{2m+1} (-1)^{i+1} \binom{2m+1}{i} I_{2m-i+1}(a)}{2^i}.$$

In particular,

$$I_1(a) = \tfrac{1}{2}I_0(a) = \tfrac{1}{2}\pi^{1/2}$$

and

$$I_3(a) = \tfrac{3}{2}I_2(a) - \tfrac{3}{4}I_1(a) + \tfrac{1}{8}I_0(a) = \tfrac{3}{2}I_2(a) - \tfrac{1}{4}I_0(a).$$

Differentiating (3.2.2) with respect to a, we obtain for $n = 2$

$$(2\pi)^{1/2}I_2'(a) = \int_{-\infty}^{\infty} \Phi(ax) \cdot 2xe^{-\frac{1}{2}x^2(a^2+2)} \, dx,$$

and an integration by parts gives

$$I_2'(a) = \frac{1}{\pi^{\frac{1}{2}}} \frac{a}{(a^2 + 2)(a^2 + 1)^{\frac{1}{2}}},$$

so that

$$I_2(a) = \frac{1}{\pi^{\frac{1}{2}}} \arctan [(a^2 + 1)^{\frac{1}{2}}],$$

and

$$I_3(a) = \frac{3}{2\pi^{\frac{1}{2}}} \arctan [(a^2 + 1)^{\frac{1}{2}}] - \tfrac{1}{4}\pi^{\frac{1}{2}}$$

With the help of these results the ordinary moments of order statistics can be evaluated. Thus for $n = 5$

$$\mathscr{E} X_{5:5} = 5 \int_{-\infty}^{\infty} \Phi^4(x) \cdot x\phi(x) \, dx$$

$$= 5 \int_{-\infty}^{\infty} 4\Phi^3(x) \phi(x) \cdot \phi(x) \, dx$$

by an integration by parts, using $\phi'(x) = -x\phi(x)$. Hence

$$\mu_{5:5} = \frac{10}{\pi} I_3(1)$$

$$= \frac{15}{\pi^{\frac{3}{2}}} \arctan \sqrt{2} - \frac{5}{2\pi^{\frac{1}{2}}}$$

$$= \frac{5}{4\pi^{\frac{1}{2}}} + \frac{15}{2\pi^{\frac{3}{2}}} \arcsin \tfrac{1}{3} = 1.16296.$$

Similarly

$$\mu_{4:5} = \frac{5}{2\pi^{\frac{1}{2}}} - \frac{15}{\pi^{\frac{3}{2}}} \arcsin \tfrac{1}{3} = 0.49502$$

and

$$\mu_{3:5} = 0, \quad \mu_{2:5} = -\mu_{4:5}, \quad \mu_{1:5} = -\mu_{5:5}.$$

A more general although somewhat tedious approach, which can be applied equally to product moments, consists in expressing all integrals in terms of

$$J_n = \int_0^{\infty} \cdots \int_0^{\infty} e^{-Q} \, dx_1 \cdots dx_n,$$

where Q is a quadratic form in the x's. For $n \leq 3$, J_n can be evaluated in terms of elementary functions. Thus we have (Godwin, 1949)

$$n = 1: \quad Q = ax_1{}^2$$

$$J_1 = \frac{\frac{1}{2}\pi^{\frac{1}{2}}}{a}$$

$$n = 2: \quad Q = ax_1{}^2 + 2hx_1x_2 + bx_2{}^2$$

$$J_2 = \frac{1}{(ab - h^2)^{\frac{1}{2}}}\left(\frac{\pi}{2} - \arctan\frac{h}{(ab - h^2)^{\frac{1}{2}}}\right)$$

$$n = 3: \quad Q = ax_1{}^2 + bx_2{}^2 + cx_3{}^2 + 2fx_2x_3 + 2gx_3x_1 + 2hx_1x_2$$

$$J_3 = \frac{1}{4}\left(\frac{\pi}{\Delta}\right)^{\frac{1}{2}}\left(\frac{\pi}{2} + \arctan\frac{gh - af}{(a\Delta)^{\frac{1}{2}}} + \arctan\frac{hf - bg}{(b\Delta)^{\frac{1}{2}}}\right.$$

$$\left. + \arctan\frac{fg - ch}{(c\Delta)^{\frac{1}{2}}}\right)$$

where $\Delta = abc + fgh - af^2 - bg^2 - ch^2$.

Example 3.2.1. We shall indicate the steps involved in the above procedure by considering $\mathscr{E}(X_{3:5}X_{5:5})$, which is given by

$$\mathscr{E}(X_{3:5}X_{5:5}) = \frac{5!}{2!1!0!}\int_{-\infty}^{\infty}\int_{-\infty}^{y} xy\Phi^2(x)\phi(x)[\Phi(y) - \Phi(x)]\phi(y)\, dx\, dy. \quad \text{(A)}$$

Now

$$\int_{-\infty}^{y}\Phi^2(x)[\Phi(y) - \Phi(x)] \cdot x\phi(x)\, dx = \int_{-\infty}^{y}[2\Phi(x)\Phi(y) - 3\Phi^2(x)]\phi^2(x)\, dx.$$

Substituting this in (A) and reversing the order of integration, we have

$$\mathscr{E}(X_{3:5}X_{5:5}) = 60\int_{-\infty}^{\infty}\phi^2(x)\int_{x}^{\infty}[2\Phi(x)\Phi(y) - 3\Phi^2(x)] \cdot y\phi(y)\, dy\, dx.$$

The inner integral equals

$$-\Phi^2(x)\phi(x) + \int_{x}^{\infty} 2\Phi(x)\phi^2(y)\, dy.$$

We are therefore left with the evaluation of

$$-60\int_{-\infty}^{\infty}\Phi^2(x)\phi^3(x)\, dx = -\frac{10\sqrt{3}}{\pi^{3/2}}I_2[(\tfrac{2}{3})^{\frac{1}{2}}]$$

and of

$$120 \int_{-\infty}^{\infty} \int_{x}^{\infty} \frac{e^{-x^2}}{2\pi} \cdot \frac{e^{-y^2}}{2\pi} \left(1 - \int_{x}^{\infty} \frac{e^{-\frac{1}{2}z^2}}{(2\pi)^{\frac{1}{2}}} dz \right) dy \, dx$$

$$= \frac{15}{\pi} - \frac{15\sqrt{2}}{\pi^{\frac{5}{2}}} \int_{-\infty}^{\infty} \int_{x}^{\infty} \int_{x}^{\infty} \exp -(x^2 + y^2 + \tfrac{1}{2}z^2) \, dz \, dy \, dx.$$

With $y' = y - x$, $z' = z - x$, the triple integral reduces to

$$\int_{0}^{\infty} \int_{0}^{\infty} \int_{0}^{\infty} [\exp -(\tfrac{5}{2}x^2 + y^2 + \tfrac{1}{2}z^2 + 2xy + xz)$$

$$+ \exp -(\tfrac{5}{2}x^2 + y^2 + \tfrac{1}{2}z^2 - 2xy - xz)] \, dz \, dy \, dx,$$

so that finally

$$\mathscr{E}(X_{3:5} X_{5:5}) = -\frac{10\sqrt{3}}{\pi^2} \arctan \left(\frac{5}{3}\right)^{\frac{1}{2}} + \frac{15}{\pi} - \frac{15\sqrt{2}}{\pi^{\frac{5}{2}}} \cdot \tfrac{1}{4}(2\pi)^{\frac{1}{2}}$$

$$\cdot \left(\pi + 2 \arctan \frac{1}{\sqrt{5}}\right) = 0.14815.$$

The foregoing methods give all moments and product moments for $n \leq 5$ in terms of elementary functions. By repeated integration by parts many higher moments can be obtained in like manner. Thus $\mathscr{E}X_{6:6}^2$ can be handled, but the method fails for $\mathscr{E}X_{6:6}$ itself.

Ruben (1954) has provided an ingenious but involved approach by which the ordinary moments of order statistics can be expressed as linear functions of the contents of certain hyperspherical simplices (generalized spherical triangles). For more than three dimensions these contents cannot be expressed in terms of elementary functions, a result which *inter alia* implies the same for $I_4(a)$ and $\mathscr{E}X_{6:6}$. Some similar results for the mean and variance of range are given by Ruben (1956b). See also H. T. David (1963). Of course, the designation of certain functions as elementary is rather arbitrary. Watanabe *et al.* (1957) derive by lengthy but straightforward methods all first two moments and product moments for $n \leq 7$ in terms of the inverse trigonometric functions and of a few such integrals as $\int \sin^{-1} [3/(8 - \tan^2 \Psi)]^{\frac{1}{2}} d\Psi$. Moments and product moments up to the fourth degree are similarly given in Watanabe *et al.* (1958).[3]

3.3. THE DISCRETE CASE

For a discrete parent $p(x)$ ($x = 0, 1, 2, \ldots$) the kth raw moment of $X_{r:n}$ is immediately obtainable from the definition

$$\mu_{r:n}^{(k)} = \sum_{x=0}^{\infty} x^k f_r(x),$$

[3] I am indebted to Dr. Peter Nemenyi for supplying these two papers.

where $f_r(x)$ is given by (2.4.1). Somewhat more convenient formulae involving the "tails" $1 - F_r(x)$, rather than $f_r(x)$, are readily derived from general results for discrete distributions. Following Feller (1957, p. 249), let

$$q(x) = p(x + 1) + p(x + 2) + \cdots,$$

and define the generating functions

$$\mathscr{P}(s) = \sum_{x=0}^{\infty} p(x)s^x, \qquad \mathscr{Q}(s) = \sum_{x=0}^{\infty} q(x)s^x. \tag{3.3.1}$$

Clearly, for $|s| < 1$, k differentiations of $\mathscr{P}(s)$ give

$$\mathscr{P}^{(k)}(s) = \sum_{x=k}^{\infty} x(x-1)\cdots(x-k+1)p(x)s^{x-k}.$$

If the kth factorial moment $\mu_{[k]}$ of X exists, we may set $s = 1$ and have

$$\mu_{[k]} = \mathscr{P}^{(k)}(1). \tag{3.3.2}$$

Feller proves that for $|s| < 1$

$$\mathscr{Q}(s) \cdot (1 - s) = 1 - \mathscr{P}(s), \tag{3.3.3}$$

from which, on differentiating k times and using Leibnitz's theorem, we obtain

$$\mathscr{Q}^{(k)}(s)(1 - s) + k\mathscr{Q}^{(k-1)}(s)(-1) = -\mathscr{P}^{(k)}(s).$$

When $\mu_{[k]}$ exists, we deduce from (3.3.2)

$$\mu_{[k]} = k\mathscr{Q}^{(k-1)}(1). \tag{3.3.4}$$

In particular,

$$\mu_{[1]} = \mu = \sum_{x=0}^{\infty} q(x) = \sum_{x=0}^{\infty} [1 - P(x)],$$

$$\mu_{[2]} = \mathscr{E}[X(X-1)] = 2\sum_{x=0}^{\infty} xq(x) = 2\sum_{x=0}^{\infty} x[1 - P(x)],$$

from which var X follows by

$$\text{var } X = \mu_{[2]} + \mu - \mu^2.$$

To apply these results to the moments of $X_{r:n}$ we need only replace $P(x)$ by $F_r(x)$. In view of (2.1.5)

$$\mu_{r:n} = \sum_{x=0}^{\infty} [1 - I_{P(x)}(r, n - r + 1)],$$

$$\text{var } X_{r:n} = 2\sum_{x=0}^{\infty} x[1 - I_{P(x)}(r, n - r + 1)] + \mu_{r:n} - \mu_{r:n}^2. \tag{3.3.5}$$

In particular, (3.3.5) gives for the moments of the extremes

$$\mu_{n:n} = \sum_{x=0}^{\infty} [1 - P^n(x)], \qquad \mu_{1:n} = \sum_{x=0}^{\infty} [1 - P(x)]^n,$$

and hence

$$\mathscr{E} W_n = \sum_{x=0}^{\infty} \{1 - P^n(x) - [1 - P(x)]^n\},$$

in direct analogy to results for a distribution continuous in $(0, \infty)$.

Actually these formulae follow also as limiting cases of the corresponding results for non-negative continuous variates if we take

$$F_r(x) = I_{P(i)}(r, n - r + 1) \qquad i \le x < i + 1.$$

Then

$$\begin{aligned}
\mu_{r:n} &= \int_0^{\infty} [1 - F_r(x)] \, dx \\
&= \sum_{i=0}^{\infty} [1 - F_r(i)] \int_i^{i+1} dx \\
&= \sum_{x=0}^{\infty} [1 - F_r(x)].
\end{aligned}$$

Again

$$\begin{aligned}
\mathscr{E} X_{r:n}^2 &= \int_0^{\infty} 2x[1 - F_r(x)] \, dx \\
&= \sum_{i=0}^{\infty} [1 - F_r(i)] \int_i^{i+1} 2x \, dx \\
&= \sum_{i=0}^{\infty} (2i + 1)[1 - F_r(i)] \\
&= 2 \sum_{x=0}^{\infty} x[1 - F_r(x)] + \mu_{r:n}
\end{aligned}$$

etc.

By (2.4.5) we have also

$$\mu_{rs:n} = C_{rs} \sum_{x=0}^{\infty} \sum_{y=x}^{\infty} xy \iint w^{r-1}(v - w)^{s-r-1}(1 - v)^{n-s} \, dv \, dw, \qquad (3.3.6)$$

the integration being over $w \le v$, $P(x - 1) \le w \le P(x)$, $P(y - 1) \le v \le P(y)$.

The mean and variance of the smaller of two binomial variates are explicitly considered by Craig (1962) and Shah (1966a).

3.4. RECURRENCE RELATIONS

Many authors (see, e.g., Govindarajulu, 1963a) have studied recurrence relations between the moments of order statistics, usually with the principal aim of reducing the number of independent calculations required for the evaluation of the moments. It may be noted in passing that equations (3.1.12)–(3.1.17) can also be used to this end, although they are perhaps best kept for checking purposes. In the derivation of recurrence relations it has nearly always been assumed that the parent distribution is continuous. As will become clear, most results in fact continue to hold for a discrete parent.[4]

Relation 1. For an arbitrary distribution

$$(n - r)\mu_{r:n}^{(k)} + r\mu_{r+1:n}^{(k)} = n\mu_{r:n-1}^{(k)},$$

where $r = 1, 2, \ldots, n - 1$, and $k = 1, 2, 3, \ldots$.

This result has been obtained by Cole (1951) in the continuous and by Melnick (1964) in the discrete case. A common proof is possible since we may write, respectively,

$$\mu_{r:n}^{(k)} = \int_{-\infty}^{\infty} x^k \frac{d}{dx} I_{P(x)}(r, n - r + 1) \, dx \qquad (3.4.1)$$

and

$$\mu_{r:n}^{(k)} = \sum_{x=0}^{\infty} x^k \Delta I_{P(x)}(r, n - r + 1), \qquad (3.4.2)$$

where $\Delta I_{P(x)}(a, b) = I_{P(x)}(a, b) - I_{P(x-1)}(a, b)$.

From the well-known recurrence formula for the incomplete B function

$$aI_y(a + 1, b) + bI_y(a, b + 1) = (a + b)I_y(a, b)$$

we have, on taking $a = r$, $b = n - r$, $y = P(x)$,

$$(n - r)I_{P(x)}(r, n - r + 1) + rI_{P(x)}(r + 1, n - r) = nI_{P(x)}(\varsigma, n - r).$$

Relation 1 follows at once.

Corollary 1A. For n even,

$$\tfrac{1}{2}(\mu_{\frac{1}{2}n+1:n}^{(k)} + \mu_{\frac{1}{2}n:n}^{(k)}) = \mu_{\frac{1}{2}n:n-1}^{(k)}.$$

Proof. Put $r = \tfrac{1}{2}n$ in Relation 1. ▶

Taking $k = 1$, we see that the expected values of the median in samples of n (even) and $n - 1$ are equal.

Corollary 1B. If the parent distribution is symmetric about the origin and n is even,

$$\mu_{\frac{1}{2}n:n}^{(k)} = \mu_{\frac{1}{2}n:n-1}^{(k)} \qquad k \text{ even,}$$
$$= 0 \qquad k \text{ odd.}$$

---

[4] In fact, all recurrence relations depending on no specific distributional assumption, such as normality, remain valid for exchangeable variates—see Section 5.5.

Proof.   In Corollary 1A substitute

$$\mu^{(k)}_{\frac12 n+1:n} = (-1)^k \mu_{\frac12 n:n}. \qquad \blacktriangleright$$

Comment. Since the proof of Relation 1 depends only on a property of the incomplete B function it is clear that the same recurrence relation also links the pdf's, cdf's, and in fact the expected values (if these exist) of any function $g(X_{r:n})$. Thus we have (Srikantan, 1962)

$$(n - r)\mathscr{E}g(X_{r:n}) + r\mathscr{E}g(X_{r+1:n}) = n\mathscr{E}g(X_{r:n-1}).$$

Similar generalizations apply to the next two relations, which are again stated in terms of moments, the most important case.

Relation 2.   For an arbitrary distribution

$$\mu^{(k)}_{r:n} = \sum_{i=r}^{n} \binom{i-1}{r-1}\binom{n}{i}(-1)^{i-r}\mu^{(k)}_{i:i}. \qquad (3.4.3)$$

Thus the moments of $X_{r:n}$ are expressible in terms of the simpler moments of the largest in samples of $r, r+1, \ldots, n$.

Proof.   In the equation

$$I_{P(x)}(r, n-r+1) = \int_0^{P(x)} \frac{n!}{(r-1)!(n-r)!} t^{r-1}(1-t)^{n-r}\,dt$$

expand $(1 - t)^{n-r}$. The integrand on the RHS then becomes

$$\frac{n!}{(r-1)!(n-r)!} \sum_{j=0}^{n-r} \binom{n-r}{j}(-1)^j t^{r-1+j},$$

which, on putting $i = j + r$, equals

$$\sum_{i=r}^{n} \binom{i-1}{r-1}\binom{n}{i}(-1)^{i-r}\cdot it^{i-1}.$$

Substituting this in (3.4.1) or (3.4.2), we obtain Relation 2. $\blacktriangleright$

A corresponding result in terms of $\mu_{1:i}$ is

$$\mu^{(k)}_{r:n} = \sum_{i=n-r+1}^{n} \binom{i-1}{n-r}\binom{n}{i}(-1)^{i-n+r-1}\mu^{(k)}_{1:i}.$$

Comment. In the use of (3.4.3) rounding errors become important as $n - r$ increases. See also Srikantan (1962).

Relation 3.   For an arbitrary distribution and $1 \le r < s \le n$

$$(r-1)\mu_{rs:n} + (s-r)\mu_{r-1,s:n} + (n-s+1)\mu_{r-1,s-1:n} = n\mu_{r-1,s-1:n-1}.$$

Proof. (Govindarajulu, 1963a). In the formula for $n\mu_{r-1,s-1:n-1}$, namely,

$$n\mu_{r-1,s-1:n-1} = \frac{n!}{(r-2)!(s-r-1)!(n-s)!} \int_{-\infty}^{\infty}\int_{-\infty}^{y} xy$$

$$\cdot [P(x)]^{r-2}[P(y) - P(x)]^{s-r-1}[1 - P(y)]^{n-s}\,dP(x)\,dP(y)$$

split up the integral as the sum of three similar integrals according to the partition

$$1 = P(x) + [P(y) - P(x)] + [(1 - P(y)].$$

Relation 3 follows. With the help of (3.3.6) the same argument holds also in the discrete case. Note that Relation 1 is the special case $r = 1, s = r + 1$.

▶

Some further relations are given as exercises. See also Krishnaiah and Rizvi (1966).

EXERCISES

3.1.1. For a random sample of n from the exponential distribution with pdf

$$p(x) = e^{-x} \quad x \geq 0,$$
$$= 0 \quad x < 0,$$

show that

$$\mu_{r:n} = \sum_{i=n-r+1}^{n} i^{-1},$$

and that for $r < s$

$$\sigma_{rs:n} = \sigma_{r:n}^2 = \sum_{i=n-r+1}^{n} i^{-2}.$$

3.1.2. For a random sample of n from the power-function distribution with pdf

$$p(x) = va^{-v}x^{v-1} \quad 0 \leq x \leq a, a > 0, v > 0,$$

show that

$$\mu_{r:n}^{(k)} = \frac{\Gamma(n+1)\Gamma(k/v + r)a^k}{\Gamma(r)\Gamma(n + k/v + 1)},$$

and that for $r < s$

$$\mu_{rs:n} = \frac{\Gamma(n+1)\Gamma(1/v + r)\Gamma(2/v + s)a^2}{\Gamma(r)\Gamma(s + 1/v)\Gamma(n + 2/v + 1)}.$$

(Malik, 1967)

3.1.3. For random samples from a unit normal parent show that

$$\mu_{2:2} = \pi^{-1/2}, \quad \mu_{3:3} = \tfrac{3}{2}\pi^{-1/2}.$$

3.1.4. Show that for a random sample of n from a continuous distribution with cdf $P(x)$

$$\mathscr{E}(X_{r+1:n} - X_{r:n}) = \binom{n}{r}\int_{-\infty}^{\infty} [P(x)]^r[1 - P(x)]^{n-r}\,dx \quad r = 1, 2, \ldots, n - 1.$$

(Galton, 1902; Pearson, 1902)

3.1.5. Show that in random samples from a continuous distribution with cdf $P(x)$

(a) $\mathscr{E}[X_{s:n}P(X_{r:n})] = \dfrac{r}{n+1} \mu_{s+1:n+1}$ $r \leq s,$

(b) $\mathscr{E}[X_{r:n}P(X_{s:n})] = \mu_{r:n} - \dfrac{n+1-s}{n+1} \mu_{r:n+1}$ $r < s.$

(Govindarajulu, 1968a)

3.1.6. Prove that in random samples of n from a continuous parent distribution

$$\sum_{r=1}^{n-1} \sum_{s=r+1}^{n} \mathscr{E}(X_{r:n}^k X_{s:n}^l) = \binom{n}{2} \mathscr{E}(X_{1:2}^k X_{2:2}^l).$$

(Govindarajulu, 1963a)

3.1.7. Show that the pdf of the median in samples of $n = 2k + 1$ from the Cauchy distribution

$$p(x) = \frac{1}{\pi[1 + (x - \theta)^2]} \qquad -\infty < x < \infty$$

is

$$f_{k+1,n}(x) = \frac{n!}{(k!)^2\pi} \left[\frac{1}{4} - \frac{1}{\pi^2} \text{arc tan}^2\,(x - \theta) \right]^k \frac{1}{1 + (x - \theta)^2},$$

and that the variance of the median is given by

$$\frac{2(n!)}{(k!)^2\pi^n} \int_0^{\frac{1}{2}\pi} (\pi - y)^k y^k \cot^2 y \, dy.$$

(Note that this is finite for $k \geq 2$.)

(Rider, 1960)

3.1.8. In the Cauchy distribution

$$p(x) = \frac{1}{\pi(1 + x^2)} \qquad -\infty < x < \infty$$

show that for $r = 3, 4, \ldots, n - 2$

$$\sigma_{r:n}^2 = \frac{n}{\pi} (\mu_{r:n-1} - \mu_{r-1:n-1}) - 1 - \mu_{r:n}^2.$$

(Barnett, 1966)

3.1.9. Let the distribution of X be symmetric about zero. Then the distribution of Y, obtained by folding the distribution of X at zero, has cdf $P^*(x) = 2P(x) - 1$ ($x \geq 0$). If $\mu_{r:n}^{*(k)}$ denotes the kth moment of Y, show that

$$\mu_{r:n}^{(k)} = 2^{-n} \left[\sum_{i=0}^{r-1} \binom{n}{i} \mu_{r-i:n-i}^{*(k)} + (-1)^k \sum_{i=r}^{n} \binom{n}{i} \mu_{i-r+1:i}^{*(k)} \right].$$

(Govindarajulu, 1963b)

3.1.10. The statistic $T(X_1, X_2, \ldots, X_n)$ is an *odd location statistic* if for all $x_1, x_2, \ldots, x_n,$ and every $h,$

$$T(x_1 + h, x_2 + h, \ldots, x_n + h) = T(x_1, x_2, \ldots, x_n) + h,$$

and

$$T(-x_1, -x_2, \ldots, -x_n) = -T(x_1, x_2, \ldots, x_n).$$

Likewise $S(X_1, X_2, \ldots, X_n)$ is an *even location-free statistic* if for all $x_1, x_2, \ldots, x_n$, and every h,

$$S(x_1 + h, x_2 + h, \ldots, x_n + h) = S(x_1, x_2, \ldots, x_n),$$

and

$$S(-x_1, -x_2, \ldots, -x_n) = S(x_1, x_2, \ldots, x_n).$$

Prove that for a random sample of n from any symmetric parent distribution T and S are uncorrelated.

(Hogg, 1960)

3.1.11. Let $X_1, X_2, \ldots, X_n$ be a random sample from a population with cdf $P(x)$ and pdf $p(x)$, the latter being continuous and strictly positive on $\{x \mid 0 < P(x) < 1\}$. Suppose that $\mathscr{E}X_{i:n}^2 + \mathscr{E}X_{j:n}^2 < \infty$. Show that

$$\operatorname{cov}(X_{i:n}, X_{j:n}) \geq 0. \tag{A}$$

[It is sufficient to show that $\mathscr{E}(X_{j:n} \mid X_{i:n})$ is a continuous monotone increasing function of $X_{i:n}$ in view of the following lemma: *If X and Y are random variables such that $\mathscr{E}X^2 + \mathscr{E}Y^2 < \infty$ and $\mathscr{E}(Y \mid X)$ is a monotone increasing function of X, then* $\operatorname{cov}(X, Y) \geq 0$.]

(Bickel, 1967; cf. Tukey, 1958)

3.2.1. If (X, Y) is an observation from a bivariate normal $N(0, 0, 1, 1, \rho)$ population, show that the expected value of max (X, Y) is $[(1 - \rho)/\pi]^{1/2}$.

3.2.2. Annual fleece weights of father and daughter sheep of a certain flock may be assumed to follow a bivariate normal distribution with correlation coefficient ρ. In one season only, the best 3 out of 10 available rams were used for breeding purposes. If these have, respectively, n_1, n_2, n_3 female offspring, find the expected rise in mean fleece weights, expressed in standard deviation units, of these daughter sheep. Verify that for $n_1 = n_2 = n_3$ and $\rho = 0.6$ the answer is 0.64.

3.2.3. Let (X, Y) have a bivariate normal $N(\mu_x, \mu_y, \sigma_x^2, \sigma_y^2, \rho)$ distribution. When in a random sample of n pairs the x's are arranged in ascending order, namely,

$$x_{(1)} \leq x_{(2)} \leq \cdots \leq x_{(n)},$$

denote the associated y's (which are not necessarily in ascending order) by

$$y_{[1]}, y_{[2]}, \ldots, y_{[n]}.$$

Show that for $r, s = 1, 2, \ldots, n$

$$\mathscr{E}Y_{[r]} = \mu_y + \rho\sigma_y\mu_{r:n},$$
$$\operatorname{var} Y_{[r]} = (1 - \rho^2)\sigma_y^2 + \rho^2\sigma_y^2\sigma_{r:n}^2$$
$$\operatorname{cov}(X_{(r)}, Y_{[s]}) = \rho\sigma_x\sigma_y\sigma_{rs:n},$$
$$\operatorname{cov}(Y_{[r]}, Y_{[s]}) = \rho^2\sigma_y^2\sigma_{rs:n} \qquad r \neq s,$$

where $\mu_{r:n}$, $\sigma_{rs:n}$ refer to a unit normal parent population.

(Watterson, 1959)

3.2.4. The independent normal variates X and Y have respective means μ_x, μ_y and common variance σ^2. Show that

$$\mathscr{E}(X \mid X < Y) = \mu_x - \frac{\sigma}{\sqrt{2}} A_\xi,$$

$$\operatorname{var}(X \mid X < Y) = \sigma^2(1 + \tfrac{1}{2}\xi A_\xi - \tfrac{1}{2}A_\xi^2),$$

where

$$\xi = \frac{\mu_x - \mu_y}{\sqrt{2}\,\sigma},$$

and

$$A_\xi = \frac{e^{-\frac{1}{2}\xi^2}}{\displaystyle\int_\xi^\infty e^{-\frac{1}{2}t^2}\,dt}.$$

(David, 1957)

3.2.5. (Youden's Angel Problem). Suppose that n observations are drawn at random from a normal population with unit variance. A benevolent angel tells us which is nearest the true mean, and the others are rejected. Show that the variance v_n of the retained member is given by

$$v_n = n\left(\frac{2}{\pi}\right)^{\frac{1}{2}n}\int_0^\infty x^2 e^{-\frac{1}{2}x^2}\left(\int_x^\infty e^{-\frac{1}{2}t^2}\,dt\right)^{n-1}dx$$

and find

$$v_2 = 1 - \frac{2}{\pi}, \qquad v_3 = 1 - \frac{2}{\pi}(3 - \sqrt{3}),$$

$$v_4 = 1 - \frac{12}{\pi} + \frac{16}{\pi\sqrt{3}}, \qquad v_5 = 1 - \frac{20}{\pi} + \frac{240}{\pi^2\sqrt{3}}\left[\tan^{-1}\left(\tfrac{5}{3}\right)^{\frac{1}{2}} - \frac{\pi}{6}\right]$$

(Kendall, 1954)

3.2.6. Show that for any distribution with cdf $P(x)$ possessing a mean, $\mu_{r:n}$ may be expressed as

$$\mu_{r:n} = \sum_{j=0}^{n-1} c_j \phi_j(r:n) \qquad r = 1, 2, \ldots, n,$$

where

$$\phi_j(r:n) = \frac{(j!)^3}{(2j!)}\sum_{i=0}^j (-1)^i \binom{r-1}{j-i}\binom{2j-1}{j}\binom{n-j+i-1}{i}$$

and

$$c_j = \frac{n!}{(n+j)!}\,\frac{(2j+1)!}{(j!)^2}\int_{-\infty}^\infty L_j\,[2P(x) - 1]x\,dP(x),\qquad\text{(A)}$$

$L_j(z)$ being the jth order Legendre polynomial in z.
Hence establish that

$$\frac{1}{n}\sum_{r=1}^n \mu_{r:n}^2 = \sum_{j=0}^{n-1}\frac{n!(n-1)!}{(n+j)!(n-1-j)!}(2j+1)I_j^2,$$

where I_j is the integral in (A).

(Saw and Chow, 1966; cf. Ruben 1956a)

3.3.1. Show that, if the kth moment of a discrete variate X exists, then so does $\mu_{r:n}^{(k)}$.

3.3.2. Show that the range W_n in samples of n from a continuous parent with cdf $P(x)$

has variance

$$\text{var } W_n = 2 \int_{-\infty}^{\infty} \int_{-\infty}^{y} \{1 - P^n(y) - [1 - P(x)]^n + [P(y) - P(x)]^n\} \, dx \, dy - (\mathscr{E} W_n)^2,$$

and that the corresponding formula in the discrete case ($x = 0, 1, 2, \ldots, \infty$) is

$$\text{var } W_n = 2 \sum_{y=0}^{\infty} \sum_{x=0}^{y} \{1 - P^n(y) - [1 - P(x)]^n + [P(y) - P(x)]^n\} - \mathscr{E} W_n(1 + \mathscr{E} W_n).$$

(Tippett, 1925; Siotani, 1957).

3.3.3. Prove that for random samples of n from any parent distribution

$$\sum_{\substack{r=1 \\ r \neq s}}^{n} \sum_{s=1}^{n} \mathscr{E}(X_{r:n}^k X_{s:n}^l) = n(n - 1)\mathscr{E}(X^k)\mathscr{E}(X^l)$$

and hence that

$$\sum_{r=1}^{n-1} \sum_{s=r+1}^{n} \mathscr{E}(X_{r:n}^k X_{s:n}^k) = \binom{n}{2}[\mathscr{E}(X^k)]^2.$$

3.4.1. By repeated application of Relation 1 show that for any arbitrary distribution

$$(n - r)_m \mu_{r:n}^{(k)} = \sum_{i=0}^{m} (-r)_i (n)_{m-i} \binom{m}{i} \mu_{r+i:n-m+i}^{(k)},$$

where, for example, $(n - r)_m$ denotes $(n - r)(n - r - 1) \cdots (n - r - m + 1)$.
Hence obtain Relation 2 as the special case $m = n - r$.

3.4.2. Prove that for an arbitrary distribution and $n \geq m$

$$\binom{n}{m} \mu_{r:m} = \sum_{i=0}^{n-m} \binom{n-r-i}{m-r} \binom{r+i-1}{i} \mu_{r+i:n}.$$

(Sillitto, 1964)

3.4.3. By direct use of the integral definitions of $\mu_{r:n}$ and $\mu_{rs:n}$ for a continuous parent, show that for an arbitrary distribution

$$\sum_{i=1}^{n} (i - 1)^{(k)} (n - i)^{(l)} \mu_{i:n} = k! l! \binom{n}{k + l + 1} \mu_{k+1:k+l+1},$$

$$\sum_{i=1}^{n} (i - 1)^{(k)} (n - i)^{(l)} \mu_{ii:n} = k! l! \binom{n}{k + l + 1} \mu_{k+1,k+1:k+l+1},$$

$$\sum_{i<j} (i - 1)^{(k)} (n - j)^{(l)} \mu_{ij:n} = k! l! \binom{n}{k + l + 2} \mu_{k+1,k+2:k+l+2}.$$

(Note that the first identity is equivalent to the result in Ex. 3.4.2.)

(Downton, 1966b)

3.4.4. Let $\chi_{n,r} = \mathscr{E}(X_{r+1:n} - X_{r:n})$, $\omega_n = \mathscr{E} W_n$. Show that for an arbitrary distribution

$$\omega_n = \frac{1}{n} (\chi_{n,1} + \chi_{n,n-1}) + \omega_{n-1}.$$

Deduce that

$$\mathscr{E}(X_{n-1:n} - X_{2:n}) = n\omega_{n-1} - (n-1)\omega_n.$$

(Sillitto, 1951; Cadwell, 1953a)

3.4.5. Prove that for an arbitrary distribution

$$n\chi_{n-1,r-1} - (n-r+1)\chi_{n,r-1} = r\chi_{n,r},$$

and, by repeated application of this result, that for $v \leq r-1$

$$\chi_{n,r} = \frac{(n)_v}{(r)_v} \sum_{i=0}^{v} (-1)^i \binom{v}{i} \frac{(n-r+i)_i}{(n-v+i)_i} \chi_{n-v+i,r-v}.$$

(Sillitto, 1951)

3.4.6. Show that for an arbitrary distribution

$$\binom{n}{r} \sum_{i=0}^{r} (-1)^{i+1} \binom{r}{i} \omega_{n-r+i} = \chi_{n,n-r} + \chi_{n,r}.$$

Hence prove that the value of the mean range for n odd can be deduced from the values for smaller n by

$$2\omega_n = \omega_{n-1} + \sum_{i=1}^{n-2} (-1)^{i-1} \binom{n-1}{i} \omega_{n-i}, \tag{A}$$

a result which includes the special case

$$\omega_3 = \tfrac{3}{2}\omega_2.$$

Directly from (3.1.11) obtain also the following two formulae equivalent to (A):

$$2\omega_n = \sum_{i=1}^{n-2} (-1)^{i-1} \binom{n}{i} \omega_{n-i} \qquad n \text{ odd}$$

and

$$\lambda_{2m} = \sum_{i=0}^{m-1} (-1)^{m+i+1} \binom{m}{i} \lambda_{m+i} \qquad m = 1, 2, 3, \ldots$$

where $\lambda_r = \omega_{r+1}/(2r+2)$.

(Sillitto, 1951; cf. Romanovsky, 1933)

CHAPTER 4

Bounds and Approximations for Moments of Order Statistics

4.1. INTRODUCTION

In this chapter we consider a number of general approaches, some of decided mathematical interest, giving bounds and approximations to the moments of order statistics. Sections 4.2 and 4.3 deal with the use of Schwarz's inequality and some of its generalizations. If a variate X has a finite variance, the expected values of the extremes $X_{(n)}$ and $X_{(1)}$ (and a fortiori of other order statistics) cannot be arbitrarily large even if the range of X is unbounded. A bound can be found which, in the case of the extremes, is attainable for a certain class of cdf's. Better bounds can be obtained for symmetrical cdf's. In the case of order statistics other than the extremes, bounds obtained in this manner are not attainable but can be improved by the use of a generalized Schwarz inequality. Improvement of a different kind gives approximations with known error bounds for the expected values of all order statistics drawn from a parent distribution for which the expected value of the largest is available in small samples.

It has long been known that the expected value of an order statistic can be approximated by the appropriate population quantile, especially in large samples. In Section 4.4 we consider conditions under which it can be stated whether such an approximation is an overestimate or an underestimate, thus allowing large-sample approximations to be replaced by inequalities which are valid for all sample sizes. Some related inequalities are also derived.

Sections 4.2–4.4 are mainly of theoretical interest. However, Section 4.5 deals with a simple device, based on a Taylor expansion in powers of $1/n$, which frequently provides reasonable approximations to the means, variances, and covariances of order statistics. The first term of such a series is then the asymptotic result. In the case of $\mathscr{E}X_{(r)}$ this is simply the quantile approximation mentioned above; later terms provide (under suitable conditions) successive improvement when n is finite. These later terms are less tractable, and modifications of the quantile approximation for use in finite samples are also considered.

4.2. DISTRIBUTION-FREE BOUNDS FOR THE MOMENTS OF ORDER STATISTICS AND OF THE RANGE

We begin by considering the expected value of the largest order statistic in random samples of n from a parent with continuous strictly increasing cdf $P(x)$. In place of

$$\mathscr{E}X_{(n)} = \int_{-\infty}^{\infty} nx[P(x)]^{n-1} \, dP(x)$$

it is convenient to use the alternative form obtained by the probability integral transformation $u = P(x)$, namely,

$$\mathscr{E}X_{(n)} = \int_{0}^{1} nx(u)u^{n-1} \, du, \tag{4.2.1}$$

where $x(u)$ indicates that x is now expressed as a function of u. Let us standardize the variate X to have mean 0 and variance 1, that is,

$$\int_{0}^{1} x(u) \, du = 0, \qquad \int_{0}^{1} [x(u)]^{2} \, du = 1. \tag{4.2.2}$$

This entails no loss of generality provided the parent distribution possesses a second moment. Then it turns out that $\mathscr{E}X_{(n)}$ is bounded, whatever the form of $P(x)$ (Gumbel, 1954; Hartley and David, 1954). From the calculus of variations we can find the extremal $x(u)$ giving stationary values of (4.2.1) subject to (4.2.2) by first obtaining the unconditional extremal for

$$\int_{0}^{1} (nxu^{n-1} - ax - \tfrac{1}{2}bx^2) \, du,$$

and then determining the constants a and b so as to satisfy (4.2.2). The stationary solution is given by

$$\frac{\partial}{\partial x} (nxu^{n-1} - ax - \tfrac{1}{2}bx^2) = 0,$$

that is,

$$bx = nu^{n-1} - a,$$

where

$$\int_{0}^{1} (nu^{n-1} - a) \, du = 0, \qquad \int_{0}^{1} (nu^{n-1} - a)^2 \, du = b^2.$$

Thus $a = 1$, $b = (n - 1)/(2n - 1)^{1/2}$,

$$x(u) = \frac{(2n - 1)^{1/2}(nu^{n-1} - 1)}{n - 1}, \tag{4.2.3}$$

and for this extremal

$$\mathscr{E} X_{(n)} = \frac{n(2n - 1)^{1/2}}{n - 1} \int_0^1 u^{n-1}(nu^{n-1} - 1) \, du$$

$$= \frac{n - 1}{(2n - 1)^{1/2}}.$$

The calculus of variations is useful in suggesting the form of the solution; it is not sufficient or even necessary. To show that (4.2.3) leads to a maximum of $\mathscr{E} X_{(n)}$—and not merely to a stationary value—we use Schwarz's inequality:

$$\int fg \, du \leq (\int f^2 \, du \int g^2 \, du)^{1/2}$$

with

$$f = x, \qquad g = nu^{n-1} - 1.$$

This gives

$$\mathscr{E} X_{(n)} \leq \left[1 \cdot \int_0^1 (n^2 u^{2n-2} - 2nu^{n-1} + 1) \, du \right]^{1/2}$$

and hence

$$\mathscr{E} X_{(n)} \leq \frac{n - 1}{(2n - 1)^{1/2}}. \tag{4.2.4}$$

Equality occurs when $x(u)$ is as in (4.2.3) or, on inverting, when

$$u = P(x) = \left(\frac{1 + bx}{n} \right)^{1/(n-1)} - \frac{(2n - 1)^{1/2}}{n - 1} \leq x \leq (2n - 1)^{1/2}. \tag{4.2.5}$$

Although in this argument beginning with (4.2.2) we have not specifically required $P(x)$ to be a cdf, we see from (4.2.5) that it satisfies all the conditions. The corresponding pdf

$$p(x) = \frac{b}{n} \left(\frac{1 + bx}{n} \right)^{1/(n-1)-1} - \frac{(2n - 1)^{1/2}}{n - 1} \leq x \leq (2n - 1)^{1/2},$$

is shown in Fig. 4.2.1 for various n. If the mean and variance of the parent are μ and σ^2 (rather than 0 and 1), (4.2.4) simply becomes

$$\mathscr{E} X_{(n)} \leq \mu + \frac{(n - 1)\sigma}{(2n - 1)^{1/2}}, \tag{4.2.6}$$

and likewise

$$\mathscr{E} X_{(1)} \geq \mu - \frac{(n - 1)\sigma}{(2n - 1)^{1/2}}.$$

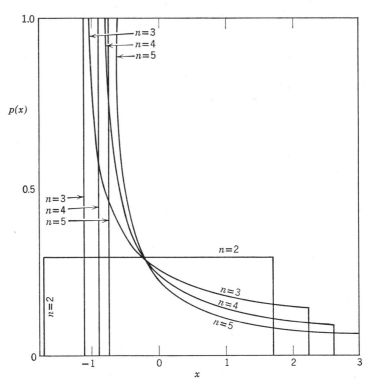

Fig. 4.2.1. Extremal pdf (general parent) for sample size n maximizing $\mathscr{E} X_{(n)}$ (from Gumbel, 1954, with permission of the Editor of the *Annals of Mathematical Statistics*).

For the class of symmetric parent distributions inequality (4.2.4) can be sharpened. Since (after taking $\mu = 0$) we have $P(x) = 1 - P(-x)$, it follows that

$$\mathscr{E} X_{(n)} = \int_0^\infty nx\{[P(x)]^{n-1} - [1 - P(x)]^{n-1}\}\, dP(x)$$

$$= \int_{1/2}^1 nx[u^{n-1} - (1 - u)^{n-1}]\, du. \qquad (4.2.7)$$

The same approach as before applied to (4.2.7) now yields the extremal

$$cx(u) = u^{n-1} - (1 - u)^{n-1}, \qquad (4.2.8)$$

where

$$c = \left\{ \frac{2\left[1 - 1 \middle/ \binom{2n-2}{n-1}\right]}{2n-1} \right\}^{1/2},$$

and the inequality

$$\mathscr{E}X_{(n)} \leq \tfrac{1}{2}nc. \tag{4.2.9}$$

We see directly from (4.2.8) that x is confined to the range $(-1/c, 1/c)$, that is, $P(x)$ is again a finite-range distribution. It is of some interest to note that by (4.2.5) and (4.2.8) the uniform distribution over $(-\sqrt{3}, \sqrt{3})$ is the extremal distribution for $n = 2$ and, in the class of symmetrical cdf's, also for $n = 3$. Figure 4.2.2 shows the symmetrical extremal pdf for various n. In Table 4.2 the two upper bounds of $\mathscr{E}X_{(n)}$ are given for $n \leq 20$ and are compared with the expected value of a unit normal variate. It will be noted that the symmetrical upper bound is not much above the normal-theory values.

Since for any continuous parent we have

$$\mathscr{E}W_n = \int_0^1 nx[u^{n-1} - (1 - u)^{n-1}]\, du,$$

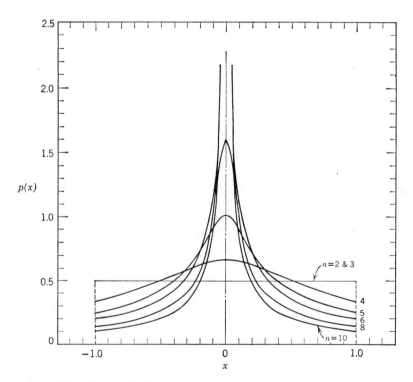

Fig. 4.2.2. Extremal pdf (symmetrical parent) maximizing $\mathscr{E}X_{(n)}$ (from Moriguti, 1951, with permission of the author and the Editor of the *Annals of Mathematical Statistics*).

Table 4.2.   *Comparison of two upper bounds of $\mathscr{E}(X_{(n)} - \mu)/\sigma$ with exact values in normal and uniform parents*

n	Upper Bound (any Parent)	Upper Bound (Symmetrical Parent)	Normal Parent	Uniform Parent
2	.5774	.5774	.5642	.5774
3	.8944	.8660	.8463	.8660
4	1.1339	1.0420	1.0294	1.0392
5	1.3333	1.1701	1.1630	1.1547
6	1.5076	1.2767	1.2672	1.2372
7	1.6641	1.3721	1.3522	1.2990
8	1.8074	1.4604	1.4236	1.3472
9	1.9403	1.5434	1.4850	1.3856
10	2.0647	1.6222	1.5388	1.4171
12	2.2937	1.7693	1.6292	1.4656
15	2.5997	1.9696	1.7359	1.5155
20	3.0424	2.2645	1.8673	1.5671
50	4.9247	3.5533	2.2491	1.6641
100	7.0179	5.0125	2.5076	1.6978
1000	22.3439	15.8153	3.2414	1.7286

(From Moriguti, 1951; Hartley and David, 1954; and Tippett, 1925)

it follows on comparison with (4.2.7) that the extremal distribution leading to a maximum of $\mathscr{E}W_n$ is (4.2.8), whether the parent is symmetrical or not, and that

$$\mathscr{E}W_n \leq nc.$$

This was actually the first bound obtained in the above sequence (Plackett, 1947).

Our derivations have started with the assumption of a strictly increasing continuous cdf. It is clear, however, that the extremals and bounds continue to hold for all cdf's possessing a variance, even if we admit discrete parents, for all cdf's can be approximated with arbitrary precision by strictly increasing continuous cdf's.

We turn now to some extensions. Moriguti (1951, 1954) considers the extreme and the range for a symmetrical parent. In both cases he derives the upper bound for the expectation, and lower bounds for the variance and the coefficient of variation.

All the foregoing results involve extreme order statistics only. Since

$$\mathscr{E}X_{(r)} = \int_0^1 x(u)i_u \, du, \tag{4.2.10}$$

where

$$i_u = \frac{d}{du} I_u(r, n - r + 1) = \frac{n!}{(r-1)!(n-r)!} u^{r-1}(1-u)^{n-r}, \quad (4.2.11)$$

we find as before, on applying conditions (4.2.2),

$$|\mathcal{E}(X_{(r)})| \leq \left[n \frac{\binom{2n-2r}{n-r}\binom{2r-2}{r-1}}{\binom{2n-1}{n-1}} - 1 \right]^{\frac{1}{2}}, \quad (4.2.12)$$

and also, for $s > r$ (Ludwig, 1960),

$$\mathcal{E}(X_{(s)} - X_{(r)}) \leq \left\{ \frac{2n}{\binom{2n}{n}} \left[\binom{2s-2}{s-1}\binom{2n-2s}{n-s} + \binom{2r-2}{r-1}\binom{2n-2r}{n-r} \right. \right.$$

$$\left. \left. - 2\binom{r+s-2}{r-1}\binom{2n-r-s}{n-s} \right] \right\}^{\frac{1}{2}}. \quad (4.2.13)$$

The last result follows also directly from Schwarz's inequality with

$$f = x, \quad g = i_u(s, n - s + 1) - i_u(r, n - r + 1).$$

Numerical values of the bounds in (4.2.12) and (4.2.13) for $n \leq 10$ are given by Ludwig (1959). However, as was first pointed out by Moriguti (1953a), these bounds are sharp ($=$ attainable) only for the extremes (i.e., $r = n$ or 1 in (4.2.12) and $r = 1$, $s = n$ in (4.2.13)). The reason is easy to see. For example, (4.2.10) attains its upper bound only when $x(u) \propto i_u - 1$. But if $u = P(x)$ is to be a cdf, then $x(u)$ and hence i_u must be monotonic in u, which is true only for $r = 1$ or n.

In the course of a more general study Moriguti shows that sharp bounds for $\mathcal{E}X_{(r)}$ result if $i_u - 1$ is replaced by a closely related *nondecreasing* function before the application of Schwarz's inequality. More precisely, write (4.2.10) in the form

$$\mathcal{E}X_{(r)} = \int_0^1 x(u) \, dI_u(r, n - r + 1).$$

Now replace I_u by $\bar{I}_u$, its "greatest convex minorant" in the interval $(0, 1)$; that is, $\bar{I}_u$ is the supremum of all convex functions dominated by I_u throughout $0 \leq u \leq 1$. (A convex function is characterized by the fact that any chord of its graph lies on or above the graph.) It can be shown that $\bar{I}_u$ is continuous and has a right-hand derivative $\bar{i}_u$ which, in addition to being nondecreasing by the convexity of $\bar{I}_u$, is continuous, except possibly at a denumerable set of values of u. Since $\bar{I}_u$ is clearly nondecreasing and $\bar{I}_0 = 0$,

$\bar{I}_1 = 1$, it is a distribution function—in fact, the cdf of a variate that is stochastically larger than $X_{(r)}$. It follows that

$$\mathscr{E} X_{(r)} \leq \int_0^1 x(u) \, d\bar{I}_u = \int_0^1 x(u)\bar{i}_u \, du. \qquad (4.2.14)$$

Since on integration by parts

$$\mathscr{E} X_{(r)} - \int_0^1 x(u) \, d\bar{I}_u = \int_0^1 x(u) \, d(I_u - \bar{I}_u)$$

$$= -\int_0^1 (I_u - \bar{I}_u) \, dx(u),$$

it follows that equality holds in (4.2.14) only if $x(u)$ is constant whenever $I_u > \bar{I}_u$, and that

$$\mathscr{E}(X_{(r)} - \mu) \leq \int_0^1 x(u)(\bar{i}_u - 1) \, du = \int_0^1 [x(u) - \mu](\bar{i}_u - 1) \, du$$

$$\leq \left\{ \int_0^1 [x(u) - \mu]^2 \, du \int_0^1 (\bar{i}_u - 1)^2 \, du \right\}^{1/2}. \quad (4.2.15)$$

Thus

$$\frac{\mathscr{E}(X_{(x)} - \mu)}{\sigma} \leq \left\{ \int_0^1 (\bar{i}_u - 1)^2 \, du \right\}^{1/2}. \qquad (4.2.16)$$

To determine $\bar{I}_u$ (and hence $\bar{i}_u$) consider Fig. 4.2.3, in which the solid curve

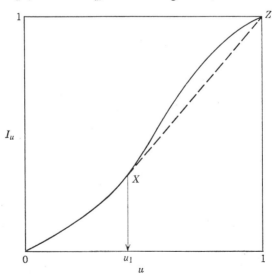

Fig. 4.2.3

represents I_u as a function of u. The greatest convex minorant $\bar{I}_u$ initially is equal to I_u and then continues as the dotted line XZ, the tangent from Z to I_u. The value u_1 of u at the point of contact X clearly satisfies

$$1 - I_{u_1} = i_{u_1}(1 - u_1),$$

that is,

$$1 - I_{u_1} = \frac{n!}{(r-1)!(n-r)!} u_1^{r-1}(1 - u_1)^{n-r+1}, \qquad (4.2.17)$$

which can be solved numerically. Correspondingly, $\bar{i}_u$ is given by

$$\begin{aligned} \bar{i}_u &= i_u & 0 \le u < u_1, \\ &= i_{u_1} & u_1 \le u \le 1, \end{aligned} \qquad (4.2.18)$$

so that the upper bound in (4.2.14) can be calculated.

Now (4.2.15) is an equality when

$$x(u) - \mu \propto \bar{i}_u - 1,$$

so that $x(u)$ is constant for $u_1 \le u \le 1$, which in turn yields equality in (4.2.14). Thus bound (4.2.16) is attained for $x(u) - \mu = c(\bar{i}_u - 1)$, where c is a constant, and we see that the maximizing $P(x)$ is the cdf of a variate continuous in $(\mu - c, \mu - c + ci_{u_1})$, with the remaining probability concentrated at $x = \mu - c + ci_{u_1}$.

Example 4.2. Consider the simple case of the median in samples of 3. Then u_1 is given by

$$1 - I_{u_1}(2, 2) = 6u_1(1 - u_1)^2,$$

which reduces to

$$4u_1^3 - 9u_1^2 + 6u_1 - 1 = 0,$$

so that $u_1 = \frac{1}{4}$. Thus

$$\begin{aligned} \bar{i}_u &= 6u(1 - u) & 0 \le u < \tfrac{1}{4}, \\ &= \tfrac{9}{8} & \tfrac{1}{4} \le u \le 1, \end{aligned}$$

and by (4.2.14) we obtain

$$\frac{\mathscr{E}(X_{(2)} - \mu)}{\sigma} \le 0.271.$$

This upper bound is considerably lower than the value 0.447 given by (4.2.12).

Moriguti (1953a) compares the two bounds for the expected value of the sample median for odd n up to $n = 19$ and finds that the discrepancy increases with n, the corresponding figures for $n = 19$ being 0.598 and 1.242. Of course, one would expect the simple Schwarz inequality to lose steadily in sharpness as it is applied to order statistics $X_{(r)}$ increasingly removed from the extremes.

No equally general results are possible for *lower* bounds of $\mathscr{E}(X_{(r)} - \mu)/\sigma$ ($r > [\tfrac{1}{2}n]$) or $\mathscr{E}(X_{(s)} - X_{(r)})/\sigma$ ($s > r$). Since for the existence of the numerators the existence of μ suffices, it is clear that the lower bounds may be made arbitrarily close to zero by choice of a parent cdf $P(x)$ with sufficiently large σ. More worthwhile lower bounds can be obtained only by imposing suitable conditions on $P(x)$. One possible condition which reflects a frequently occurring practical situation is to limit the range of X, say $a \leq X \leq b$, with a and b finite. For this restriction the case of the range has been investigated in detail by Hartley and David (1954), who find that the minimizing distribution is a two-point distribution. With $a = -c$, $b = c$ they obtain and provide a short table of the lower bound, namely,

$$\mathscr{E}\left(\frac{W_n}{\sigma}\right) \geq \min \begin{cases} 2[1 - (\tfrac{1}{2})^{n-1}], \\ (1 - p^n - q^n)/(pq)^{1/2}, \end{cases}$$

where $p = c^2/(1 + c^2)$ and $q = 1 - p$. It may be noted that as $c \to \infty$ the distribution becomes increasingly skew and the lower bound tends to 0.

A more general discussion of bounds (lower and upper) in the case $-c \leq X \leq c$ has been given by Rustagi (1957). See also Karlin and Studden (1966, Chapter 14).

4.3. BOUNDS AND APPROXIMATIONS BY ORTHOGONAL INVERSE EXPANSION

An interesting approach which provides both bounds and approximations for the means, variances, and covariances of order statistics has been given by Sugiura (1962, 1964). Let $\{\psi_k(u)\}$ ($k = 0, 1, 2, \dots$) be an orthonormal system over the interval $(0, 1)$, that is, $\psi_0(u) = 1$ and for all positive integral k, k' ($k' \neq k$)

$$\int_0^1 \psi_k(u)\,du = 0, \quad \int_0^1 \psi_k^2(u)\,du = 1, \quad \int_0^1 \psi_k(u)\psi_{k'}(u)\,du = 0.$$

Write

$$a_k = \int_0^1 f(u)\psi_k(u)\,du, \quad b_k = \int_0^1 g(u)\psi_k(u)\,du,$$

where f, g are square-integrable functions over $(0, 1)$. Then by Schwarz's inequality we have, in slightly simplified notation,

$$\left| \int \left(f - \sum_{k=0}^m p_k\psi_k\right)\left(g - \sum_{k=0}^m g_k\psi_k\right) du \right|$$
$$\leq \left\{ \int \left(f - \sum_{k=0}^m a_k\psi_k\right)^2 du \cdot \int \left(g - \sum_{k=0}^m b_k\psi_k\right)^2 du \right\}^{1/2},$$

which immediately reduces to the basic inequality

$$\left| \int fg \, du - \sum_{k=0}^{m} a_k b_k \right| \leq \left(\int f^2 \, du - \sum_{k=0}^{m} a_k^2 \right)^{\frac{1}{2}} \left(\int g^2 \, du - \sum_{k=0}^{m} b_k^2 \right)^{\frac{1}{2}}. \quad (4.3.1)$$

Equality holds if

$$f - \sum_{k=0}^{m} a_k \psi_k \propto g - \sum_{k=0}^{m} b_k \psi_k.$$

Thus $\sum_{k=0}^{m} a_k b_k$ provides an approximation to $\int fg \, du$ whose maximum error is a function of only the coefficients a_k, b_k ($k = 0, 1, 2, \ldots, m$) used in the approximation. If, moreover, $\{\psi_k\}$ is a complete orthonormal system, then $\sum_{k=0}^{\infty} a_k^2 = \int f^2 \, du$ and $\sum_{k=0}^{\infty} b_k^2 = \int g^2 \, du$. The right side of (4.3.1) therefore converges to 0 as $m \to \infty$, and the approximation may be made as accurate as we please.

Theorem 4.3.1   *Let $u = P(x)$ be the (strictly increasing) continuous cdf of a standardized variate and let $\psi_0 = 1$, $\psi_1, \ldots, \psi_m$ be any orthonormal system over $(0, 1)$. Put*

$$a_k = \int_0^1 x(u)\psi_k(u) \, du, \quad b_k = \frac{1}{B(r, n - r + 1)} \int_0^1 u^{r-1}(1 - u)^{n-r}\psi_k(u) \, du;$$

then

$$\left| \mathscr{E}X_{r:n} - \sum_{k=1}^{m} a_k b_k \right| \leq \left(1 - \sum_{k=1}^{m} a_k^2 \right)^{\frac{1}{2}} \left\{ \frac{B(2r - 1, 2n - 2r + 1)}{[B(r, n - r + 1)]^2} - 1 - \sum_{k=1}^{m} b_k^2 \right\}^{\frac{1}{2}}.$$
$$(4.3.2)$$

Proof. In (4.3.1) take

$$f = x(u), \quad g = \frac{1}{B(r, n - r + 1)} u^{r-1}(1 - u)^{n-r}. \quad (4.3.3)$$

The theorem follows at once since $a_0 = 0$, $b_0 = 1$, and

$$\int fg \, du = \mathscr{E}X_{r:n}, \quad \int f^2 \, du = 1, \quad \int g^2 \, du = \frac{B(2r - 1, 2n - 2r + 1)}{[B(r, n - r + 1)]^2}. \quad (4.3.4)$$
▶

An example of a complete orthonormal system over $(0, 1)$ is the sequence of Legendre polynomials [adjusted to $(0, 1)$]

$$L_k(u) = \frac{(2k + 1)^{\frac{1}{2}}}{k!} \frac{d^k}{du^k} u^k(u - 1)^k \quad k = 0, 1, 2, \ldots.$$

If the orthonormal system is taken to be just ψ_0, the present approach clearly reduces to the simple use of Schwarz's inequality, and (4.3.2) becomes (4.2.12). It is by taking suitable additional members of the orthonormal system that further improvements in the bound may be possible, while at the same time $\sum_{k=1}^{m} a_k b_k$ gives a manageable approximation to $\mathscr{E}X_{r:n}$. This process will be illustrated more fully in the case when $P(x)$ is the cdf of a symmetric standardized variate. We will need the known results that in the class of (i) even and (ii) odd functions, square-integrable in $(0, 1)$, complete orthonormal systems are given respectively by the Legendre functions

$$
\begin{aligned}
&\text{(i) } \{L_{2k}(u)\}, \\
&\text{(ii) } \{L_{2k+1}(u)\},
\end{aligned}
\quad k = 0, 1, 2, \ldots. \tag{4.3.5}
$$

We now obtain, corresponding to (4.3.2),

$$
\left| \mathscr{E}X_{r:n} - \sum_{k=0}^{m} a_{2k+1} b_{2k+1} \right| \leq \left(1 - \sum_{k=0}^{m} a_{2k+1}^2 \right)^{\!1/2}
$$
$$
\times \left[\frac{\mathrm{B}(2r - 1, 2n - 2r + 1) - \mathrm{B}(n, n)}{2[\mathrm{B}(r, n - r + 1)]^2} - \sum_{k=0}^{m} b_{2k+1}^2 \right]^{\!1/2}. \tag{4.3.6}
$$

Proof. By our assumptions on $u = P(x)$, the inverse function $x(u)$ is odd and square-integrable over $(0, 1)$. Then, since $L_k(1 - u) = (-1)^k L_k(u)$,

$$
\begin{aligned}
a_{2k} &= \int_0^1 x(u) L_{2k}(u)\, du \\
&= \int_0^1 -x(1 - u) L_{2k}(1 - u)\, du \\
&= -\int_0^1 x(v) L_{2k}(v)\, dv \qquad v = 1 - u \\
&= -a_{2k}.
\end{aligned}
$$

Thus

$$
a_{2k} = 0.
$$

With f and g as in (4.3.3), apply (4.3.1) with $k = 1, 3, \ldots, 2m + 1, 0, 2, 4, 6, \ldots$, giving

$$
\left| \int fg\, du - \sum_{k=0}^{m} a_{2k+1} b_{2k+1} \right|
$$
$$
\leq \left(\int f^2\, du - \sum_{k=0}^{m} a_{2k+1}^2 \right)^{\!1/2} \left(\int g^2\, du - \sum_{k=0}^{m} b_{2k+1}^2 - \sum_{k=0}^{\infty} b_{2k}^2 \right)^{\!1/2}. \tag{4.3.7}
$$

To evaluate the term Σb_{2k}^2, representing the reduction in the upper bound due to symmetry, put

$$
g^*(u) = \frac{1}{\mathrm{B}(r, n - r + 1)} u^{n-r}(1 - u)^{r-1}.
$$

Then

$$
b_{2k} = \int g L_{2k}\, du = \tfrac{1}{2} \int (g + g^*) L_{2k}\, du.
$$

Since $g(u) + g^*(u)$ is an even square-integrable function over $(0, 1)$, it follows from (i) of (4.3.5) that

$$\sum_{k=0}^{\infty} b_{2k}^2 = \tfrac{1}{4} \int [g(u) + g^*(u)]^2 \, du$$

$$= \frac{1}{2[(B(r, n - r + 1)]^2} \, [B(2r - 1, 2n - 2r + 1) + B(n, n)].$$

Substituting this and (4.3.4) in (4.3.7) yields (4.3.6).   ▶

For the calculation of the approximation $\sum_{k=0}^{m} a_{2k+1} b_{2k+1}$ and the associated error, note that the Legendre functions are

$$L_0(u) = 1, \quad L_1(u) = \sqrt{3}(2u - 1), \quad L_2(u) = \sqrt{5}(6u^2 - 6u + 1),$$

$$L_3(u) = \sqrt{7}(20u^3 - 30u^2 + 12u - 1), \quad \text{etc.}$$

In general, put

$$L_k(u) = \sum_{i=0}^{k} a_{k,i} u^i.$$

Then the coefficients a_k and b_k are given by

$$a_k = \sum_{i=0}^{k} a_{k,i} \int_0^1 u^i x(u) \, du$$

$$= \sum_{i=0}^{k} a_{k,i} \frac{\mathscr{E} X_{i+1:i+1}}{i + 1},$$

$$b_k = \sum_{i=0}^{k} a_{k,i} \frac{r(r + 1) \cdots (r + i - 1)}{(n + 1)(n + 2) \cdots (n + i)}.$$

The approximation therefore requires evaluation of the expected values of the maximum in small samples. In the normal case, $a_1 b_1 + a_3 b_3 + a_5 b_5$ gives reasonable results for $n = 10$ and 50, as seen from the column headed "3rd Bound" in Table 4.3. As is to be expected, bounds of the same degree are sharper for the smaller n. No easy pattern is discernible in the behavior of the associated errors for given n as r varies. See also Joshi (1969).

4.4. BOUNDS FOR THE EXPECTED VALUES OF ORDER STATISTICS IN TERMS OF QUANTILES OF THE PARENT DISTRIBUTION

It is intuitively obvious and has long been known that for sufficiently large n an approximation to $\mathscr{E}(X_{r:n})$ is provided by the value of x satisfying

$$P(x) = \frac{r}{n + 1}.$$

Table 4.3.   Upper and lower bounds for $\mathscr{E}X_{r:n}$
(unit normal parent)

r	1st Bound	2nd Bound	3rd Bound	Exact Value
		$n = 10$		
6	.15 ± .06	.113 ± .016	.1246 ± .0032	.12267
7	.46 ± .13	.357 ± .028	.3775 ± .0024	.37576
8	.77 ± .14	.651 ± .008	.6527 ± .0048	.65606
9	1.08 ± .09	1.030 ± .035	1.0032 ± .0026	1.00136
10	1.38 ± .17	1.527 ± .015	1.5384 ± .0005	1.53875
		$n = 50$		
26	.03 ± .04	.020 ± .020	.0278 ± .0112	.02496
30	.30 ± .27	.189 ± .118	.2468 ± .0586	.22653
35	.63 ± .28	.444 ± .092	.5036 ± .0321	.49354
40	.96 ± .26	.798 ± .091	.7770 ± .0564	.80225
45	1.29 ± .26	1.302 ± .133	1.2180 ± .0479	1.21846
50	1.63 ± .67	2.007 ± .253	2.1556 ± .1044	2.24907

(Results for $n = 10$ from Sugiura, 1962, with permission of the author and the Editor of the *Osaka Journal of Mathematics*.)

If for greater clarity we now denote the inverse function $x(P)$ by $Q(P)$, that is, $Q[P(x)] = x$, we have asymptotically

$$\mathscr{E}(X_{r:n}) \sim Q\left(\frac{r}{n+1}\right). \tag{4.4.1}$$

This kind of quantile approximation is discussed in the next section. Here, following van Zwet (1964), we establish several inequalities closely related to (4.4.1) but applicable even in small samples. The following definitions are needed.

Elaborating on the definition in Section 4.2, we shall call the real-valued function $g(x)$, defined on some nondegenerate interval I, *convex* on I if for all x_1, x_2 in I and $0 \leq \lambda \leq 1$

$$g[\lambda x_1 + (1 - \lambda)x_2] \leq \lambda g(x_1) + (1 - \lambda)g(x_2). \tag{4.4.2}$$

Clearly, if $g(x)$ is convex, then $g'(x)$ is nondecreasing and $g''(x) \geq 0$, provided these derivatives exist. Also, for every interior point x_0 of I, there exists a straight line L having $L(x_0) = g(x_0)$ and lying wholly on or below the graph of g. We speak of L as a *line of support* of g at $x = x_0$. If (4.4.2) holds with the inequality sign reversed, g is *concave* on I. Note that a linear function is both convex and concave. A function g on I will be called *antisymmetrical* on I

about x_0 if

$$g(x_0 + x) + g(x_0 - x) = 2g(x_0)$$

for some x_0 in I and for all x with both $x_0 - x$ and $x_0 + x$ in I; x_0 will be called a *central point* of g. An antisymmetrical function g on I will be called *concave-convex* on I if g is concave for $x \leq x_0$ and convex for $x \geq x_0$, x in I.

The various definitions ensure that g is continuous except possibly at the end points of I and, in the case of a concave-convex function, at x_0. If we suppose g to be nondecreasing, continuity at x_0 follows.

All the above definitions are purely mathematical. Now take I as the largest interval for which the random variable X has cdf $P(x)$ such that $0 < P(x) < 1$. (The case X constant is excluded by the nondegeneracy of I.) We now prove the important

Jensen's Inequality.   *If g is convex on I, then*

$$g(\mathscr{E}X) \leq \mathscr{E}[g(X)]$$

provided both expectations exist. Equality holds if and only if g is linear on I.

Proof.   Let L be a line of support of g at $x = \mathscr{E}X$. Since $L(x) \leq g(x)$ on I and L is linear, we have

$$\mathscr{E}[g(X)] \geq \mathscr{E}[L(X)] = L(\mathscr{E}X) = g(\mathscr{E}X).$$

If g is linear on I, equality holds trivially. Conversely, for equality, $g(x) = L(x)$ must clearly hold almost everywhere. By convexity, g is continuous on I and must therefore be linear throughout I.   ▶

Now consider the class $\mathscr{P}$ of cdf's $P(x)$ which have a positive continuous derivative $p(x)$ over some interval I.[1] Then $Q(P)$ is uniquely defined for $0 < P < 1$ and has a positive continuous derivative in this range.

Lemma A.   *For any pair of cdf's P, P^* in $\mathscr{P}$ there exists a strictly increasing function g on I such that, if X has cdf P, then $g(X)$ has cdf P^*. The function g is uniquely defined on I by $g(x) = Q^*[P(x)]$ and has a continuous derivative on I.*

Proof.   It is necessary and sufficient that g should satisfy

$$P^*[g(x)] = \Pr\{g(X) \leq g(x)\} = \Pr\{X \leq x\} = P(x)$$

for all x in I, and be strictly increasing on I. Obviously $g(x) = Q^*[P(x)]$ is the only function meeting the first requirement. Since Q^* is strictly increasing in its argument $P(x)$, itself strictly increasing in x, $g(x)$ is strictly increasing on I. Also $g(x)$ has a continuous derivative on I by the remark preceding Lemma A.   ▶

---

[1] Van Zwet lists further restrictions not needed for our purposes.

We now define an ordering relation for cdf's in $\mathscr{P}$: If P, P^* are in $\mathscr{P}$, then $P \underset{c}{<} P^*$ if and only if $Q^*(P)$ is convex on I. The letter c stands for convex and P is said to *c-precede* P^*. From the lemma it is clear that $P \underset{c}{<} P^*$ if and only if a variate with cdf P may be transformed into one with cdf P^* by an increasing convex transformation. Obviously $P \underset{c}{<} P$. Since an increasing, convex function of a convex function is again convex, $P \underset{c}{<} P^* \underset{c}{<} P^{**}$ yields $P \underset{c}{<} P^{**}$. The relation $\underset{c}{<}$ is thus a weak ordering on $\mathscr{P}$, and we speak of *c-ordering* or *c-comparison*. An equivalence relation $\sim$ may now be defined: If P, P^* are in $\mathscr{P}$, then $P \sim P^*$ if and only if $P \underset{c}{<} P^*$ and $P^* \underset{c}{<} P$.

Lemma B. *If P, P^* are in $\mathscr{P}$, then $P \sim P^*$ if and only if $P(x) = P^*(ax + b)$ for some constants $a > 0$ and b.*

Proof. $P \sim P^*$ if and only if both $Q^*(P)$ and $Q(P^*)$ are convex on I. But the convexity of $Q(P^*)$ is equivalent to the concavity of $Q^*(P)$, which, being both convex and concave, must be linear. Thus $Q^*[P(x)] = ax + b$ or $P(x) = P^*(ax + b)$ with $a > 0$, since P^* is increasing in x. ▶

This lemma asserts that c-ordering is independent of location and scale parameters. Attention may therefore be confined to standardized cdf's. In order to establish that $P \underset{c}{<} P^*$ it is useful to have convenient criteria for the convexity of $Q^*(P)$.

Lemma C. *If P, P^* are in $\mathscr{P}$, then $P \underset{c}{<} P^*$ if and only if $Q^{*\prime}(y)/Q'(y)$ is nondecreasing for $0 < y < 1$.*

Proof. From $g(x) = Q^*[P(x)]$ we have, on setting $x = Q(y)$, that $Q^*(y) = g[Q(y)]$ for $0 < y < 1$. Hence $P \underset{c}{<} P^*$ if and only if $Q^*(y)$ is a convex function of $Q(y)$. Differentiating $Q^*(y)$ with respect to $Q(y)$ gives the lemma. ▶

We are now able to use these results to obtain inequalities for the expected values of order statistics.

Theorem 4.4.1. *If P, P^* are in $\mathscr{P}$, then $P \underset{c}{<} P^*$ implies that*

$$P(\mathscr{E}X_{r:n}) \le P^*(\mathscr{E}X_{r:n}^*) \qquad (4.4.3)$$

for all r ($r = 1, 2, \ldots, n$) and all n, for which $\mathscr{E}X_{r:n}$ and $\mathscr{E}X_{r:n}^*$ exist.

Proof. The convex transformation $g(x) = Q^*[P(x)]$ takes X with cdf P into $X^* = g(X)$ with cdf P^*. Since g is strictly increasing, it will also take

$X_{r:n}$ with cdf F_r into $X_{r:n}^*$ with cdf F_r^*. By Jensen's inequality we now have

$$g(\mathscr{E}X_{r:n}) \le \mathscr{E}[g(X_{r:n})] = \mathscr{E}X_{r:n}^*.$$

Thus

$$Q^*[P(\mathscr{E}X_{r:n})] \le \mathscr{E}X_{r:n}^*$$

or

$$P(\mathscr{E}X_{r:n}) \le P^*(\mathscr{E}X_{r:n}^*). \qquad \blacktriangleright$$

c-Comparison with the Uniform Distribution

Take $P^*(x) = x \, (0 < x < 1)$; then $Q^*(y) = y \, (0 < y < 1)$. Since $Q^*(P) = P$, any convex P c-precedes P^*. Now

$$P^*(\mathscr{E}X_{r:n}^*) = \frac{r}{n+1}, \qquad (4.4.4)$$

so that for any convex P

$$P(\mathscr{E}X_{r:n}) \le \frac{r}{n+1}. \qquad (4.4.5)$$

For any concave P, the inequality is reversed.

c-Comparison with $P^*(x) = -1/x$ and $P^*(x) = (x-1)/x$

For $P^*(x) = -1/x \, (-\infty < x < -1)$ or $Q^*(y) = -1/y$, we find

$$\mathscr{E}X_{r:n}^* = \frac{-n}{r-1} \qquad \text{for } r > 1.$$

Thus, if $1/P(x)$ is concave on I, so that $Q^*(P) = -1/P(x)$ is convex, we have

$$P(\mathscr{E}X_{r:n}) \le \frac{r-1}{n} \qquad r > 1. \qquad (4.4.6)$$

If $1/P(x)$ is convex, the inequality is reversed.

Similarly, c-comparison with $P^*(x) = (x-1)/x$ gives

$$P(\mathscr{E}X_{r:n}) \le \frac{r}{n} \qquad r < n \qquad (4.4.7)$$

if $1/[1 - P(x)]$ is convex. The inequality is reversed if $1/[1 - P(x)]$ is concave.

For the normal distribution $p(x) = (2\pi)^{-1/2}e^{-1/2x^2} \, (-\infty < x < \infty)$ it is easy to show that both $1/P(x)$ and $1/[1 - P(x)]$ are convex. For example, writing P for $P(x)$, etc., we have

$$\frac{d^2}{dx^2}P^{-1} = \frac{d}{dx}(-pP^{-2})$$

$$= P^{-3}(2p^2 + Pxp)$$

$$= pP^{-3}(2p + Px) > 0,$$

where the inequality holds since $2p + Px$ increases for $x < 0$ from 0 at $x = -\infty$, and is obviously positive for $x > 0$.

It follows that

$$\frac{r-1}{n} \leq P(\mathscr{E}X_{r:n}) \leq \frac{r}{n}. \tag{4.4.8}$$

c-Comparison with the Exponential Distribution

Here $P^*(x) = 1 - e^{-x}$ $(0 < x < \infty)$, and hence

$$g(x) = Q^*[P(x)] = -\log[1 - P(x)].$$

Thus convexity of $g(x)$ requires that

$$g'(x) = \frac{p(x)}{1 - P(x)}$$

be nondecreasing. It is interesting to note that $g'(x)$ is the *hazard rate* $h(x)$ much used in life testing, namely, the conditional pdf of lifetime X of an item with cdf $P(x)$, given that it has survived to time x. Since $h(x)\,dx$ is the corresponding conditional probability of failure in $(x, x + dx)$, it is often reasonable to suppose that the hazard rate is monotonic increasing (see, e.g., Barlow *et al.*, 1963) or equivalently that $P \underset{c}{<} P^*$. Now

$$\mathscr{E}X^*_{r:n} = \sum_{i=0}^{r-1} \frac{1}{n-i} < \int_{n-r+\frac12}^{n+\frac12} \frac{1}{x}\,dx = \log\frac{n+\frac12}{n-r+\frac12},$$

so that, if $P(x)$ is a distribution with increasing hazard rate, then

$$P(\mathscr{E}X_{r:n}) \leq 1 - \exp\left(-\sum_{i=0}^{r-1} \frac{1}{n-i}\right) < \frac{r}{n+\frac12}. \tag{4.4.9}$$

See also Barlow (1965).

Stronger inequalities are possible if we confine ourselves to the subclass $\mathscr{S}$ of distributions in $\mathscr{P}$ which are symmetric. Formally, this additional requirement, which we shall impose for the remainder of this section, is that

$$P(x_0 - x) + P(x_0 + x) = 1 \qquad \text{for some } x_0 \text{ and all } x,$$

or equivalently

$$Q(y) + Q(1 - y) = 2x_0 \qquad \text{for all } y \text{ in } 0 < y < 1. \tag{4.4.10}$$

Applying (4.4.10) to the cdf P^* in $\mathscr{S}$, with $y = P(x_0 - x)$, we have

$$Q^*[P(x_0 - x)] + Q^*[P(x_0 + x)] = 2x_0^*,$$

where x_0^* is the point of symmetry of P^*. This means that $Q^*(P)$ is antisymmetrical (on I) about x_0. Consequently convexity (concavity) of $Q^*(P)$

for $x > x_0$ implies concavity (convexity) of $Q^*(P)$ for $x < x_0$. It follows that, if P, P^* are in $\mathscr{S}$, then $P \underset{c}{<} P^*$ implies $P \sim P^*$, so that symmetric distributions are not c-comparable unless they are equivalent. We therefore define *s-ordering* or *s-comparison*: If P, P^* are in $\mathscr{S}$, then $P \underset{s}{<} P^*$ if and only if $Q^*(P)$ is convex for $x > x_0$. The letter s stands for symmetry, and P is said to *s-precede P**.

It is now easy to parallel for s-ordering the previous development for c-ordering, and in particular to show that the relation $\underset{s}{<}$ is a weak ordering on $\mathscr{S}$. Lemma A holds with $\mathscr{P}$ replaced by $\mathscr{S}$; the transformation $g(x) = Q^*[P(x)]$ is now in addition antisymmetrical concave-convex on I. In Lemma B we simply substitute $\mathscr{S}$ for $\mathscr{P}$. We have, corresponding to Lemma C,

Lemma C'. *If P, P^* are in $\mathscr{S}$, then $P \underset{s}{<} P^*$ if and only if $Q^{*\prime}(y)/Q'(y)$ is nondecreasing for $\frac{1}{2} < y < 1$.*

Finally the counterpart of Theorem 4.4.1 is

Theorem 4.4.2. *If P, P^* are in $\mathscr{S}$, then $P \underset{s}{<} P^*$ implies that*

$$P(\mathscr{E}X_{r:n}) \leq P^*(\mathscr{E}X_{r:n}^*) \qquad (4.4.11)$$

for all r in $\frac{1}{2}(n+1) \leq r \leq n$ and all n for which $\mathscr{E}X_{r:n}^*$ exists.

For a proof we refer the reader to van Zwet (1964, p. 67).

s-Comparison with the Uniform Distribution

Let P^* be the uniform cdf. Then $Q^*(P) = P$, and any concave-convex P in $\mathscr{S}$ s-precedes P^*. Consider distributions in $\mathscr{S}$ having pdf's which are either U-shaped or unimodal (single maximum). Then P is concave-convex and convex-concave, respectively. It follows that for P in $\mathscr{S}$ and $r \geq \frac{1}{2}(n+1)$:

For a symmetric U-shaped distribution, $P(\mathscr{E}X_{r:n}) \leq r/(n+1)$,
for a symmetric unimodal distribution, $P(\mathscr{E}X_{r:n}) \geq r/(n+1)$. $\qquad (4.4.12)$

These results may be compared with (4.4.5). A direct proof of (4.4.12) is given by Ali and Chan (1965). See Ex. 4.4.3.

s-Comparison of Normal and Logistic Distributions

Let $P(x)$ be the unit normal cdf $\Phi(x)$, and let

$$P^*(x) = \frac{1}{1 + e^{-x}} \qquad -\infty < x < \infty.$$

Clearly P, P^* are in $\mathscr{S}$. Also the function

$$Q^*[P(x)] = \log P(x) - \log [1 - P(x)]$$

is easily shown to be convex for $x \geq 0$, so that $P \underset{s}{<} P^*$.

Table 4.4. Bounds and approximations to $\mathscr{E} X_{r:n}$ for $n = 10$

r	(1)	(2)	(3)	(4)	(5)	(6)	(7)
6	.178	.180	.125	.126	0	.123	.123
7	.428	.431	.384	.385	.253	.375	.376
8	.708	.712	.671	.674	.524	.655	.656
9	1.057	1.067	1.027	1.036	.842	1.000	1.001
10	1.612	1.669	1.591	1.645	1.282	1.547	1.539

Now for $r \geq \frac{1}{2}(n + 1)$

$$\mathscr{E}(X_{r:n}^*) = \sum_{i=n+1-r}^{r-1} \frac{1}{i} < \log \frac{r - \frac{1}{2}}{n - r + \frac{1}{2}},$$

and hence we have from Theorem 4.4.2

$$P(\mathscr{E} X_{r:n}) \leq \frac{1}{1 + \exp\left(-\sum\limits_{i=n+1-r}^{r-1} \frac{1}{i}\right)} < \frac{r - \frac{1}{2}}{n}. \qquad (4.4.13)$$

These two inequalities are respectively stronger than those of (4.4.9) applied to the normal case. Table 4.4 provides a numerical comparison for $n = 10$ and $r = 6(1)10$ of the four upper bounds:

$$(1) \qquad \Phi^{-1}\left[1 - \exp\left(-\sum_{i=n+1-r}^{n} \frac{1}{i}\right)\right],$$

$$(2) \qquad \Phi^{-1}\left(\frac{r}{n + \frac{1}{2}}\right)$$

$$(3) \qquad \Phi^{-1}\left[\frac{1}{1 + \exp\left(-\sum\limits_{i=n+1-r}^{r-1} \frac{1}{i}\right)}\right],$$

$$(4) \qquad \Phi^{-1}\left(\frac{r - \frac{1}{2}}{n}\right),$$

as well as of the lower bound (from 4.4.8):

$$(5) \qquad \Phi^{-1}\left(\frac{r - 1}{n}\right),$$

of an approximation suggested by Blom (1958), which is further discussed in

Section 4.5:

$$(6) \quad \Phi^{-1}\left(\frac{r - \frac{3}{8}}{n + \frac{1}{4}}\right),$$

and of the exact value of

$$(7) \quad \mathscr{E}X_{r:n}.$$

Of the upper bounds, (3) is best but (4) is almost as good. The lower bound (5) is rather poor. The results, especially the lower bounds, do not generally compare favorably with the more complicated 3rd bounds of Table 4.3.

See also van Zwet (1967).

4.5. APPROXIMATIONS TO MOMENTS IN TERMS OF THE INVERSE CDF AND ITS DERIVATIVES

As we have seen, the probability integral transformation $u = P(x)$ takes the continuous order statistic $X_{(r)}$ into the rth order statistic $U_{(r)}$ in a sample of n from a rectangular $R(0, 1)$ distribution. We now invert the relation $U_{(r)} = P(X_{(r)})$, writing $X_{(r)} = Q(U_{(r)})$, and expand $Q(U_{(r)})$ in a Taylor series about

$$\mathscr{E}U_{(r)} = \frac{r}{n + 1} = p_r. \tag{4.5.1}$$

This gives

$$X_{(r)} = Q(p_r) + (U_{(r)} - p_r)Q'(p_r) + \tfrac{1}{2}(U_{(r)} - p_r)^2 Q''(p_r)$$
$$+ \tfrac{1}{6}(U_{(r)} - p_r)^3 Q'''(p_r) + \cdots. \tag{4.5.2}$$

Replacing $Q(p_r)$ by Q_r, etc., and setting $q_r = 1 - p_r$, we obtain with the help of (3.1.7) to order $(n + 2)^{-2}$

$$\mathscr{E}X_{(r)} = Q_r + \frac{p_r q_r}{2(n + 2)}Q_r'' + \frac{p_r q_r}{(n + 2)^2}[\tfrac{1}{3}(q_r - p_r)Q_r'''$$
$$+ \tfrac{1}{8}p_r q_r Q_r''''], \tag{4.5.3}$$

$$\operatorname{var} X_{(r)} = \frac{p_r q_r}{n + 2}Q_r'^2 + \frac{p_r q_r}{(n + 2)^2}[2(q_r - p_r)Q_r'Q_r''$$
$$+ p_r q_r(Q_r'Q_r''' + \tfrac{1}{2}Q_r''^2)], \tag{4.5.4}$$

$$\operatorname{cov}(X_{(r)}, X_{(s)}) = \frac{p_r q_s}{n + 2}Q_r'Q_s' + \frac{p_r q_s}{(n + 2)^2}[(q_r - p_r)Q_r''Q_s'$$
$$+ (q_s - p_s)Q_r'Q_s'' + \tfrac{1}{2}p_r q_r Q_r'''Q_s'$$
$$+ \tfrac{1}{2}p_s q_s Q_r'Q_s''' + \tfrac{1}{2}p_r q_s Q_r''Q_s'']. \tag{4.5.5}$$

Note that, since $p_r = P(Q_r)$, we have

$$Q_r' = \frac{1}{dp_r/dQ_r} = \frac{1}{p(Q_r)},$$

where $p(Q_r)$ is the pdf of X evaluated at Q_r. This approach, essentially due to K. and M. V. Pearson (1931), has been systematically pursued by F. N. David and Johnson (1954), who present results to order $(n + 2)^{-3}$ for all the first four cumulants and cross cumulants. The expansions need not be in inverse powers of $n + 2$ (see Clark and Williams, 1958), but David and Johnson have found this advantageous. Conditions for the validity of the approach are given by Blom (1958, Chapter 5) and van Zwet (1964, Chapter 3). Saw (1960) has obtained bounds for the remainder term when the expansion of $\mathscr{E}X_{(r)}$ is terminated after an even number of terms. From the practical point of view, the most important feature of the expansion is that convergence may be slow or even nonexistent if r/n is too close to 0 or 1.

Example 4.5. For a standard normal parent with cdf $\Phi(x)$ and pdf $\phi(x)$ we have $Q(p_r) = \Phi^{-1}(p_r)$ and $Q'(p_r) = 1/\phi(Q)$. Then

$$Q''(p_r) = \frac{d}{d\Phi(Q)}\left(\frac{1}{\phi(Q)}\right) = \frac{d}{dQ}\left(\frac{1}{\phi(Q)}\right)\frac{dQ}{d\Phi(Q)}$$

$$= \frac{Q}{\phi^2(Q)},$$

since $d\phi(Q)/dQ = -Q\phi(Q)$. Continuing, we also find

$$Q'''(p_r) = \frac{1 + 2Q^2}{\phi^3(Q)}, \quad Q''''(p_r) = \frac{Q(7 + 6Q^2)}{\phi^4(Q)}.$$

A different approach based on the logistic rather than the uniform distribution has been developed by Plackett (1958) (but see Chan, 1967b). Although this is less convenient, there are indications in the normal case that for the same number of terms Plackett's series for $\mathscr{E}X_{(r)}$ is somewhat more accurate than that of David and Johnson (Saw, 1960).

Formulae (4.5.3)–(4.5.5) are rather tedious to apply, especially for distributions that, unlike the normal, do not allow $dp(Q)/dQ$ to be simply expressed. From mean-value theorem considerations Blom (1958, Chapter 6) has suggested semi-empirical "α,β-corrections" and writes

$$\mathscr{E}X_{(r)} = Q(\pi_r) + R,$$

where $\pi_r = (r - \alpha_r)/(n + 1 - \alpha_r - \beta_r)$, and R is of order $1/n$. By suitable choice of α_r and β_r (which generally also depend on n) it should be possible to make the remainder R sufficiently small, so that $Q(\pi_r)$ may be used as an approximation to $\mathscr{E}X_{(r)}$. This approach simplifies in the case of a symmetric parent; for, taking the mean to be zero, we then want

$$Q(\pi_r) = -Q(\pi_{n-r+1}) \quad \text{or} \quad \pi_r + \pi_{n-r+1} = 1.$$

In turn, this suggests taking $\alpha_r = \beta_r$ and hence

$$\pi_r = \frac{r - \alpha_r}{n + 1 - 2\alpha_r}.$$

Solving for α_r in $\mathscr{E}X_{(r)} = Q(\pi_r)$, we have

$$\alpha_r = \frac{r - (n + 1)P(\mathscr{E}X_{(r)})}{1 - 2P(\mathscr{E}X_{(r)})}.$$

In the normal case Blom finds α_r to be remarkably stable for $n \leq 20$ and all r, its smallest and largest values being 0.33 and 0.39. He therefore suggests $\alpha = \frac{3}{8}$ as a convenient general value. This is the approximation in column (6) of Table 4.4. The approximation is quite good. However, calculations for larger n (≤ 400) carried out by Harter (1961a) indicate that $\alpha = 0.4$ is better in the range $50 \leq n \leq 400$. For more detailed recommendations see Harter's paper.

EXERCISES

4.2.1. For any distribution with cdf $u = P(x)$, symmetric about zero, show that

$$\text{var } X_{(n)} \geq \lambda_n \sigma^2,$$

and that the lower bound is achieved if

$$x \propto \frac{n[u^{n-1} - (1 - u)^{n-1}]}{n[u^{n-1} + (1 - u)^{n-1}] - 2\lambda_n},$$

where λ_n is the only root of the following equation for λ:

$$\int_{\frac{1}{2}}^{1} \frac{n^2[u^{n-2} - (1 - u)^{n-1}]^2}{n[u^{n-1} + (1 - u)^{n-1}] - 2\lambda} \, du = 1$$

in the interval $0 \leq \lambda \leq n/2^{n-1}$.

(Moriguti, 1951)

4.2.2. By setting

$$f = xn^{\frac{1}{2}}[u^{n-1} + (1 - u)^{n-1}]^{\frac{1}{2}},$$

$$g = \frac{n^{\frac{1}{2}}[u^{n-1} - (1 - u)^{n-1}]}{[u^{n-1} + (1 - u)^{n-1}]^{\frac{1}{2}}},$$

in Schwarz's inequality, show that, for any distribution symmetric about zero and having a variance,

$$\frac{\text{var } X_{(n)}}{(\mathscr{E}X_{(n)})^2} \geq \frac{1}{M_n} - 1, \tag{A}$$

here

$$M_n = \int_{\frac{1}{2}}^{1} \frac{n[u^{n-1} - (1-u)^{n-1}]^2}{u^{n-1} + (1-u)^{n-1}} \, du,$$

and that equality in (A) is attained if and only if

$$x \propto \frac{u^{n-1} - (1-u)^{n-1}}{u^{n-1} + (1-u)^{n-1}}.$$

(Moriguti, 1951)

4.2.3. Show that for any parent with finite variance σ^2

$$\mathscr{E}(X_{(r)} - X_{(n-r+1)}) \leq \sigma \left\{ \int_0^1 [\bar{\Psi}(u)]^2 \, du \right\}^{\frac{1}{2}} \qquad r > [\tfrac{1}{2}n],$$

where $\bar{\Psi}(u) = -\Psi(u_1) \qquad 0 \leq u < 1 - u_1,$

$$= \Psi(u) \qquad 1 - u_1 \leq u \leq u_1,$$

$$= \Psi(u_1) \qquad u_1 \leq u < 1,$$

$$\Psi(u) = \frac{n!}{(r-1)!(n-r)!} [u^{r-1}(1-u)^{n-r} - u^{n-r}(1-u)^{r-1}],$$

and u_1 is determined by

$$(1 - u_1)\Psi(u_1) = \int_{u_1}^1 \Psi(u) \, du \qquad \tfrac{1}{2} < u_1 < 1.$$

(Moriguti, 1953a)

4.3.1. If X is a variate from any standardized symmetric distribution with cdf $u = P(x)$, show that

$$|\mathscr{E}X_{(r)}| \leq \frac{1}{\sqrt{2}B(r, n-r+1)} [B(2r-1, 2n-2r+1) - B(n, n)]^{\frac{1}{2}},$$

with equality holding if and only if

$$x(u) = \pm \frac{1}{\sqrt{2}} [B(2r-1, 2n-2r+1) - B(n,n)]^{-\frac{1}{2}}$$

$$\cdot [u^{r-1}(1-u)^{n-r} - u^{n-r}(1-u)^{r-1}].$$

(Sugiura, 1962)

4.4.1. For the gamma distribution with pdf

$$P'(x) = \frac{1}{\Gamma(\alpha)} e^{-x} x^{\alpha-1} \qquad \alpha > 0, 0 < x < \infty,$$

show with the help of (4.4.4)–(4.4.7) that

$$\frac{r-1}{n} \leq P(\mathscr{E}X_{r:n}) \leq \frac{r}{n} \qquad \alpha > 1,$$

$$\frac{r}{n+1} \leq P(\mathscr{E}X_{r:n}) \leq \frac{r}{n} \qquad \alpha = 1,$$

$$\frac{r}{n+1} \leq P(\mathscr{E}X_{r:n}) \qquad \alpha < 1.$$

(van Zwet, 1964, p. 56)

4.4.2. For the beta distribution with pdf

$$P'(x) = \frac{1}{B(\alpha, \beta)} x^{\alpha-1}(1 - x)^{\beta-1} \qquad \alpha, \beta > 0, 0 < x < 1,$$

show that

$$\frac{r - 1}{n} \leq P(\mathscr{E}X_{r:n}) \leq \frac{r}{n} \qquad \alpha > 1, \beta > 1,$$

$$\frac{r - 1}{n} \leq P(\mathscr{E}X_{r:n}) \leq \frac{r}{n + 1} \qquad \alpha > 1, \beta = 1,$$

$$\frac{r}{n + 1} \leq P(\mathscr{E}X_{r:n}) \leq \frac{r}{n} \qquad \alpha = 1, \beta > 1,$$

$$P(\mathscr{E}X_{r:n}) \leq \frac{r}{n + 1} \qquad \alpha \geq 1, \beta < 1,$$

$$\frac{r}{n + 1} \leq P(\mathscr{E}X_{r:n}) \qquad \alpha < 1, \beta \geq 1.$$

(van Zwet, 1964, p. 57)

4.4.3. Let $P(x)$ be the continuous, strictly increasing cdf of a variate X, symmetric about zero. With i_u as defined in (4.2.11), write

$$C = \int_{1/2}^{1} (i_u - i_{1-u}) \, du.$$

Show that, for $r > \frac{1}{2}(n + 1)$ and $Q = P^{-1}$,

(a) $0 < C < 1,$

(b) $\dfrac{\mathscr{E}X_{r:n}}{C} \geq Q\left\{ \int_{1/2}^{1} \left[\dfrac{u(i_u - i_{1-u})}{C} \right] du \right\},$

(c) $\mathscr{E}X_{r:n} \geq CQ\left[\dfrac{1}{2} + \dfrac{1}{C}\left(\dfrac{r}{n + 1} - \dfrac{1}{2} \right) \right] \geq Q\left(\dfrac{r}{n + 1} \right).$

(Ali and Chan, 1965)

4.5.1. Show that for a symmetric pdf $p(x)$ with mean μ and variance σ^2 the efficiency of the median M relative to the mean in samples of n is, to terms of order $1/n$,

$$4f^2\sigma^2\left(1 + \frac{g + 8}{4n}\right) \qquad n \text{ odd},$$

$$4f^2\sigma^2\left(1 + \frac{g + 12}{n}\right) \qquad n \text{ even},$$

where

$$f = p(\mu) \quad \text{and} \quad g = \frac{p''(\mu)}{p^3(\mu)}.$$

Show also that to order $1/n^2$

$$\text{var } M = \frac{1}{8mf^2}\left(1 - \frac{g + 12}{8mf^2}\right)$$

both for $n = 2m$ and $n = 2m + 1$ (m integral).

(Thus up to the accuracy of this approximation it does not pay to base the median on an odd number of observations; the next smaller even number provides a median which is just as accurate.)

(Hodges and Lehmann, 1967)

CHAPTER 5

Further Distribution Theory

5.1. INTRODUCTION

In Chapter 2 we considered the basic distribution theory of order statistics and of simple systematic statistics. It is the purpose of the present chapter to deal with more complicated quantities involving order statistics. Suppose first that we have a random sample $X_1, X_2, \ldots, X_n$ from a $N(\mu, \sigma^2)$ parent. Then the important class of studentized statistics has the form of a linear function $\sum_{i=1}^{n} a_i X_{(i)}$ of order statistics divided by an independent root-mean-square (rms) estimator S_ν of σ having ν DF (i.e., $\nu S_\nu^2/\sigma^2$ is distributed as χ^2 with ν DF). Most important in this class is the studentized range W_n/S_ν, useful in the problem of ranking "treatment" means in an an analysis of variance. For tests of normality and the presence of outlying observations (Chapter 8) it is of interest to consider statistics of the form $\sum_{i=1}^{n} a_i X_{(i)}/S$, where S without subscript denotes the rms estimator obtained from the sample at hand, that is, $(n-1)S^2 = \Sigma (X_i - \bar{X})^2$. In this case we may speak of internal studentization in contrast to the first, more familiar process, which is external. When external information on σ is available, we may wish to supplement it with internal information, suggesting as a divisor the pooled estimator $S^{(P)}$, where

$$(n - 1 + \nu)(S^{(P)})^2 = (n - 1)S^2 + \nu S_\nu^2.$$

Use of $S^{(P)}$ leads to yet another kind of studentization.

Many statistics are expressible as maxima. Indeed, the studentized range is the largest of the $n(n - 1)$ differences $(X_i - X_j)/S_\nu$, a property closely related to its major role in problems of ranking and multiple comparisons. In Section 5.3 we present an approach frequently useful for obtaining exact or approximate upper percentage points of such statistics. The method does not work well for the studentized range but is effective for such outlier statistics as $X_{(n)} - \bar{X}$, $\max\limits_{i=1,2,\ldots,n} |X_i - \bar{X}|$, and their studentized versions, as well

70

as for many other statistics not necessarily assuming underlying normality. Another application of this approach is to the distribution of the largest subinterval created by the random division of the unit interval.

Although in the foregoing examples the X_i are independent, all the statistics considered are maxima of correlated variates; for example, $X_{(n)} - \bar{X}$ is the maximum of the n correlated deviates $Y_i = X_i - \bar{X}$. Section 5.5 deals more generally with the distribution of order statistics for dependent variates.

5.2. STUDENTIZATION

We shall illustrate various general methods of handling studentized statistics by dealing in detail with the studentized forms of the range.

For the (externally) studentized range $Q_{n,\nu} = W_n/S_\nu$ it follows at once from the independence of W_n and S_ν that the kth raw moment is

$$\mathcal{E}(Q_{n,\nu}^k) = \mathcal{E}(S_\nu^{-k})\mathcal{E}(W_n^k) \qquad 0 \leq k < \nu$$

$$= \frac{\nu^{\frac{1}{2}k}\Gamma\left(\dfrac{\nu - k}{2}\right)\mathcal{E}\left(\dfrac{W_n}{\sigma}\right)^k}{2^{\frac{1}{2}k}\Gamma(\frac{1}{2}\nu)}. \tag{5.2.1}$$

Thus the kth raw moment of Q can be found from that of W_n/σ, and Pearson or other types of curves fitted to give the approximate distribution of Q. Another approach is through Hartley's (1944) process of studentization, which allows the cdf of $Q_{n,\nu}$ to be expressed in terms of the cdf of W_n and a series in powers of $1/\nu$, namely,

$$\Pr\{Q_{n,\nu} < q\} = \Pr\{W_n < q\} + a_1\nu^{-1} + a_2\nu^{-2} + \cdots, \tag{5.2.2}$$

where a_1, a_2 are functions of n and q which have been tabulated by Pearson and Hartley (1943) for $n \leq 20$ and $\nu \geq 10$. The representation of the LHS by only three terms is not altogether satisfactory, especially for $\nu \leq 20$. See also Moriguti (1953b), Kudô (1956c), and Chambers (1967). Harter et al. (1959) in their definitive tables have gone back (essentially) to the simple relation

$$\Pr\{Q_{n,\nu} < q\} = \int_0^\infty \Pr\{W_n < s_\nu q\}f(s_\nu)\,ds_\nu$$

$$= \frac{2(\frac{1}{2}\nu)^{\frac{1}{2}\nu}}{\Gamma(\frac{1}{2}\nu)}\int_0^\infty s^{\nu-1}e^{-\frac{1}{2}\nu s^2}\Pr\{W_n < sq\}\,ds. \tag{5.2.3}$$

Study of the internally studentized range W_n/S is facilitated by the independence of W_n/S and S in normal samples. This result follows immediately from the fact that W_n/S, having a distribution free of μ and σ, is independent of the complete sufficient statistic $(\bar{X}, S)$ (Basu, 1955). A more elementary proof is indicated in Ex. 5.2.1. We now have

$$\mathscr{E}(W_n{}^k) = \mathscr{E}\left[\left(\frac{W_n}{S}\right)^k \cdot S^k\right] = \mathscr{E}\left(\frac{W_n}{S}\right)^k \cdot \mathscr{E}S^k, \qquad (5.2.4)$$

so that the kth raw moment of W_n/S is given by

$$\mathscr{E}\left(\frac{W_n}{S}\right)^k = \frac{\mathscr{E}(W_n{}^k)}{\mathscr{E}S^k} = [\tfrac{1}{2}(n-1)]^{\frac{1}{2}k} \frac{\Gamma[\frac{1}{2}(n-1)]}{\Gamma[\frac{1}{2}(n-1+k)]} \mathscr{E}\left(\frac{W_n}{\sigma}\right)^k. \quad (5.2.5)$$

Approximate distributions are therefore again obtainable by curve fitting. Some exact upper percentage points can also be found by the method of the next section (David $et\ al.$, 1954).

The ratio of range to the pooled rms estimator $S^{(P)}$ of σ can be handled similarly; in fact, (5.2.4) holds with S replaced by $S^{(P)}$. To see this, suppose without essential loss of generality that $S_\nu{}^2$ is derived from a sample of $\nu + 1$ from a $N(\mu_1, \sigma^2)$ parent. Then the mean $\bar{X}$ of the sample at hand, the mean $\bar{X}_1$ of the sample of $\nu + 1$, and $S^{(P)}$ are jointly complete sufficient statistics for μ, μ_1, and σ^2. Since the ratio $W_n/S^{(P)}$ has a distribution not involving these parameters, it is independent of $S^{(P)}$, etc.

5.3. STATISTICS EXPRESSIBLE AS MAXIMA

The most basic of the statistics to be considered is the extreme deviate (from the sample mean), $X_{(n)} - \bar{X}$. We find its distribution when the X_i are independent normal $N(\mu, 1)$ variates (Nair, 1948; Grubbs, 1950). To this end, transform from $x_{(i)}$ to x_i' to y_i by means of the relations

$$y_1 = n^{\frac{1}{2}}x_1' = \sum_{i=1}^{n}(x_{(i)} - \mu) = n(\bar{x} - \mu),$$

$$y_2 = (2 \cdot 1)^{\frac{1}{2}}x_2' = -x_{(1)} + x_{(2)} = 2\left(x_{(2)} - \frac{x_{(1)} + x_{(2)}}{2}\right),$$

$$y_3 = (3 \cdot 2)^{\frac{1}{2}}x_3' = -x_{(1)} - x_{(2)} + 2x_{(3)} = 3\left(x_{(3)} - \frac{x_{(1)} + x_{(2)} + x_{(3)}}{3}\right),$$

$$\cdot$$
$$\cdot$$
$$\cdot$$

$$y_n = [n(n-1)]^{\frac{1}{2}}x_n' = -x_{(1)} - x_{(2)} \cdots - x_{(n-1)} + (n-1)x_{(n)} = n(x_{(n)} - \bar{x}).$$

Since the transformation from $x_{(i)}$ to x_i' is orthogonal,

$$f(x_1', x_2', \ldots, x_n') = \frac{n!}{(2\pi)^{n/2}} \exp\left(-\frac{1}{2}\sum_{i=1}^{n} x_i'^2\right),$$

and hence

$$f(y_1, y_2, \ldots, y_n) = \frac{1}{(2\pi)^{n/2}} \exp\left[-\frac{1}{2}\left(\frac{y_1^2}{n} + \sum_{i=2}^{n} \frac{y_i^2}{i(i-1)}\right)\right],$$

$$f(y_2, \ldots, y_n) = \frac{n^{1/2}}{(2\pi)^{(n-1)/2}} \exp\left[-\frac{1}{2}\sum_{i=2}^{n} \frac{y_i^2}{i(i-1)}\right]. \qquad (5.3.1)$$

From $y_i - y_{i-1} = (i - 1)(x_{(i)} - x_{(i-1)}) \geq 0 \ (i = 3, 4, \ldots, n)$ we see that (5.3.1) holds over the region $0 \leq y_2 \leq \cdots \leq y_n$.

For $n = 2, 3$ we have, introducing the function H_n,

$$\Pr\{X_{(2)} - \bar{X} < c\} = \sqrt{2} \int_0^{2c} \frac{1}{(2\pi)^{1/2}} \exp\left(-\frac{1}{2}\frac{y_2^2}{2 \cdot 1}\right) dy_2 = H_2(2c),$$

$$\Pr\{X_{(3)} - \bar{X} < c\} = \left(\frac{3}{2}\right)^{1/2} \int_0^{3c} H_2(y_3) \frac{1}{(2\pi)^{1/2}} \exp\left(-\frac{1}{2}\frac{y_3^2}{3 \cdot 2}\right) dy_3 = H_3(3c),$$

and so, by successive integration of (5.3.1),

$$\Pr\{X_{(n)} - \bar{X} < c\} = \left(\frac{n}{n-1}\right)^{1/2} \int_0^{nc} H_{n-1}(y_n) \frac{1}{(2\pi)^{1/2}}$$

$$\times \exp\left[-\frac{1}{2}\frac{y_n^2}{n(n-1)}\right] dy_n = H_n(nc).$$

Grubbs used this relation to tabulate the cdf of $X_{(n)} - \bar{X}$ for $n \leq 25$.

General Method of Approximating to Upper Percentage Points of Statistics Expressible as Maxima

Although we have just shown how the cdf of $X_{(n)} - \bar{X}$ was successfully tabulated *in toto* for a normal parent distribution, there are very few systematic statistics for which such a table is available. Nor is there a crying need for detailed tables of this sort, since usually upper percentage points for a few α levels suffice. We now describe a method which often provides these points $y_{n,\alpha}$ approximately—and sometimes exactly. Lower percentage points of statistics expressible as minima may, of course, be obtained in the same manner.

Let $A_1, A_2, \ldots, A_n$ be n events. Then the principle of inclusion and exclusion gives the well-known Boole formula for the probability of occurrence of at least one of the A_i:

$$\Pr\left\{\bigcup_{i=1}^{n} A_i\right\} = \sum_i \Pr\{A_i\} - \sum\sum_{i<j} \Pr\{A_i A_j\}$$

$$+ \cdots (-1)^{n-1} \Pr\{A_1 A_2 \cdots A_n\}. \qquad (5.3.2)$$

Moreover, the sum of an odd number of terms on the RHS provides an upper bound and the sum of an even number a lower bound to the LHS, the bounds increasing in sharpness with the number of terms included. Thus we have a sequence of inequalities (sometimes attributed to Bonferroni) of which the first are

$$\sum_i \Pr\{A_i\} - \sum_{i<j}\sum \Pr\{A_iA_j\} \leq \Pr\{\bigcup A_i\} \leq \Sigma \Pr\{A_i\}.^1 \quad (5.3.3)$$

Now identify A_i with the event $Y_i > y$, where the Y_i are any random variables. Then we have $A_i: Y_i > y$, $A_iA_j: Y_i > y$, $Y_j > y$, etc., and $\bigcup A_i$: $Y_{(n)} > y$. If, in addition, the joint distribution of the Y_i is symmetrical in the Y_i, then (5.3.2) becomes

$$\Pr\{Y_{(n)} > y\} = n \Pr\{Y_1 > y\} - \binom{n}{2}\Pr\{Y_1 > y, Y_2 > y\}$$
$$+ \cdots (-1)^{n-1} \Pr\{Y_1 > y, Y_2 > y, \ldots, Y_n > y\}. \quad (5.3.4)$$

The usefulness of this result lies in the often rapid decline of the terms on the right for y sufficiently large, say exceeding $y_{n,0,1}$, the upper 10% point of $Y_{(n)}$. Then the upper bound $y^{(1)}$ to $y_{n,\alpha}$, obtained by solving

$$n \Pr\{Y_1 > y\} = \alpha \quad (5.3.5)$$

for y, may also serve as a good first approximation. Thus $y^{(1)}$ is simply the upper α/n significance point of Y_1. With $Y_i = X_i - \bar{X}$ as in the beginning of this section, we have

$$y^{(1)} = \left(\frac{n-1}{n}\right)^{1/2}\Phi^{-1}\left(1 - \frac{\alpha}{n}\right).$$

To gauge the accuracy of $y^{(1)}$ as a general approximation to $y_{n,\alpha}$ note that (5.3.3) gives

$$\alpha - \binom{n}{2}\Pr\{Y_1 > y^{(1)}, Y_2 > y^{(1)}\} \leq \Pr\{Y_{(n)} > y^{(1)}\} \leq \alpha. \quad (5.3.6)$$

Now if, for $y \geq y^{(1)}$,

$$\Pr\{Y_1 > y, Y_2 > y\} \leq [\Pr\{Y_1 > y\}]^2,$$

or equivalently $\qquad\qquad\qquad\qquad\qquad\qquad\qquad\qquad\qquad (5.3.7)$

$$\Pr\{Y_1 < y, Y_2 < y\} \leq [\Pr\{Y_1 < y\}]^2,$$

a condition which holds for the negatively correlated normal variates $Y_1 = X_1 - \bar{X}$, $Y_2 = X_2 - \bar{X}$ and in many other cases of interest (Doornbos and Prins, 1956; Doornbos, 1966; Hume, 1965; and, especially, Lehmann,

---

[1] Since $\Sigma \Pr\{A_i\}$ can exceed 1, the upper bound is strictly min $(\Sigma \Pr\{A_i\}, 1)$.

1966),[2] then (5.3.6) implies

$$\alpha - \tfrac{1}{2}\alpha^2 < \alpha - \frac{\tfrac{1}{2}(n-1)\alpha^2}{n} \le \Pr\{Y_{(n)} > y^{(1)}\} \le \alpha. \qquad (5.3.8)$$

A second approximation $y^{(2)}$, a lower bound to $y_{n,\alpha}$, is the solution for y of

$$n\Pr\{Y_1 > y\} - \binom{n}{2}\Pr\{Y_1 > y, Y_2 > y\} = \alpha.$$

This is not very convenient, but if (5.3.7) holds the second term may be replaced by $\tfrac{1}{2}(n-1)\alpha^2/n$ to yield a simple and usually only slightly less sharp lower bound.

An interesting sharpening of the *upper* bound in (5.3.3) has apparently only recently been discovered, in spite of its simplicity (Kounias, 1968). Clearly

$$\Pr\{\bigcup A_i\} \le \Pr\{A_i\} + \sum_j{}' \Pr\{\bar{A}_i A_j\},$$

where $\bar{A}_i$ is the event complementary to A_i and $\sum_j{}'$ denotes summation over $j = 1, 2, \ldots, n$ with $j \ne i$. Hence

$$\Pr\{\bigcup A_i\} \le \Pr\{A_i\} + \sum_j{}'(\Pr\{A_j\} - \Pr\{A_i A_j\}) = \sum_{i=1}^{n}\Pr\{A_i\} - \sum_j{}'\Pr\{A_i A_j\}$$

so that

$$\Pr\{\bigcup A_i\} \le \min_{i=1,2,\ldots,n}\left(\Sigma\Pr\{A_i\} - \sum_j{}'\Pr\{A_i A_j\}\right).$$

In particular, if $\Pr\{A_i\} = \Pr\{A_1\}$ and $\Pr\{A_i A_j\} = \Pr\{A_1 A_2\}$ for all i, j ($i \ne j$), then

$$\Pr\{\bigcup A_i\} \le n\Pr\{A_1\} - (n-1)\Pr\{A_1 A_2\}.$$

Generalizations of (5.3.4) to order statistics other than the extremes are possible, although less useful. Let $P_{i_1 i_2 \cdots i_m}$ denote the joint probability of m ($\le n$) events $A_{i_1}, A_{i_2}, \ldots, A_{i_m}$, and S_m the sum of the $\binom{n}{m}$ P's with m different subscripts. Then the probability $p_{r,n}$ of the realization of at least r ($\le n$) events out of n is given by

$$p_{r,n} = \sum_{m=r}^{n}(-1)^{m-r}\binom{m-1}{r-1}S_m. \qquad (5.3.9)$$

When A_i is the event $Y_i > y$, as before, $p_{r,n}$ becomes the probability of $Y_{(n-r+1)} > y$. Writing $P_{12\cdots m}(y)$ for $\Pr\{Y_1 > y, Y_2 > y, \ldots, Y_m > y\}$,

[2] More generally, the inequality $\Pr\{Y_1 > y_1, Y_2 > y_2\} \le \Pr\{Y_1 > y_1\}\Pr\{Y_2 > y_2\}$ also holds in many instances (see, e.g., Doornbos, 1966; Mallows, 1968).

we therefore obtain the following generalization of (5.3.4):

$$\Pr\{Y_{(n-r+1)} > y\} = \sum_{m=r}^{n} (-1)^{m-r} \binom{m-1}{r-1} \binom{n}{m} P_{12\cdots m}(y). \quad (5.3.10)$$

An interesting alternative formulation of these results, essentially due to Fréchet, is used extensively by Barton and F. N. David (1959) in the study of combinatorial extreme-value distributions. Let $R = \sum_{i=1}^{n} \alpha_i$, where $\alpha_i = 1$ or 0 according as the event A_i occurs or not. Then $\Pr\{R = r\}$ is the probability that exactly r events occur. Now for a discrete variate, such as R, ranging over $0, 1, 2, \ldots, n$ it is easily verified that

$$\Pr\{R = r\} = \frac{1}{r!} \sum_{i=0}^{n-r} \frac{(-1)^i}{i!} \mu_{[r+i]},$$

where $\mu_{[m]}$ is the mth factorial moment of R, that is, $\mu_{[m]} = \mathscr{E}R^{(m)}$. By the multinomial generalization of Vandermonde's theorem we have

$$R^{(m)} = \sum \frac{m!}{m_1! m_2! \cdots m_n!} \alpha_1^{(m_1)} \alpha_2^{(m_2)} \cdots \alpha_n^{(m_n)},$$

where the summation extends over all compositions $(m_1, m_2, \ldots, m_n)$ of m, including zero parts. Since $\alpha_i^{(m_i)} = 0$ for $m_i \geq 2$, it follows that

$$R^{(m)} = m! \sum_{i_1 < i_2 < \cdots < i_m} \alpha_{i_1} \alpha_{i_2} \cdots \alpha_{i_m},$$

the summation being over the $\binom{n}{m}$ selections of $i_1, i_2, \ldots, i_m$ from $1, 2, \ldots, n$. Thus

$$\mu_{[m]} = m! \sum_{i_1 < i_2 < \cdots < i_m} \mathscr{E}(\alpha_{i_1} \alpha_{i_2} \cdots \alpha_{i_m})$$

$$= m! \sum_{i_1 < i_2 < \cdots < i_m} \Pr\{A_{i_1} A_{i_2} \cdots A_{i_m}\} = m! S_m.$$

Hence

$$p_{r,n} = \Pr\{R \geq r\} = \sum_{j=r}^{n} \Pr\{R = j\}$$

$$= \sum_{j=r}^{n} \frac{1}{j!} \sum_{i=0}^{n-j} \frac{(-1)^i}{i!} = S_{j+i}(j+i)!$$

$$= \sum_{j=r}^{n} \frac{1}{j!} \sum_{m=j}^{n} \frac{(-1)^{m-j}}{(m-j)!} S_m m!$$

$$= \sum_{m=r}^{n} S_m \sum_{j=r}^{m} (-1)^{m-j} \binom{m}{j}$$

$$= \sum_{m=r}^{n} S_m (-1)^{m-r} \binom{m-1}{r-1},$$

which is (5.3.9). See also Takács (1967), which contains a historical review of the method of inclusion and exclusion.

A generalization in a multivariate direction has been considered by Siotani (1959). Let $\mathbf{y}_j' = (\mathbf{y}_{1j}, \mathbf{y}_{2j}, \ldots, \mathbf{y}_{pj})$ $(j = 1, 2, \ldots, n)$ be n p-variate vectors with mean vector $\mathbf{0}$ and covariance matrix $\gamma \mathbf{\Lambda}$ $(\gamma > 0)$, and let the covariance matrix of $\mathbf{y}_\alpha$ and $\mathbf{y}_\beta$ $(\alpha \neq \beta)$ be $\delta \mathbf{\Lambda}$, where $\mathbf{\Lambda}$ is a positive definite symmetric matrix and $\gamma > |\delta|$. Then

$$_j\chi^2 = \frac{1}{\gamma} \mathbf{y}_j' \mathbf{\Lambda}^{-1} \mathbf{y}_j \qquad (5.3.11)$$

may be called the *generalized distance* of $\mathbf{y}_j$ from the origin. A studentized form is obtained on replacing $\mathbf{\Lambda}$ by $\mathbf{L}$, where the elements of $\mathbf{L}$ are the usual unbiased estimators of covariance based on ν DF and independent of the $\mathbf{y}_j$. Siotani applies (5.3.2)–(5.3.4) to deal with the distributions of

$$\gamma \chi^2_{\max} = \max_j (\mathbf{y}_j' \mathbf{\Lambda}^{-1} \mathbf{y}_j) \quad \text{and} \quad \gamma T^2_{\max} = \max_j (\mathbf{y}_j' \mathbf{L}^{-1} \mathbf{y}_j),$$

specifically when $\mathbf{y}_j$ is multivariate normal and $\mathbf{L}$ has the Wishart distribution with ν DF. In this case, $_j\chi^2$ has a chi-square distribution with p DF, and

$$_jT^2 = \left(\frac{1}{\gamma}\right) \mathbf{y}_j' \mathbf{L}^{-1} \mathbf{y}_j$$

has a Hotelling distribution with ν DF, that is, the pdf of $_jT^2/\nu$ is

$$f\left(\frac{_jT^2}{\nu}\right) = \frac{1}{B[\frac{1}{2}(\nu + 1 - p), \frac{1}{2}p]} \left(\frac{_jT^2}{\nu}\right)^{\frac{1}{2}p - 1} \left(1 + \frac{_jT^2}{\nu}\right)^{-\frac{1}{2}(\nu+1)} .$$

Thus we have the first-term approximations

$$\Pr\{\gamma \chi^2_{\max} > a^2\} \doteq n \Pr\{\gamma_1 \chi^2 > a^2\}$$

and

$$= n \int_{a^2/\gamma}^{\infty} \frac{x^{\frac{1}{2}p - 1} e^{-\frac{1}{2}x}}{2^{\frac{1}{2}p} \Gamma(\frac{1}{2}p)} \, dx,$$

$$\Pr\{\gamma T^2_{\max} > b^2\} \doteq n \Pr\left\{\frac{_1T^2}{\nu} > \frac{b^2}{\nu\gamma}\right\}$$

$$= \frac{n}{B[\frac{1}{2}(\nu + 1 - p), \frac{1}{2}p]} \int_{b^2/\nu\gamma}^{\infty} x^{\frac{1}{2}p - 1}(1 + x)^{-\frac{1}{2}(\nu+1)} \, dx$$

$$= n I_{\nu\gamma/(\nu\gamma + b^2)}[\frac{1}{2}(\nu + 1 - p), \frac{1}{2}p].$$

Siotani also examines two-term approximations and uses them to tabulate upper percentage points for small p of the multivariate extreme deviate from the sample mean:

$$\max_j [(\mathbf{x}_j - \bar{\mathbf{x}})' \mathbf{\Lambda}^{-1} (\mathbf{x}_j - \bar{\mathbf{x}})] = \chi^2_{\max \, D},$$

and of the multivariate studentized extreme deviate:

$$\max_j [(\mathbf{x}_j - \bar{\mathbf{x}})' \mathbf{L}^{-1} (\mathbf{x}_j - \bar{\mathbf{x}})] = T^2_{\max \mathrm{D}}.$$

Here the $\mathbf{x}_j'$ are n independent p-variate vectors with mean vector $\boldsymbol{\mu}'$ and covariance matrix $\boldsymbol{\Lambda}$, corresponding to taking $\gamma = (n - 1)/n$ and $\delta = -1/n$ in (5.3.11).

An Alternative Approach Assuming Independence

To illustrate this method let us consider its application to the joint distribution of the n variance ratios S_j^2/S_0^2, where $\nu_j S_j^2/\sigma^2 \frown \chi_{\nu j}^2$ ($j = 0, 1, \ldots, n$) and all the S_j^2 are independent. A by-product will be the distribution of the largest variance ratio

$$F^*_{(n)} = \max_{i=1,2,\ldots,n} (S_i^2/S_0^2),$$

relevant when $\nu_i = \nu$ for $i = 1, 2, \ldots, n$. The dependence among the ratios S_i^2/S_0^2 is here due entirely to their common denominator and may be expected to be weak if ν_0 is large. As an approximation one may therefore simply ignore the dependence, thus obtaining (Hartley, 1938; Finney, 1941)

$$\Pr\{S_i^2/S_0^2 \leq y_i; i = 1, 2, \ldots, n\} \doteq \prod_{i=1}^n \Pr\{S_i^2/S_0^2 \leq y_i\} \quad (5.3.12)$$

and

$$\Pr\{F^*_{(n)} \leq y\} \doteq [\Pr\{F_{\nu,\nu_0} \leq y\}]^n. \quad (5.3.13)$$

The accuracy of (5.3.13) and related approximations has also been investigated by Hartley (1955). We now show that in (5.3.12) and (5.3.13) it is always possible to replace $\doteq$ by $\geq$. To do this we need the easily proved result (e.g., Kimball, 1951; Esary et al., 1967) that for any n non-negative increasing functions $g_i(X)$ of a random variable X

$$\mathscr{E}\left[\prod_{i=1}^n g_i(X)\right] \geq \prod_{i=1}^n \mathscr{E}[g_i(X)].$$

Taking $g_i(x) = \Pr\{S_i^2 < x\}$, we have

$$\Pr\{S_i^2/S_0^2 \leq y_i; i = 1, 2, \ldots, n\}$$

$$= \int_0^\infty \Pr\{S_i^2 < s_0^2 y_i; i = 1, 2, \ldots, n\} f(s_0^2)\, ds_0^2$$

$$= \mathscr{E}\left[\prod_{i=1}^n g_i(S_0^2 y_i)\right] \geq \prod_{i=1}^n \mathscr{E}[g_i(S_0^2 y_i)]$$

$$= \prod_{i=1}^n \Pr\{S_i^2/S_0^2 \leq y_i\}.$$

If $\Pr\{S_i^2/S_0^2 > y_i\} = \beta_i$, then, as was to be shown,

$$\Pr\{S_i^2/S_0^2 \leq y_i; i = 1, 2, \ldots, n\} \geq \prod_{i=1}^{n} (1 - \beta_i). \qquad (5.3.14)$$

It may be noted that this is a stronger result than the first Bonferroni inequality (Dunnett and Sobel, 1955):

$$\Pr\{S_i^2/S_0^2 \leq y_i; i = 1, 2, \ldots, n\} \geq 1 - \sum_{i=1}^{n} \beta_i$$

since $\Pi (1 - \beta_i) > 1 - \Sigma \beta_i$ for $0 < \beta_i < 1$ $(i = 1, 2, \ldots, n)$.
If $\nu_i = \nu$, $\beta_i = \beta$ for all i, then (5.3.14) reduces to

$$\Pr\{F_{(n)}^* \leq y\} \geq (1 - \beta)^n.$$

Setting this equal to $1 - \alpha$, we see that an upper bound to $F_{n,\alpha}^*$, the upper α significance point of $F_{(n)}^*$, is given by the upper β significance point of F_{ν,ν_0}, where $\beta = 1 - (1 - \alpha)^{1/n}$. If α is small, we may approximate to β by α/n, which takes us right back to the use of the always valid but here slightly less sharp first Bonferroni bound in (5.3.5).

Interesting multivariate versions of the above inequalities have been developed by a series of authors beginning with Dunn (1958) and culminating in Šidák (1968). The latter's results include the following as a special case: If $(X_1, X_2, \ldots, X_k)$ is multivariate normal with means 0 and an arbitrary correlation matrix, and if Z is a positive variate independent of $(X_1, X_2, \ldots, X_k)$, then

$$\Pr\{|X_1|/Z < c_1, \ldots, |X_k|/Z < c_k\} \geq \prod_{i=1}^{k} \Pr\{|X_i|/Z < c_i\}.$$

In the one-sided case Slepian (1962) has established that $\Pr\{X_1/Z < c_1, \ldots, X_k/Z < c_k\}$ is a nondecreasing function of the correlations.

5.4. RANDOM DIVISION OF AN INTERVAL

Suppose that $n - 1$ points are dropped at random on the unit interval $(0, 1)$. As indicated in Fig. 5.4, denote the ordered distances of these points from the origin by $u_{(i)}$ $(i = 1, 2, \ldots, n - 1)$ and let $y_i = u_{(i)} - u_{(i-1)}$ $(u_{(0)} = 0)$. Then the random variables $U_{(1)}, U_{(2)}, \ldots, U_{(n-1)}$

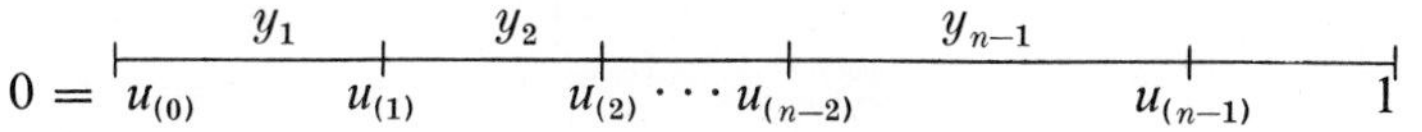

Fig. 5.4

are distributed as $n - 1$ order statistics from a uniform $R(0, 1)$ parent, that is, with joint pdf $(n - 1)!$ over the simplex

$$0 \leq u_{(1)} \leq u_{(2)} \leq \cdots \leq u_{(n-1)} \leq 1.$$

Correspondingly, the pdf of the Y_i is

$$f(y_1, y_2, \ldots, y_{n-1}) = (n - 1)! \qquad y_i \geq 0, \sum_{i=1}^{n-1} y_i \leq 1 \qquad (5.4.1)$$

The distribution is completely symmetrical in the y_i. Indeed, if we define

$$y_n = 1 - \sum_{i=1}^{n-1} y_i,$$

we have the (degenerate) joint pdf $(j = 1, 2, \ldots, n)$

$$f(y_1, y_2, \ldots, y_{n-1}, y_n) = (n - 1)! \qquad y_j \geq 0, \sum_{j=1}^{n} y_j = 1, \qquad (5.4.2)$$

still symmetrical in all y_j. It follows that the joint distribution of any r of the Y_j $(r = 1, 2, \ldots, n - 1)$ is the same as that of the first r, and in particular that the distribution of the sum of any r of the Y_j is that of

$$U_{(r)} = Y_1 + Y_2 + \cdots + Y_r,$$

namely,

$$f_r(u) = \frac{1}{B(r, n - r)} u^{r-1}(1 - u)^{n-r-1} \qquad 0 \leq u \leq 1.$$

If $P(x)$ is the cdf of a continuous variate X, then, in view of the probability integral transformation, $P(X_{(j)})$ is distributed as $U_{(j)}$ and $P(X_{(j)}) - P(X_{(j-1)})$ as Y_j. In this context the Y_j have been called *elementary coverages* by Wilks (1948, 1962), who shows them to play an important part in the theory of nonparametric statistics.

Again, if the X_j have a common exponential distribution and $T = \sum_{j=1}^{n} X_j$, it is easy to show that the ratios X_j/T have the same joint pdf (5.4.2) as the Y_j. The random division of the unit interval may in fact originate from a Poisson process with $n - 1$ events in some time interval which for convenience is scaled down to unit length.

Because of the two major fields of application just outlined, the random division of an interval has received considerable study (see, e.g., Darling, 1953). We confine ourselves to finding the distribution of $Y_{(n)}$, the length of the longest interval. From (5.4.1) the joint pdf of $Y_1, Y_2, \ldots, Y_r$ is, for $\sum_{i=1}^{r} y_i \leq 1$,

$$f(y_1, y_2, \ldots, y_r) = (n - 1)! \int_0^{1-y_1-\cdots-y_r} \cdots \int_0^{1-y_1-\cdots-y_{n-2}} dy_{n-1} \cdots dy_{r+1}$$

$$= \frac{(n - 1)!}{(n - r - 1)!} (1 - y_1 - \cdots - y_r)^{n-r-1}$$

$$r = 1, 2, \ldots, n - 1.$$

Hence for constants $c_i \geq 0$ $(i = 1, 2, \ldots, r)$ with $\sum_{i=1}^{r} c_i \leq 1$

$$\Pr \{ Y_1 > c_1, \, Y_2 > c_2, \ldots, Y_r > c_r \} = (1 - c_1 - c_2 - \cdots - c_r)^{n-1}.$$
$$(5.4.3)$$

Taking $c_1 = c_2 = \cdots = c_r = y$, we now have from (5.3.4)

$$\Pr \{ Y_{(n)} > y \} = n(1 - y)^{n-1} - \binom{n}{2}(1 - 2y)^{n-1}$$

$$+ \cdots (-1)^{i-1} \binom{n}{i}(1 - iy)^{n-1} \cdots, \quad (5.4.4)$$

where the series continues as long as $1 - iy > 0$. This result was first obtained by Fisher (1929) through an ingenious geometric argument and applied by him to harmonic analysis, where a *specified* harmonic may be tested by a statistic of the form X_j/T defined above. In practice, one will usually wish to test first the largest of n such ratios, which may be done with the help of (5.4.4). Fisher (1950, p. 16.59a) provides upper percentage points for which use of the first term on the right of (5.4.4) often suffices.[3]

The distribution of $Y_{(n-1)}$ is readily handled as a special case of (5.3.9) (Fisher, 1940). In this last paper Fisher also points out an interesting connection with a result in geometric probability due to Stevens (1939). Suppose that n arcs of equal length y are marked off at random on the circumference of a circle of unit perimeter. What is the probability that the arcs will cover the entire circumference and, more generally, that there will be at most r breaks? The answer to the first part is just $\Pr \{ Y_{(n)} < y \}$, the probability complementary to (5.4.4), and the general answer is $\Pr \{ Y_{(n-r)} < y \}$. To see this, note that the midpoints of the n arcs divide the unit circle into n intervals of length Y_i with joint pdf (5.4.2). If $Y_{(n)} < y$, there will be no break, and if $Y_{(n-r)} < y$, at most r breaks.

Cochran (1941) has extended some of Fisher's results in order to treat the ratio $\max_j S^2 / \sum_{j=1}^{n} {}_j S^2$, where ${}_j S^2 \nu/\sigma^2 \frown \chi_\nu^2$ and ν is even (cf. Ex. 5.4.4). Upper percentage points of this statistic, providing a possible test for equality of variances of n normal populations, have been tabulated by Eisenhart and Solomon (1947). Bliss *et al.* (1956) give upper 5% points of the corresponding short-cut criterion $\max_j W / \sum_j {}_j W$, ${}_j W$ being the range in the jth sample.

Many of the foregoing topics, as well as the more general distribution theory for differences (or spacings) between successive order statistics when the underlying population has any continuous form, are reviewed by Pyke (1965). See also Naus (1966). Although Pyke's main emphasis is on non-parametric tests of goodness of fit based on suitable functions of the spacings,

---

[3] As an alternative test of significance in harmonic analysis, Hartley (1949) uses the ratio of $X_{(n)}$ to an *independent* mean square error. He also discusses the approximate power of his test.

the subject has applications to the distribution of circular serial correlation coefficients, since these are expressible as linear functions of spacings (cf. Dempster and Kleyle, 1968). The reader is referred also to the elegant account given by Feller (1966, Chapter I and III.3).

We may also mention the discrete analogue to the random division of an interval. Consider a line of N elements broken at $n - 1$ randomly chosen places. What is the distribution of the longest interval (Ex. 5.4.7)? Or—a closely related problem—when black and white balls are arranged in a line, what is the distribution of the longest run of white balls? Treatment of these problems by combinatorial methods goes back at least to Whitworth (see Barton and F. N. David, 1959).

5.5. ORDER STATISTICS FOR DEPENDENT VARIATES

We begin by noting that the distribution of ordered dependent variates Y_i ($i = 1, 2, \ldots, n$) has already been considered in the general approach of Section 5.3. The results there developed are most useful for bounding, approximating, or evaluating $\Pr\{Y_{r:n} > y\}$ for y large, especially when $r = n$. To supplement these results let us write the joint cdf of $Y_1, Y_2, \ldots, Y_n$ as $P_n(y_1, y_2, \ldots, y_n)$. Clearly we have

$$\Pr\{Y_{n:n} \leq y\} = P_n(y, y, \ldots, y). \tag{5.5.1}$$

Consider now (5.3.9), but let A_i be the event $Y_i \leq y$. Then

$$p_{r,n} = \Pr\{Y_{r:n} \leq y\} = \sum_{m=r}^{n} (-1)^{m-r} \binom{m-1}{r-1} S_m, \tag{5.5.2}$$

where S_m is the sum of the $\binom{n}{m}$ probabilities $\Pr\{Y_{i_1} \leq y, Y_{i_2} \leq y, \ldots, Y_{i_m} \leq y\}$ with $i_1 < i_2 < \cdots < i_m$. In the important special case where the Y_i are exchangeable variates (i.e., P_n is symmetrical in $y_1, y_2, \ldots, y_n$) equation (5.5.2) reduces to

$$\Pr\{Y_{r:n} \leq y\} = \sum_{m=r}^{n} (-1)^{m-r} \binom{m-1}{r-1} \binom{n}{m} \Pr\{Y_{m:m} \leq y\}. \tag{5.5.3}$$

This relation links the cdf $F_{r:n}(y)$ of $Y_{r:n}$ with the simpler cdf's of the maxima in samples of $r, r + 1, \ldots, n$. Differentiating or differencing, multiplying by e^{ity}, and integrating or summing, we obtain similar relations between pdf's, characteristic functions, and hence raw moments. Thus (5.5.3) is a generalization of (3.4.3) from independent to exchangeable random variables. With the help of (5.5.3) we can derive the basic recurrence relation

$$(n - r)F_{r:n}(y) + rF_{r+1:n}(y) = nF_{r:n-1}(y). \tag{5.5.4}$$

This formula, usually stated in terms of moments (Relation 1 of Section 3.4), follows on applying (5.5.3) to each term of (5.5.4). We now give a more direct proof of this result which lends itself to further extensions.

Of the n order statistics $Y_{i:n}$ obtained by rearrangement from n exchangeable variates, drop one at random. Then the resulting $n - 1$ variates will be the order statistics in a sample of $n - 1$ exchangeable variates. If $Y_{i:n}$ is dropped ($i = 1, 2, \ldots, r$), the rth largest variate in the set of $n - 1$ was the $(r + 1)$th largest out of n, that is,

$$Y_{r:n-1} = Y_{r+1:n}. \tag{A}$$

Likewise, if $Y_{i:n}$ is dropped ($i = r + 1, r + 2, \ldots, n$), then

$$Y_{r:n-1} = Y_{r:n}. \tag{B}$$

Since (A) and (B) have respective probabilities r/n and $(n - r)/n$, it follows that for any y

$$\Pr\{Y_{r:n-1} \leq y\} = \frac{r}{n} \Pr\{Y_{r+1:n} \leq y\} + \frac{n - r}{n} \Pr\{Y_{r:n} \leq y\},$$

which is (5.5.4).

Consider now $Y_{r:n-1}$ and $Y_{s:n-1}$ ($1 \leq r < s \leq n - 1$) with joint cdf $F_{rs:n-1}(y, z)$. Corresponding to the random dropping of one of the (C) first r, (D) next $s - r$, (E) last $n - s$, of the $Y_{i:n}$, we have

$$Y_{r:n-1} = Y_{r+1:n} \qquad Y_{s:n-1} = Y_{s+1:n}, \tag{C}$$

$$Y_{r:n-1} = Y_{r:n} \qquad Y_{s:n-1} = Y_{s+1:n}, \tag{D}$$

$$Y_{r:n-1} = Y_{r:n} \qquad Y_{s:n-1} = Y_{s:n}. \tag{E}$$

Since events (C), (D), and (E) have respective probabilities r/n, $(s - r)/n$, and $(n - s)/n$, we have, for any y, z ($y \leq z$),

$$nF_{rs:n-1}(y, z) = rF_{r+1,s+1:n}(y, z)$$
$$+ (s - r)F_{r,s+1:n}(y, z) + (n - s)F_{rs:n}(y, z). \tag{5.5.5}$$

As before, this result can be converted into one linking the corresponding product moments of any order, to give in particular Relation 3 of Section 3.4.

Subject to the caution regarding rounding errors, issued in connection with (3.4.3), equation (5.5.3) permits the evaluation of the cdf of $Y_{r:n}$ whenever the cdf's of the extreme in samples up to size n are available. This includes the following cases, where the Y_i are standard multinormal variates with equal correlation coefficient ρ:

(i) Gupta (1963a) gives $\Pr\{Y_{n:n} \leq y\}$ for $n = 1(1)12$ and many positive values of ρ.

(ii) Krishnaiah and Armitage (1965a) tabulate $\Pr\{Y^2_{n:n} \leq y\}$ in great detail for $n = 1(1)10$.

For the important case $r = n - 1$ (5.5.3) reduces to

$$\Pr\{Y_{n-1:n} \leq y\} = n \Pr\{Y_{n-1:n-1} \leq y\} - (n - 1)\Pr\{Y_{n:n} \leq y\}. \quad (5.5.6)$$

As pointed out in Section 5.4, Fisher (1940) gave upper 5% and 1% points of $X_{n-1:n}/\sum_{i=1}^{n} X_i$, where the X_i are iid exponential variates. Such tables are needed when the test on $X_{n:n}/\Sigma X_i$ is inconclusive or unreliable. With similar motivation Young (1967) and David and Joshi (1968) have used (5.5.6) to obtain upper percentage points of $Y_{n-1:n}$ from Gupta's tables.

It should be noted that (5.5.3) does not give the cdf of

$$Y_{r:n} = \frac{1}{\sigma}\left(X_{r:n} - \frac{1}{n}\sum_{i=1}^{n} X_i\right)$$

where the X_i are independent $N(\mu, \sigma^2)$ variates, from tables of the cdf of the extreme deviate (from the sample mean) $Y_{n:n}$. With $Y_{n:n}$ so defined, $Y_{m:m}$ in (5.5.3) is the maximum of m equicorrelated normal variates with $\rho = -1/(n - 1)$ and hence not the extreme deviate in samples of m for $m < n$ (cf. Bland and Owen, 1966). An extension of Gupta's tables to $\rho < 0$ would therefore be needed. The distribution of $Y_{r:n}$ can, however, be handled approximately, since the cumulants of $Y_{r:n}$ can be expressed in terms of those of $X_{r:n}/\sigma$ (Ex. 5.3.1). These in turn can be found from the cumulants of $X_{n:n}/\sigma$ by means of the relation (3.4.3) between the corresponding moments. Kendall (1954) has used these methods to solve Youden's "Demon Problem": Given a (small) sample $X_1, X_2, \ldots, X_n$ from a normal population, what is the probability that $\bar{X}$ lies between $X_{n-1:n}$ and $X_{n:n}$? See also H. T. David (1962, 1963), Sarkadi et al. (1962).

Upper percentage points are available not only for cases (i) and (ii) above but also for the following related statistics: $Y_{n:n}/S_v$ (the studentized maximum) and the studentized largest and smallest chi square.

We will now consider in more detail the case where the Y_i are identically distributed equicorrelated multinormal variates. Without loss of generality the Y_i may be taken as unit normal. Since

$$0 \leq \text{var}\left(\sum_{i=1}^{n} Y_i\right) = n \text{ var } Y_i + n(n - 1)\text{ cov}(Y_i, Y_j) \qquad i \neq j,$$

it follows that the common correlation coefficient ρ must satisfy $\rho \geq -1/(n - 1)$. It is easy to verify that the Y_i may be generated from random variables X_i as follows (e.g., Gupta et al., 1964):

$$Y_i = \rho^{1/2}X_0 + (1 - \rho)^{1/2}X_i \qquad \rho \geq 0; i = 1, 2, \ldots, n, \quad (5.5.7)$$

where $X_0, X_1, \ldots, X_n$ are independent $N(0, 1)$ variates, and

$$Y_i = (-\rho)^{\frac{1}{2}}Z_0 + (1 - \rho)^{\frac{1}{2}}Z_i \qquad \rho < 0; i = 1, 2, \ldots, n,$$

where $Z_1, Z_2, \ldots, Z_n$ are independent $N(0, 1)$, Z_0 is also $N(0, 1)$, and

$$\mathscr{E}(Z_0 Z_i) = \frac{-(-\rho)^{\frac{1}{2}}}{(1 - \rho)^{\frac{1}{2}}}.$$

Thus, if $Y = \sum_{i=1}^{n} a_i Y_{i:n}$, we have for $\rho \geq 0$

$$\Pr\{Y \leq y\} = \Pr\left\{\sum_{i=1}^{n} a_i X_{i:n} \leq -\left(\frac{\rho}{1 - \rho}\right)^{\frac{1}{2}}(\Sigma a_i)X_0 + \frac{y}{(1 - \rho)^{\frac{1}{2}}}\right\},$$

and for $\rho < 0$

$$\Pr\{Y \leq y\} = \Pr\left\{\sum_{i=1}^{n} a_i Z_{i:n} \leq -\left(\frac{-\rho}{1 - \rho}\right)^{\frac{1}{2}}(\Sigma a_i)Z_0 + \frac{y}{(1 - \rho)^{\frac{1}{2}}}\right\}.$$

In the former case it follows that

$$\Pr\{Y \leq y\} = H(y; \rho) = \int_{-\infty}^{\infty} H\left[-\left(\frac{\rho}{1 - \rho}\right)^{\frac{1}{2}} x(\Sigma a_i) + \frac{y}{(1 - \rho)^{\frac{1}{2}}}; 0\right] d\Phi(x).$$

If $\Sigma a_i = 0$, then for all ρ we see that

$$H(y; \rho) = H\left[\frac{y}{(1 - \rho)^{\frac{1}{2}}}; 0\right],$$

a well-known result showing in particular (Hartley, 1950a) that the range of the Y_i is distributed as the range of iid normal variates with variance $1 - \rho$. The ratio $Y^{(1)}/Y^{(2)}$, where

$$Y^{(1)} = \sum_{i=1}^{n} a_i^{(1)} Y_{i:n},$$

etc., is clearly distributed independently of ρ.

Approximations to the distribution of the extreme in a general multivariate normal sample have been studied by Greig (1967). See also Exs. 5.5.2–5.5.5.

The distributions of the maximum and the range of partial sums

$$S_r = \sum_{i=1}^{r} X_i \qquad r = 1, 2, \ldots, n$$

of n iid variates X_i are of interest in storage problems, X_i being the input in the ith year. Lower moments of these distributions can be found by special methods when the X_i are unit normal; see, for example, Feller (1951), Anis and Lloyd (1953), Anis (1955, 1956), Solari and Anis (1957), and Moran (1964).

EXERCISES

5.2.1. Let $X_1, X_2, \ldots, X_n$ be a random sample from a $N(\mu, \sigma^2)$ distribution. It is well known that there exist orthogonal transformations from $x_1, x_2, \ldots, x_n$ to $y_1, y_2, \ldots, y_n$ such that $y_n = \sqrt{n}\,\bar{x}$. Applying a generalized spherical polar transformation to $y_1, y_2, \ldots, y_{n-1}$, namely,

$$y_1 = R \cos \theta_1 \cos \theta_2 \cdots \cos \theta_{n-3} \cos \theta_{n-2},$$
$$y_2 = R \cos \theta_1 \cos \theta_2 \cdots \cos \theta_{n-3} \sin \theta_{n-2},$$
$$y_3 = R \cos \theta_1 \cos \theta_2 \cdots \sin \theta_{n-3},$$

$$\vdots$$

$$y_{n-1} = R \sin \theta_1,$$

show that the ratio of any linear function

$$\sum_{i=1}^{n} c_i X_{(i)} \qquad \text{with} \qquad \sum c_i = 0$$

divided by S is independent of S.

5.3.1. Let X_i $(i = 1, 2, \ldots, n)$ be a random sample from a normal $N(\mu, \sigma^2)$ population. Show that the cumulants K_k' of $X_{(r)} - \bar{X}$ are related to the cumulants K_k of $X_{(r)}$ by

$$K_1' = K_1 - \mu,$$
$$K_2' = K_2 - \sigma^2/n,$$
$$K_k' = K_k \qquad k > 2.$$

Hence indicate how to find the first four moments of the statistics

    (a) $(X_{(n)} - \bar{X})/S_\nu$,

    (b) $(X_{(n)} - \bar{X})/S^{(P)}$.

                                        (McKay, 1935; Ruben, 1954)

[See also Borenius (1959, 1966) for a detailed study of $(X_{(n)} - \bar{X})/[\Sigma (X_i - \bar{X})^2/n]^{1/2}$].

5.3.2. Let $X_1, X_2, \ldots, X_n$ be a random sample from a $N(\mu, \sigma^2)$ parent, and let S_ν^2 be an independent estimator of σ^2 such that $\nu S_\nu^2/\sigma^2 \frown \chi_\nu^2$. Show that a first approximation to the upper α significance point of $(X_{(n)} - \bar{X})/S_\nu$ is given by

$$\left(\frac{n-1}{n}\right)^{1/2} t_\nu\left(\frac{\alpha}{n}\right),$$

where $t_\nu(\alpha/n)$ denotes the upper α/n significance point of a t variate with ν DF.

                                                (David, 1956)

5.3.3. Let $X_1, X_2, \ldots, X_n$ be a random sample from a $N(\mu, \sigma^2)$ parent, and let $S^2 = \Sigma (X_i - \bar{X})^2/(n-1)$.

(a) Show that $Y_i = (X_i - \bar{X})/S$ $(i = 1, 2, \ldots, n)$ is distributed as

$$\frac{(n-1)t_{n-2}}{[n(t_{n-2}^2 + n - 2)]^{1/2}}, \tag{A}$$

where t_{n-2} denotes a t variate with $n - 2$ DF.

(b) Noting that the Y_i are bounded, show that $Y_{(n-1)}$, the second largest of the Y_i, cannot exceed

$$y'_{n-1} = \left[\frac{\frac{1}{2}(n-1)(n-2)}{n} \right]^{\frac{1}{2}}.$$

(c) Hence prove that for $y \geq y'_{n-1}$

$$\Pr \{Y_{(n)} > y\} = n \Pr \{Y_1 > y\},$$

and that the upper α significance point of $Y_{(n)}$ is obtained by setting $t_{n-2} = t_{n-2}(\alpha/n)$ in (A). (Note: This method provides *exact* upper 5% points for $n \leq 14$ and 1% points for $n \leq 19$.)

(Pearson and Chandra Sekar, 1936)

5.3.4. If $G(X) \geq 0$, $H(X) \geq 0$ are both strictly monotonically increasing functions of a variate X with cdf $F(x)$ ($0 \leq x \leq \infty$), and if both $G(X)$ and $H(X)$ have finite expectations, then

$$\mathscr{E}[G(X)H(X)] > \mathscr{E}[G(X)]\mathscr{E}[H(X)].$$

Hence show that (5.3.7) does not hold when

$$Y_1 = \chi_1^2/\chi^2, \qquad Y_2 = \chi_2^2/\chi^2,$$

where χ_1^2, χ_2^2, χ^2 are independent χ^2 variates.

(Kimball, 1951)

[For other counter-examples to (5.3.7) see Halperin (1967) and Esary *et al.* (1967).]

5.3.5. Show that, for every choice of $i_1, i_2, \ldots, i_k$ such that $1 \leq i_1 < i_2 < \cdots < i_k \leq n$, and with $x_{i_1} < x_{i_2} < \cdots < x_{i_k}$,

$$\Pr \{X_{(i_1)} \leq x_{i_1}, \ldots, X_{(i_k)} \leq x_{i_k}\} \geq \prod_{j=1}^{k} \Pr \{X_{i_j} \leq x_{i_j}\},$$

$$\Pr \{X_{(i_1)} > x_{i_1}, \ldots, X_{(i_k)} > x_{i_k}\} \geq \prod_{j=1}^{k} \Pr \{X_{i_j} > x_{i_j}\}.$$

(Esary *et al.*, 1967)

5.3.6. Let $Y_1, Y_2, \ldots, Y_n$ have the equiprobability multinomial distribution

$$p(y_1, y_2, \ldots, y_n) = \frac{N!}{y_1! y_2! \cdots y_n!} \left(\frac{1}{n}\right)^N \qquad y_i \geq 0, \; \sum_{i=1}^{n} y_i = N.$$

Show that

$$S_1 - S_2 \leq \Pr \{Y_{(n)} > y\} \leq S_1,$$

where

$$S_1 = n \sum_{i=y+1}^{N} \binom{N}{i} \frac{(n-1)^{N-i}}{n^N},$$

$$S_2 = \binom{n}{2} \sum \frac{N!}{i! j! (N-i-j)!} \frac{(n-2)^{N-i-j}}{n^N},$$

the summation extending over

$$y < i, j \qquad i + j \le N.$$

If

$$Z_i = \frac{Y_i - N/n}{[N(n-1)/n^2]^{1/2}},$$

show that approximate upper α significance points of $Z_{(n)}$ are given by $\Phi^{-1}[1 - (\alpha/n)]$.

(F. N. David and Barton, 1962; Kozelka, 1956)

5.3.7. Write $P_i = \Pr\{A_i\}$, $P_{ij} = \Pr\{A_i A_j\}$ for $i, j = 1, 2, \ldots, n$ (so that $P_{ii} = P_i$).

Let I_i be the indicator variate of A_i. Then $\max\limits_{i=1,2,\ldots,n} I_i$ is the indicator variate of $\bigcup\limits_{i=1}^{n} A_i$.
Let

$$\mathbf{P}' = (P_1, P_2, \ldots, P_n), \quad \mathbf{Q} = (P_{ij}), \quad \mathbf{I}' = (I_1, I_2, \ldots, I_n),$$

and let $\mathbf{Q}^-$ denote the generalized inverse of the $n \times n$ matrix $\mathbf{Q}$, that is, $\mathbf{Q}\mathbf{Q}^-\mathbf{Q} = \mathbf{Q}$. By noting that for any vector $\mathbf{a}' = (a_1, a_2, \ldots, a_n)$

$$(\mathbf{a}'\mathbf{I})^2 - 2(\mathbf{a}'\mathbf{I}) + \max\limits_{i=1,2,\ldots,n} I_i \ge 0$$

and taking expectations, show that

$$\Pr\left\{\bigcup_{i=1}^{n} A_i\right\} \ge 2\mathbf{a}'\mathbf{P} - \mathbf{a}'\mathbf{Q}\mathbf{a},$$

and hence that

$$\Pr\left\{\bigcup_{i=1}^{n} A_i\right\} \ge \mathbf{P}'\mathbf{Q}^-\mathbf{P}.$$

(Kounias, 1968)

5.3.8. In Ex. 5.3.7 let A_i be the event $Y_i > y$ and put $V = \mathbf{a}'\mathbf{I}$. Noting that $\Pr\{V \ne 0\} \le \Pr\{Y_{(n)} > y\}$, obtain the inequality

$$\Pr\{Y_{(n)} > y\} \ge \frac{\mathbf{a}'\mathbf{P}\mathbf{P}'\mathbf{a}}{\mathbf{a}'\mathbf{Q}\mathbf{a}}$$

and, in particular (Chung and Erdös, Whittle),

$$\Pr\{Y_{(n)} > y\} \ge \frac{\left(\sum\limits_{i=1}^{n} P_i\right)^2}{\sum\limits_{i=1}^{n}\sum\limits_{j=1}^{n} P_{ij}}. \qquad (A)$$

Also show that (A) is stronger than the Bonferroni inequality

$$\Pr\{Y_{(n)} > y\} \ge \Sigma P_i - \sum_{i<j}\sum P_{ij}$$

iff

$$\Sigma P_i < 2\sum_{i<j}\sum P_{ij}$$

and, in particular, for identically distributed Y_i iff

$$P_1 < (n-1)P_{12}.$$

(Cf. Gallot, 1966)

5.4.1. Let $y_{(1)} \leq y_{(2)} \leq \cdots \leq y_{(n)}$ denote the ordered values of the spacings $y_j (j = 1, 2, \ldots, n)$ of (5.4.2), and let

$$z_j = (n + 1 - j)(y_{(j)} - y_{(j-1)}) \qquad y_{(0)} = 0.$$

Show that the joint distribution of the Z_j is the same as that of the Y_j.

(Durbin, 1961)

5.4.2. Show that, when $n - 1$ points randomly divide the unit interval, the probability that exactly r intervals exceed x is

$$\binom{n}{r} \left\{ (1 - rx)^{n-1} - (n - 1)[1 - (r + 1)x]^{n-1} + \cdots \pm \frac{(n - 1)!}{(k - r)!(n - k)!} (1 - kx)^{n-1} \right\},$$

where k is the largest integer less than $1/x$.

(Stevens, 1939; Fisher, 1940)

5.4.3. With the help of (5.4.3) show that

(a) $\Pr\{Y_{(1)} > c\} = (1 - nc)^{n-1} \qquad 0 \leq c \leq 1/n.$

(b) $\Pr\{Y_{(1)} > c_1, Y_{(2)} > c_2\} = n[1 - c_1 - (n - 1)c_2]^{n-1} - (n - 1)(1 - nc_1)^{n-1}$

$$0 \leq c_1 \leq c_2; c_1 + (n - 1)c_2 \leq 1.$$

(c) $\Pr\{Y_{(1)} > c_1, \ldots, Y_{(r)} > c_r\} = \binom{n}{r-1}[1 - c_1 - \cdots - c_{r-1} - (n - r + 1)c_r]^{n-1}$

minus terms involving fewer than r of the c_i's.

(d) The joint pdf of $Y_{(1)}, Y_{(2)}, \ldots, Y_{(r)}$ is

$$\binom{n - 1}{r - 1} n^{(r+1)} [1 - y_1 - \cdots - y_{r-1} - (n - r + 1)y_r]^{n-r-1}$$

$$0 \leq y_1 \leq \cdots \leq y_r; y_1 + \cdots + y_{r-1} + (n - r + 1)y_r \leq 1.$$

(Cf. Barton and F. N. David, 1956)

5.4.4. Suppose that n random samples of m have been drawn independently from normal $N(\mu_i, \sigma_i^2)$ populations. Let $_iS^2$ be the unbiased mean-square estimator of σ_i^2, and write

$$Y_{(n)} = \frac{\max {}_iS^2}{\sum_{i=1}^{n} {}_iS^2} .$$

Show that on the null hypothesis

$$H_0: \sigma_1^2 = \sigma_2^2 = \cdots = \sigma_n^2$$

upper α significance points y_α of $Y_{(n)}$ are given approximately by

$$I_{y_\alpha}\left[\frac{m - 1}{2}, \frac{(n - 1)(m - 1)}{2}\right] = 1 - \frac{\alpha}{n}, \qquad (A)$$

where $I_x(a, b)$ is the incomplete beta function. Also show that (A) gives y_α *exactly* if $y_\alpha > \frac{1}{2}$.

(Cochran, 1941)

5.4.5. Let $X_1, X_2, \ldots, X_n$ be a random sample from a continuous distribution with cdf $P(x)$, and let

$$Y = X_{(i)} - X_{(i-1)}, \quad Z = X_{(j)} - X_{(j-1)} \qquad i, j = 2, 3, \ldots, n.$$

By using the Markov property of order statistics (Section 2.7), or otherwise, show that

$$f(y, z) = n! \int_{-\infty}^{\infty} \int_{t+y}^{\infty} \frac{[P(t)]^{i-2}}{(i-2)!} \frac{[P(u) - P(t+y)]^{j-i-2}}{(j-i-2)!}$$

$$\cdot \frac{[1 - P(u+z)]^{n-j}}{(n-j)!} p(t)p(t+y)p(u)p(u+z) \, du \, dt \qquad j > i+1,$$

$$= n! \int_{-\infty}^{\infty} \frac{[P(t)]^{i-2}}{(i-2)!} \frac{[1 - P(t+y+z)]^{n-i-1}}{(n-i-1)!}$$

$$\cdot p(t)p(t+y)p(t+y+z) \, dt \qquad\qquad\qquad j = i+1.$$

$$\text{(Pyke, 1965)}$$

5.4.6. Let $X_1, X_2, \ldots, X_n$ be a random sample from a population with pdf $p(x)$ ($x \geq 0$).
(a) Show that the joint pdf of $X_1, X_2, \ldots, X_n$, given $X_1 = X_{(n)}$, is

$$f(x_1, x_2, \ldots, x_n) = np(x_1)p(x_2) \cdots p(x_n) \quad \text{if } x_1 = x_{(n)},$$
$$= 0 \qquad\qquad\qquad\qquad \text{otherwise.}$$

(b) Let $Y_{(1)} = \Sigma X_i / X_{(n)}$. Writing the characteristic function $\mathscr{E}(e^{itY}{}_{(1)})$ as a n-fold integral, show that

$$\mathscr{E}(e^{itY}{}_{(1)}) = ne^{it} \int_0^{\infty} \left(\int_0^{\beta} e^{it\alpha/\beta} p(\alpha) \, d\alpha \right)^{n-1} p(\beta) \, d\beta.$$

(c) Hence prove that, when the X_i are uniform over $(0, a)$, the distribution of $Y_{(1)}$ is the same as the distribution of $1 + U_1 + U_2 + \cdots + U_{n-1}$, where the U_i are independent and uniformly distributed over $(0, 1)$.
(d) Prove the result of (c) by using the Markov property of order statistics.

$$\text{(Darling, 1952a,b)}$$

5.4.7. A line of N elements is broken at $n - 1$ randomly chosen places to give n intervals. By considering the coefficient of x^N in $(x + x^2 + \cdots + x^m)^n$, show that the cdf of the length M of the longest interval is

$$\Pr \{M \leq m\} = \frac{1}{\binom{N-1}{n-1}} \sum_{i=0}^{a} (-1)^i \binom{n}{i} \binom{N - mi - 1}{n - 1},$$

where

$$a = \min \left(n, \left[\frac{N - n}{m} \right] \right), \quad N - n + 1 \geq m \geq \left[\frac{N + n - 1}{n} \right].$$

$$\text{(Barton and F. N. David, 1959)}$$

5.5.1. Let $X_0, X_1, X_2, \ldots, X_n$ be $n + 1$ independent variates with common variance σ^2.
(a) Show that the variates Y_i defined by

$$Y_i = X_i - aX_0 \qquad i = 1, 2, \ldots, n$$

are equicorrelated and that by suitable choice of the constant a the Y_i may be made to assume any positive equal correlation.

(b) Hence prove that, for *any* set of n standardized multinormally distributed variates Y_i with equal positive correlation coefficient ρ, the cdf of their maximum $Y_{n:n}$ is given by

$$F_{n:n}(y) = \int_{-\infty}^{\infty} \Phi^n[(y + ax)(1 + a^2)^{\frac{1}{2}}] \, d\Phi[x(1 + a^2)^{\frac{1}{2}}],$$

where Φ is the unit normal cdf and $a = [\rho/(1 - \rho)]^{\frac{1}{2}}$. In particular, show that, if $\rho = \frac{1}{2}$, the Y_i are all positive with probability $1/(n + 1)$.

(c) Show also that for the Y_i as in (b) the kth cumulant $K_{k,r}^*$ of $Y_{r:n}$ is related to the kth cumulant $K_{k,r}$ of the rth normal order statistic by

$$K_{1,r}^* = (1 - \rho)^{\frac{1}{2}}K_{1,r},$$
$$K_{2,r}^* = \rho + (1 - \rho)K_{2,r},$$
$$K_{k,r}^* = (1 - \rho)^{k/2}K_{k,r} \qquad k > 2.$$

(Stuart, 1958; Owen and Steck, 1962)

5.5.2. Let Y_i $(i = 1, 2, \ldots, n)$ be standardized multinormally distributed variates with correlation matrix (ρ_{ij}) of the form $(\rho_{ij}) = (\alpha_i\alpha_j)$, where $-1 \leq \alpha_i < 1$.

(a) Show that the Y_i can be generated from $n + 1$ independent unit normal variates X_0, $X_1, \ldots, X_n$ by setting

$$Y_i = (1 - \alpha_i^2)^{\frac{1}{2}}X_i + \alpha_iX_0 \qquad i = 1, 2, \ldots, n.$$

(b) Hence show that

$$\Pr\left\{\bigcap_{i=1}^{n} (Y_i < y_i)\right\} = \int_{-\infty}^{\infty} \left[\prod_{i=1}^{n} \Phi\left(\frac{y_i - \alpha_ix_0}{(1 - \alpha_i^2)^{\frac{1}{2}}}\right) \right] d\Phi(x_0).$$

(Dunnett and Sobel, 1955; see also Curnow and Dunnett, 1962)

5.5.3. If Y_1, Y_2 are standardized bivariate normal variates with correlation coefficient ρ, show that the pdf of $Y = a_1Y_{1:2} + a_2Y_{2:2}$ is

$$f(y) = 2\xi^{-\frac{1}{2}}\phi(\xi^{-\frac{1}{2}}y)\Phi(\eta y),$$

where

$$\xi = a_1^2 + a_2^2 + 2\rho a_1a_2,$$

and

$$\eta = \frac{[2(1 - \rho)]^{\frac{1}{2}}(a_2 - a_1)}{(1 + \rho)(a_2 + a_1)^2}.$$

(Gupta and Pillai, 1965)

5.5.4. Let Z_i $(i = 1, 2, \ldots, k)$ be the minimum of variates Y_{ij} $(j = 1, 2, \ldots, n)$ in a random sample of size n from a k-variate population with joint pdf or pf $p(y_1, y_2, \ldots, y_k)$

(a) Show that

$$\Pr\{Z_1 > z_1, Z_2 > z_2, \ldots, Z_k > z_k\} = [\Pr\{Y_1 > z_1, Y_2 > z_2, \ldots, Y_k > z_k\}]^n.$$

(b) If the joint pdf is the k-variate Pareto Type I population

$$p(y_1, y_2, \ldots, y_k; \theta) = (\theta + k - 1)^{(k)} \Big/ \left(\prod_{i=1}^{k} a_i\right)\left[\left(\sum_{i=1}^{k} a_i^{-1}y_i\right) - k + 1\right]^{\theta+k}$$

$$y_i > a_i > 0; \quad \theta > 0,$$

prove that the joint distribution of the Z_i is again Pareto Type I with pdf $p(y_1, y_2, \ldots, y_k; n\theta)$.

(Mardia, 1964b)

5.5.5. The variates $X_1, X_2, \ldots, X_n, \ldots$ follow a stationary Markov process. Writing

$$P_1(x) = \Pr \{X_i \leq x\},$$
$$P_2(x, y) = \Pr \{X_i \leq x, X_{i+1} \leq y\} \qquad \text{all } i,$$

and using the Markov chain condition

$$\Pr \{X_n \leq x \mid X_1 \leq x, X_2 \leq x, \ldots, X_{n-1} \leq x\} = \Pr \{X_n \leq x \mid X_{n-1} \leq x\}$$

(for all x and all positive n), show that the cdf of the largest value in samples of n is

$$F_n(x) = [P_2(x, x)]^{n-1}/[P_1(x)]^{n-2}.$$

(Epstein, 1949b)

CHAPTER 6

Order Statistics in Estimation and Hypothesis Testing

6.1. INTRODUCTION AND BASIC RESULTS

Order statistics enter problems of estimation and hypothesis testing in a variety of ways. Most basic of these is the situation in which the range of a variate X depends at one or both terminals on parameter(s) to be estimated. Standard methods, whatever they are, then lead inevitably to estimators which involve order statistics, a prime example being provided by the various versions of the rectangular distribution.

After discussion of this area, where *all* use order statistics, we turn in Section 6.2 to an important application due to Lloyd (1952) (see also Sarhan and Greenberg, 1962) of generalized least squares, whereby linear functions of the order statistics ("linear estimators") can be found to estimate the parameters of distributions depending on location and scale only. Here the normal distribution is a case in point. The estimator of μ is as always the sample mean $\bar{X}$, which is also the mean of the $X_{(i)}$ $(i = 1, 2, \ldots, n)$, but the estimator of σ, being of the form $\Sigma a_i X_{(i)}$, is quite distinct from the usual optimal root-mean-square estimator. What can be said in favor of such an estimator? Actually very little in the normal case with all observations at hand, although the efficiency of the linear estimator is always close to unity. But for certain non-normal populations more traditional estimation procedures may be very tedious. What is perhaps more important in these days of computers, such procedures may give us estimators whose properties in small samples are little understood and possibly far from satisfactory; it is hardly enough that, as in the method of maximum likelihood, large-sample properties are often attractive. On the other hand, Lloyd's approach gives estimators which are always unbiased and also of minimum variance (for all n) in the class of unbiased linear estimators.

93

When an experiment, such as a life test, is terminated as soon as a prescribed number N ($< n$) of items has failed, the resulting data are censored in a manner ensuring that the estimators, whether maximum likelihood or Lloyd's, will involve order statistics. The latter estimators tend to be much more convenient provided the necessary tables of coefficients are available. These matters are discussed in Section 6.3 and, with emphasis on the exponential distribution, in Section 6.4.

We conclude this chapter with some remarks on an interesting and largely new subject: robust estimation. The aim here is to find estimators which are satisfactory not only when conditions are ideal but also when the distributional assumptions on the parent population are violated (within limits).

Order statistics have long played a useful role in short-cut techniques. This aspect is treated in Chapter 7. We return now to some basic results.

As has already been seen in connection with distribution-free confidence and tolerance intervals (Sections 2.5 and 2.6), order statistics are of fundamental importance in nonparametric inference. From a more theoretical standpoint it is also of interest that, if X_1, X_2, ... , X_n are independent[1] continuous variates with common cdf $P(x, \theta)$, where θ may be vector-valued, then the order statistics $(X_{(1)}, X_{(2)}, \ldots, X_{(n)}) = \mathbf{T}$ (say) are sufficient for θ. To see this, note that, given

$$\mathbf{T} = \mathbf{t} \equiv (x_{(1)}, x_{(2)}, \ldots, x_{(n)}),$$

the X_i ($i = 1, 2, \ldots, n$) are constrained to take on the values $x_{(j_i)}$, which by symmetry they must do with equal probability for each of the $n!$ permutations $(j_1, j_2, \ldots, j_n)$ of $(1, 2, \ldots, n)$. In other words, for every $(j_1, j_2, \ldots, j_n)$,

$$\Pr \{X_1 = x_{(j_1)}, X_2 = x_{(j_2)}, \ldots, X_n = x_{(j_n)} \mid \mathbf{T} = \mathbf{t}\} = 1/n!.$$

Since this probability is independent of θ, the sufficiency of $\mathbf{T}$ for θ is established. Informally this result merely says that the serial order of the X_i has no bearing on any inference concerning θ (under the null hypothesis of independent identically distributed X_i), since all serial orders give rise to the same order statistics. Although the $X_{(i)}$ may not seem to represent much of a condensation of the X_i (themselves trivially sufficient for θ), it can in fact be shown that, if no knowledge about P is available other than its absolute continuity, the $X_{(i)}$ provide the maximum condensation and do so uniquely (except for equivalent characterizations). Basic as these properties of minimal sufficiency and completeness are in the theory of nonparametric inference, we have touched on them only lightly, as they will concern us no further. The interested reader is referred to Bell *et al.* (1960) and the references there given for a more extensive and rigorous account.

---

[1] Independence may be broadened to exchangeability.

Estimation

Suppose now that the functional form of P is known. If the range of x depends on θ at one or both terminals, order statistics enter very forcibly into the estimation process. The resulting problems have become part and parcel of books on (parametric) inference. Since the difficulties are primarily inferential and not order statistical, the reader may wish to consult, for example, Hogg and Craig (1959), Kendall and Stuart (1961), or Lehmann (1959). However, we will now present some of the main results, with special emphasis on the uniform distribution. Consideration of the exponential distribution with unknown starting point is deferred to Section 6.4.

Suppose first that the range of x depends on θ (scalar) only at the lower terminal, namely, $a(\theta) \leq x \leq b$. Then, if a sufficient statistic for θ exists, (i) it must be a monotonic function of $X_{(1)}$; and (ii) the pdf must be of the form $p(x; \theta) = c(\theta)g(x)$ with $c(\theta)$, $g(x)$ both non-negative (Pitman, 1936; Davis, 1951). To see (i), we need note only that the conditional pdf $f_1(x \mid T = t)$ of $X_{(1)}$ for any statistic T cannot be independent of θ unless T uniquely determines $X_{(1)}$ in (a, b). For (ii), let X_t ($t = 1, 2$) be any two variates in the sample other than $X_{(1)}$. Then $f(x_t \mid x_{(1)})$, which equals

$$\frac{p(x_t; \theta)}{\displaystyle\int_{x_{(1)}}^{b} p(x_t; \theta)\, dx_t},$$

is independent of θ, so that $p(x_1; \theta)/p(x_2; \theta)$ is also independent of θ, which establishes (ii). The results stated hold, of course, equally for $a \leq x \leq b(\theta)$, in which case $X_{(n)}$ is sufficient for θ.

Example 6.1.1.   Let X be uniform in $(0, \theta)$, namely,

$$p(x; \theta) = 1/\theta \qquad 0 \leq x \leq \theta; \theta > 0,$$
$$= 0 \qquad \text{elsewhere.}$$

This is of the above form with $g(x) = 1$. To show that $X_{(n)}$ is indeed sufficient for θ it is convenient to absorb the dependence of x on θ into the pdf itself, rather than have it as a side restriction. To this end, define $v(x, y) = 1$ for $x \leq y$, $v(x, y) = 0$ for $x > y$. Then $p(x; \theta)$ may be written as

$$p(x; \theta) = v(x, \theta)/\theta \qquad x \geq 0; \theta > 0.$$

The likelihood function is

$$L(\theta) = \theta^{-n} \prod_{i=1}^{n} v(x_i, \theta)$$
$$= \theta^{-n} v(x_{(n)}, \theta),$$

which by the factorization criterion establishes the sufficiency of $X_{(n)}$.

Other properties of $X_{(n)}$ are easily determined from first principles. $X_{(n)}$ clearly underestimates θ; in fact,

$$\mathscr{E}X_{(n)} = n\theta/(n + 1),$$

so that $(n + 1)X_{(n)}/n$ is an unbiased estimator of θ. The distribution of $X_{(n)}$ is given by

$$\Pr\{X_{(n)} \leq x\} = (x/\theta)^n \qquad 0 \leq x \leq \theta,$$

from which it follows that the asymptotic distribution of (the maximum-likelihood estimator) $X_{(n)}$, suitably standardized, is not normal but exponential, since

$$\lim_{n \to \infty} \Pr\{n(\theta - X_{(n)}) \leq u\} = \lim_{n \to \infty}\left[1 - \left(1 - \frac{u}{n\theta}\right)^n\right]$$
$$= 1 - e^{-u/\theta} \qquad u \geq 0,$$

a simple example of Case II of the asymptotic distributions of the extreme (see Section 9.3). Finally, $X_{(n)}$ is complete for θ; for if $u(X_{(n)})$ is some function of $X_{(n)}$, then the identity $\mathscr{E}[u(X_{(n)})] = 0$ for all θ implies

$$\int_0^\theta u(x)x^{n-1}\,dx = 0 \quad \text{for all } \theta, \tag{6.1.1}$$

which in turn implies $u(x) = 0$ a.e. (Ex. 6.1.1). By definition this proves the completeness of $X_{(n)}$.

From the small-sample results among the above, it follows that $(n + 1)X_{(n)}/n$ is the unique uniformly minimum variance unbiased (UMVU) estimator of θ. Also by Basu's (1955) theorem a statistic sufficient and complete for a parameter $\boldsymbol{\theta}$ (which may be vector-valued) is distributed independently of any statistic whose distribution does not involve $\boldsymbol{\theta}$. In particular, $X_{(n)}$ is therefore statistically independent of

$$\sum_{i=1}^n X_i/X_{(n)} = Y\text{(say)}.$$

Consequently, $f(y \mid x_{(n)})$ does not depend on $x_{(n)}$, which may be taken to be unity. Given $x_{(n)} = 1$, the other $n - 1$ X_i and hence the ratios $U_i = X_i/X_{(n)}$ are independently uniformly distributed in $(0, 1)$ by the Markov property of order statistics. Thus Y is distributed as $1 + \sum_{i=1}^{n-1} U_i$. This line of proof of a result of Darling (Ex. 5.4.6) is due to Hogg and Craig (1956).

Example 6.1.2. Let X be uniform in (θ_1, θ_2):

$$p(x; \theta_1, \theta_2) = \frac{1}{\theta_2 - \theta_1} \qquad \theta_1 \leq x \leq \theta_2; \theta_2 > \theta_1.$$

We leave it to the reader to show that $X_{(1)}$ and $X_{(n)}$ are jointly sufficient and complete for θ_1 and θ_2. It is of interest to rewrite the pdf as

$$p(x; \mu, \omega) = \frac{1}{\omega} \quad \mu - \tfrac{1}{2}\omega \leq x \leq \mu + \tfrac{1}{2}\omega; \, \omega > 0,$$

$$= 0 \quad \text{elsewhere.}$$

Then

$$M = \tfrac{1}{2}(X_{(1)} + X_{(n)}) \quad \text{and} \quad W' = \frac{n+1}{n-1}(X_{(n)} - X_{(1)})$$

are unbiased estimators of μ and ω. Being functions of $X_{(1)}$ and $X_{(n)}$, which are of course also sufficient and complete for μ and ω, M and W' are unique UMVU estimators. Since var $M = \omega^2/[2(n+1)(n+2)]$ (Ex. 2.3.5), the efficiency, defined as the ratio of variances, of $\bar{X}$ relative to M is $6n/(n+1)(n+2)$ and tends to 0 as $n \to \infty$.[2] See also Ex. 6.1.2.

Example 6.1.3.   Let X be uniform in $(\theta - \tfrac{1}{2}, \theta + \tfrac{1}{2})$:

$$p(x; \theta) = 1 \quad \theta - \tfrac{1}{2} \leq x \leq \theta + \tfrac{1}{2}.$$

Here the likelihood function is a maximum (namely, 1) if both $x_{(1)}$ and $x_{(n)}$ lie in $(\theta - \tfrac{1}{2}, \theta + \tfrac{1}{2})$. Thus $X_{(1)}$ and $X_{(n)}$ are jointly sufficient for the single parameter θ. No single sufficient statistic exists, but $X_{(1)}$ and $X_{(n)}$ are not complete since

$$\mathscr{E}\left(X_{(n)} - X_{(1)} - \frac{n-1}{n+1}\right) = 0$$

for all θ. However, M is still the MVU estimator of θ, being unbiased and a function of the minimal sufficient statistics.

When the range of x depends on θ at both terminals, namely, $a(\theta) \leq x \leq b(\theta)$, it is again necessary but no longer sufficient that $p(x; \theta) = C(\theta)g(x)$. An additional requirement is now that $a(\theta)$ be monotonic increasing and $b(\theta)$ monotonic decreasing, or the reverse. In the former case the sufficient statistic is

$$\hat{\theta} = \min\{a^{-1}(X_{(1)}), b^{-1}(X_{(n)})\};$$

in the latter it is

$$\hat{\theta}' = \max\{a^{-1}(X_{(1)}), b^{-1}(X_{(n)})\},$$

where $a^{-1}(x)$, $b^{-1}(x)$ are the functions inverse to $a(x)$, $b(x)$.

Following Huzurbazar (1955), we now derive the pdf of $\hat{\theta} = Z$. From Fig. 6.1 it is seen that

$$\Pr\{z \leq \hat{\theta} \leq z + dz\} = \Pr\{a(z) \leq X_{(1)} \leq a(z + dz), a(z) \leq X_{(n)} \leq b(z)\}$$
$$+ \Pr\{a(z) \leq X_{(1)} \leq b(z), b(z + dz) \leq X_{(n)} \leq b(z)\}. \quad (6.1.2)$$

[2] Note, however, that M is not asymptotically normally distributed, so that "efficiency" is used here in a wider sense than is customary.

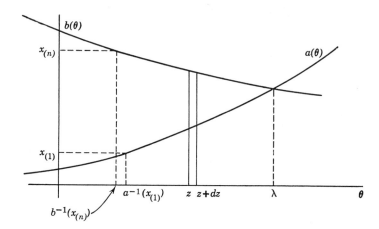

Fig. 6.1

Since, in our usual notation, with θ understood, the joint pdf of $X_{(1)}$ and $X_{(n)}$ is

$$f_{1,n}(x, y) = n(n - 1)p(x)[P(y) - P(x)]^{n-2}p(y) \qquad x \leq y,$$

the RHS of (6.1.2) can be written as

$$f(z) = n(n - 1)p[a(z)]a'(z)\int_{a(z)}^{b(z)} \{P(y) - P[a(z)]\}^{n-2}p(y)\,dy$$

$$- n(n - 1)p[b(z)]b'(z)\int_{a(z)}^{b(z)} \{P[b(z)] - P(x)\}^{n-2}p(x)\,dx, \quad (6.1.3)$$

where primes denote differentiation.

Now

$$1 = \int_{a(\theta)}^{b(\theta)} p(y)\,dy = C(\theta)\int_{a(\theta)}^{b(\theta)} g(y)\,dy,$$

so that

$$\int_{a(z)}^{b(z)} p(y)\,dy = C(\theta)\int_{a(z)}^{b(z)} g(y)\,dy = \frac{C(\theta)}{C(z)}. \qquad (6.1.4)$$

Hence

$$(n - 1)\int_{a(z)}^{b(z)} \{P(y) - P[a(z)]\}^{n-2}p(y)\,dy = \{P[b(z)] - P[a(z)]\}^{n-1}$$

$$= \left[\frac{C(\theta)}{C(z)}\right]^{n-1}$$

and (6.1.3) reduces to

$$f(z) = \frac{n\{p[a(z)]a'(z) - p[b(z)]b'(z)\}C^{n-1}(\theta)}{C^{n-1}(z)}$$

$$= \frac{n\{-(d/dz)[C(\theta)/C(z)]\}C^{n-1}(\theta)}{C^{n-1}(z)} \quad \text{by (6.1.4)}$$

$$= \frac{nC^n(\theta)C'(z)}{C^{n+1}(z)}. \quad (6.1.5)$$

The range of z is from θ to λ, the latter given by $a(\lambda) = b(\lambda)$.

Example 6.1.4. Suppose that X is uniform in $(-\theta, \theta)$:

$$p(x; \theta) = 1/(2\theta) \qquad -\theta \le x \le \theta.$$

Here $a(\theta) = -\theta$ *decreases* and $b(\theta) = \theta$ increases with θ. Thus

$$\theta' = Z = \max\{-X_{(1)}, X_{(n)}\} = \max\{|X_{(1)}|, |X_{(n)}|\},$$

and (6.1.5) holds with a minus sign on the RHS, giving

$$f(z) = -n\left(\frac{1}{2\theta}\right)^n\left(-\frac{1}{2z^2}\right)(2z)^{n+1}$$

$$= \frac{nz^{n-1}}{\theta^n} \qquad 0 \le z \le \theta,$$

as is easily seen also from first principles.

Confidence intervals for θ in the situation of (6.1.5) are readily found on noting that $V = C(\theta)/C(Z)$ has pdf $f(v) = nv^{n-1}$ in $(0, 1)$ (Ex. 6.1.4).

Hypothesis Testing

Tests of significance corresponding to the various rectangular cases considered can now be constructed, but some of the results may be a little surprising at first sight. Thus in the simplest $R(0, \theta)$ case the obvious test of $H: \theta \le \theta_0$ against $K: \theta > \theta_0$ is to reject H when $x_{(n)}$ is too large, choosing the α-level significance point $x_{n,\alpha}$ so that

$$\Pr\{X_{(n)} > x_{n,\alpha} \mid \theta = \theta_0\} = \alpha,$$

that is, $x_{n,\alpha} = \theta_0(1 - \alpha)^{1/n}$. This test is uniformly most powerful (UMP) but, as Lehmann (1959, p. 110) points out, by no means unique; in fact, any test which (i) rejects when $x_{(n)} > \theta_0$, (ii) has level α when $\theta = \theta_0$, and (iii) has rejection level $\le \alpha$ for $\theta < \theta_0$ is also UMP [e.g., the test combining (i) and rejection with probability α whenever $x_{(n)} \le \theta_0$]. However, there is a unique

UMP test of $H_0: \theta = \theta_0$ against $K: \theta \neq \theta_0$, namely,

reject H_0 if $x_{(n)} > \theta_0$ or if $x_{(n)} < \theta_0 \alpha^{1/n}$, and
accept H_0 otherwise.

Next, suppose that two independent samples $X_1, X_2, \ldots, X_{n_1}$ and Y_1, $Y_2, \ldots, Y_{n_2}$ are available from $R(0, \theta_1)$, $R(0, \theta_2)$ parents. To test $H_0: \theta_1 = \theta_2$ (or $H: \theta_1 \leq \theta_2$) against $K: \theta_1 > \theta_2$, one is led to reject for large values of $V = X_{(n_1)}/Y_{(n_2)}$. From the null distribution of V in Ex. 2.3.12, upper α significance points v_α of V are given by

$$v_\alpha(n_1, n_2) = \left[\frac{n_1}{\alpha(n_1 + n_2)} \right]^{1/n_2}.$$

For n_1, $n_2 \leq 10$ Murty (1955) tabulates $v_{0.05}$ (but the footnote to his table should be ignored). A two-sided form of this test consists in rejecting H_0 (against $K: \theta \neq \theta_0$) when

$$\text{either} \quad v > v_{\frac{1}{2}\alpha}(n_1, n_2), \quad \text{or} \quad v < \frac{1}{v_{\frac{1}{2}\alpha}(n_2, n_1)} . \tag{6.1.6}$$

This merely utilizes the fact that the null distribution of $1/V$ is the same as that of V with the degrees of freedom reversed. However, as in other similar situations, this convenient equal-tails test is biased (i.e., the probability of rejection is less than α for some $\theta_1 \neq \theta_2$) unless $n_1 = n_2$. The likelihood ratio test rejects H_0 when

$$\text{either} \quad v > \alpha^{-1/n_2}, \quad \text{or} \quad v < \alpha^{1/n_1}, \tag{6.1.7}$$

and this is UMP unbiased. For $n_1 = n_2$ tests (6.1.6) and (6.1.7) are the same and the common test is UMP. See Barr (1966) for these and further results.

To test whether the two samples are from rectangular populations with the same range (without assuming a common mean), one would use the range ratio W_1/W_2 (see Ex. 2.3.10). Tables of percentage points are given by Rider (1951) and Hyrenius (1953). The latter author also deals with tests for (a) differences in location by use of the statistic $T = (Y_{(1)} - X_{(1)})/(X_{(n_1)} - X_{(1)})$ (Ex. 2.7.1) and (b) differences in location and dispersion by use of $V = (Y_{(n_2)} - X_{(1)})/(X_{(n_1)} - X_{(1)})$. [Note that his samples are labeled so that $X_{(1)} \leq Y_{(1)}$; the range ratio $U = (Y_{(n_2)} - Y_{(1)})/(X_{(n_1)} - X_{(1)})$ is therefore not quite the same as Rider's and has little to recommend it, except possibly in conjunction with T and V.]

Extensions to some of the corresponding k-sample problems can be made by the union-intersection principle (Khatri, 1960, 1965). Also, the likelihood ratio approach can still be applied to testing equality of the θ_i when the ith pdf is of the form $C(\theta)g(x)$ with the range of x depending at one or both ends on θ. See Ex. 6.1.6 and Hogg (1956).

The estimation of parameters for a log-normal variate X, that is, $\log(X - \gamma) \frown N(\mu, \sigma^2)$, has been discussed by many authors; see Hill (1963) and Lambert (1964). Bain and Thoman (1968) provide tests for the three-parameter Weibull distribution.

When a distribution is truncated, say on the right, so that the cdf of the truncated variate X is

$$P_\theta(x) = P(x)/P(\theta) \qquad x \geq \theta,$$
$$= 0 \qquad\qquad \text{elsewhere},$$

a simple estimator of θ is $X_{(n)}$. This is, of course, an underestimate, and the question arises: Can one reduce the bias of $X_{(n)}$ for any general class of distributions? Robson and Whitlock (1964) show that this can be done by an interesting application of Quenouille's (1956) method for the successive elimination of bias of order n^{-1}, n^{-2}, etc. Quenouille noted that, if the expectation of an estimator T_n takes the form

$$\mathscr{E}T_n(X_1, \ldots, X_n) = \theta + \frac{a_1}{n} + \frac{a_2}{n^2} + \frac{a_3}{n^3} + \cdots, \qquad (6.1.8)$$

the bias term a_1/n is eliminated in the estimator

$$T_{n,i}^{(1)}(X_1, \ldots, X_n) = nT_n(X_1, \ldots, X_n)$$
$$- (n-1)T_{n-1,(i)}(X_1, \ldots, X_{i-1}, X_{i+1}, \ldots, X_n).$$

Averaging over $i = 1, 2, \ldots, n$ produces the symmetrical estimator

$$T_n^{(1)} = nT_n - \frac{n-1}{n} \sum_{i=1}^{n} T_{n-1,(i)},$$

and the process can be repeated to remove a_2/n^2, giving $T_n^{(2)}$, etc. In the present case, with $T_n = X_{(n)}$, we have

$$T_{n-1,(i)} = \max(X_1, \ldots, X_{i-1}, X_{i+1}, \ldots, X_n)$$
$$= \begin{cases} X_{(n)} & \text{if } X_i \neq X_{(n)}, \\ X_{(n-1)} & \text{if } X_i = X_{(n)}. \end{cases}$$

Thus

$$T_n^{(1)} = nX_{(n)} - \frac{n-1}{n}[(n-1)X_{(n)} + X_{(n-1)}]$$

$$= \frac{2n-1}{n} X_{(n)} - \frac{n-1}{n} X_{(n-1)}.^3$$

$T_n^{(2)}$ would involve $X_{(n-2)}$ and, although unbiased to order n^{-2}, is likely to be less efficient than $T_n^{(1)}$.

---

[3] Actually Robson and Whitlock obtain the simpler result $T_n^{(1)} = 2X_{(n)} - X_{(n-1)}$ by replacing (6.1.8) with a series in $1/n^{(r)}$ rather than $1/n^r$ $(r = 1, 2, \ldots)$.

6.2. LEAST-SQUARES ESTIMATION OF LOCATION AND SCALE PARAMETERS BY ORDER STATISTICS

Suppose that $\mathscr{P}$ is a family of continuous distributions with cdf's of the form $P(ax + b)$, where $a > 0$, b are arbitrary constants. In other words, $\mathscr{P}$ is a family of distributions depending on location and scale parameters only. We denote these parameters by μ and σ, although they need not be the mean and standard deviation. It follows that $p(x) = P'(x)$ may be written as

$$p(x) = \frac{1}{\sigma} g\left(\frac{x - \mu}{\sigma}\right) \qquad \sigma > 0,$$

and that the standardized variate $Y = (X - \mu)/\sigma$ has pdf $g(y)$, free of μ and σ. The families of normal and uniform distributions provide two important examples. In the latter case we have

$$p(x) = 1/\omega \qquad \mu - \tfrac{1}{2}\omega \leq x \leq \mu + \tfrac{1}{2}\omega.$$

Then the pdf of $Y = (X - \mu)/\omega$ is simply

$$g(y) = 1 \qquad -\tfrac{1}{2} \leq y \leq \tfrac{1}{2}.$$

Since the ordered X and Y variates (in random samples of n) are linked by

$$Y_{(r)} = (X_{(r)} - \mu)/\sigma \qquad r = 1, 2, \ldots, n,$$

the moments of the $Y_{(r)}$ depend only on the form of g, and not on μ and σ. Let

$$\mathscr{E} Y_{(r)} = \alpha_r, \quad \text{cov}\,(Y_{(r)}, Y_{(s)}) = \beta_{rs} \qquad s = 1, 2, \ldots, n.$$

Then

$$\mathscr{E} X_{(r)} = \mu + \sigma \alpha_r, \quad \text{cov}\,(X_{(r)}, X_{(s)}) = \sigma^2 \beta_{rs}, \qquad (6.2.1)$$

where the α_r, β_{rs} can be evaluated once and for all (cf. Chapter 3). Thus $\mathscr{E} X_{(r)}$ is linear in the parameters μ and σ with known coefficients, and $\text{cov}\,(X_{(r)}, X_{(s)})$ is known apart from σ^2. The Gauss-Markov least-squares (LS) theorem may therefore be applied (in its slightly generalized version, since the covariance matrix is not diagonal) to give unbiased estimators of μ and σ which have minimum variance in the class of linear unbiased estimators. To see this explicitly, write the first equation of (6.2.1) as

$$\mathscr{E}\mathbf{X} = \mu\mathbf{1} + \sigma\boldsymbol{\alpha},$$

or

$$\mathscr{E}\mathbf{X} = \mathbf{A}\boldsymbol{\theta}, \qquad (6.2.2)$$

where $\mathbf{X}$, $\boldsymbol{\alpha}$ are, respectively, the column vectors of the $X_{(r)}$, α_r; $\mathbf{1}$ is a column of n 1's, and

$$\mathbf{A} = (\mathbf{1}, \boldsymbol{\alpha}), \quad \boldsymbol{\theta}' = (\mu, \sigma).$$

Also, let the covariance matrix of the $X_{(r)}$ be $\mathscr{V}(X) = \sigma^2 B$. We have to minimize with respect to θ

$$(x - A\theta)'\Omega(x - A\theta) \qquad \text{where} \qquad \Omega = B^{-1},$$

yielding the LS estimator θ^*:

$$\theta^* = (A'\Omega A)^{-1} A'\Omega X. \tag{6.2.3}$$

The covariance matrix of θ^* is

$$(A'\Omega A)^{-1} A'\Omega \cdot \sigma^2 \Omega^{-1} \cdot \Omega A (A'\Omega A)^{-1} = \sigma^2 (A'\Omega A)^{-1}, \tag{6.2.4}$$

where

$$(A'\Omega A) = \begin{pmatrix} 1' \\ \alpha' \end{pmatrix} \Omega(1, \alpha) = \begin{pmatrix} 1'\Omega 1 & 1'\Omega\alpha \\ \alpha'\Omega 1 & \alpha'\Omega\alpha \end{pmatrix},$$

all the elements of the matrix being scalar.

It follows from (6.2.3) that with $\Delta = A'\Omega A$ and $\Delta = |\Delta|$

$$\theta^* = \frac{1}{\Delta} \begin{pmatrix} \alpha'\Omega\alpha & -\alpha'\Omega 1 \\ -1'\Omega\alpha & 1'\Omega\alpha \end{pmatrix} \begin{pmatrix} 1'\Omega \\ \alpha'\Omega \end{pmatrix} X$$

$$= \frac{1}{\Delta} \begin{pmatrix} \alpha'\Omega\alpha 1'\Omega - \alpha'\Omega 1\alpha'\Omega \\ -1'\Omega\alpha 1'\Omega + 1'\Omega 1\alpha'\Omega \end{pmatrix} X$$

or

$$\mu^* = -\alpha'\Gamma X, \quad \sigma^* = 1'\Gamma X, \tag{6.2.5}$$

where Γ is the skew-symmetric matrix

$$\Gamma = \frac{\Omega(1\alpha' - \alpha 1')\Omega}{\Delta}.$$

Also by (6.2.4)

$$\operatorname{var} \mu^* = \frac{\alpha'\Omega\alpha\sigma^2}{\Delta}, \tag{6.2.6}$$

$$\operatorname{var} \sigma^* = \frac{1'\Omega 1\sigma^2}{\Delta}, \tag{6.2.7}$$

$$\operatorname{cov}(\mu^*, \sigma^*) = \frac{-1'\Omega\alpha\sigma^2}{\Delta}. \tag{6.2.8}$$

Thus μ^* and σ^* may be expressed as linear functions of the order statistics, namely,

$$\mu^* = \sum_{i=1}^{n} \beta_i X_{(i)}, \quad \sigma^* = \sum_{i=1}^{n} \gamma_i X_{(i)}, \tag{6.2.5'}$$

with coefficients which may be tabulated once and for all (see A6.3).

Simplification for Symmetrical Populations

We now consider the important case of a symmetrical parent distribution and take μ to be the population mean. Then the distribution of $(Y_{(1)}, Y_{(2)}, \ldots, Y_{(n)})$ is the same as that of $(-Y_{(n)}, -Y_{(n-1)}, \ldots, -Y_{(1)})$. Let

$$\begin{pmatrix} -Y_{(n)} \\ -Y_{(n-1)} \\ \cdot \\ \cdot \\ \cdot \\ -Y_{(1)} \end{pmatrix} = -\mathbf{J} \begin{pmatrix} Y_{(1)} \\ Y_{(2)} \\ \cdot \\ \cdot \\ \cdot \\ Y_{(n)} \end{pmatrix} \quad \text{where } \mathbf{J} = \begin{pmatrix} 0 & & & 1 \\ & & 1 & \\ & \cdot & & \\ & \cdot & & \\ 1 & & & 0 \end{pmatrix}.$$

Note that $\mathbf{J} = \mathbf{J}' = \mathbf{J}^{-1}$, $\mathbf{J}'\mathbf{1} = \mathbf{1}$. Since $\mathbf{Y}$ and $-\mathbf{JY}$ have the same distribution, we have

$$\mathscr{E}\mathbf{Y} = \mathscr{E}(-\mathbf{JY}), \text{ that is, } \boldsymbol{\alpha} = -\mathbf{J}\boldsymbol{\alpha},$$

and

$$\mathscr{V}(\mathbf{Y}) = \Omega^{-1} = \mathscr{V}(-\mathbf{JY}),$$

that is,

$$\Omega^{-1} = -\mathbf{J}\Omega^{-1}(-\mathbf{J}) = \mathbf{J}^{-1}\Omega^{-1}\mathbf{J}^{-1},$$

or

$$\Omega = \mathbf{J}\Omega\mathbf{J}.$$

It follows that

$$\mathbf{1}'\Omega\boldsymbol{\alpha} = \mathbf{1}'(\mathbf{J}\Omega\mathbf{J})(-\mathbf{J}\boldsymbol{\alpha})$$
$$= -(\mathbf{1}'\mathbf{J})\Omega(\mathbf{J}^2)\boldsymbol{\alpha} = -\mathbf{1}'\Omega\boldsymbol{\alpha}.$$

Thus

$$\mathbf{1}'\Omega\boldsymbol{\alpha} = -\mathbf{1}'\Omega\boldsymbol{\alpha} = 0,$$

so that by (6.2.8) μ^* and σ^* are uncorrelated. Hence (6.2.5)–(6.2.7) simplify to

$$\mu^* = \frac{\boldsymbol{\alpha}'\Omega\boldsymbol{\alpha} \cdot \mathbf{1}'\Omega\mathbf{X}}{\mathbf{1}'\Omega\mathbf{1} \cdot \boldsymbol{\alpha}'\Omega\boldsymbol{\alpha}} = \frac{\mathbf{1}'\Omega\mathbf{X}}{\mathbf{1}'\Omega\mathbf{1}}, \tag{6.2.9}$$

$$\sigma^* = \frac{\boldsymbol{\alpha}'\Omega\mathbf{X}}{\boldsymbol{\alpha}'\Omega\boldsymbol{\alpha}}, \tag{6.2.10}$$

$$\text{var } \mu^* = \frac{\sigma^2}{\mathbf{1}'\Omega\mathbf{1}}, \quad \text{var } \sigma^* = \frac{\sigma^2}{\boldsymbol{\alpha}'\Omega\boldsymbol{\alpha}}. \tag{6.2.11}$$

We note that μ^* reduces to the sample mean if

$$\mathbf{1}'\Omega = \mathbf{1}' \quad \text{or equivalently if } \mathbf{B}\mathbf{1} = \mathbf{1}, \tag{6.2.12}$$

that is, if all the rows (or columns) of the covariance matrix add to unity. This is the case for a unit normal parent [equation (3.2.1')]. It may also be shown that μ^* has variance strictly smaller than σ^2/n, except when (6.2.12) is satisfied. For this result, and similar ones when the parent is not symmetrical, see Lloyd (1952), Downton (1953), and Govindarajulu (1968a).

Simplified Linear Estimates

Lloyd's procedure requires full knowledge of the expectations and the covariance matrix of the order statistics. The covariances especially may be difficult to determine. Gupta (1952) has proposed a very simple method applicable when only the expectations are known, namely, to take $\mathbf{B} = \mathbf{I}$, the unit matrix. Then $\mathbf{\Omega} = \mathbf{I}$ and previous results simplify greatly. Thus

$$\Delta = \mathbf{1}'\mathbf{\Omega}\mathbf{1} \cdot \mathbf{\alpha}'\mathbf{\Omega}\mathbf{\alpha} - (\mathbf{1}'\mathbf{\Omega}\mathbf{\alpha})^2$$
$$= n \, \Sigma \, \alpha_i^2 - (\Sigma \, \alpha_i)^2 = n \, \Sigma \, (\alpha_i - \bar{\alpha})^2,$$

and the resulting estimate of μ reduces from (6.2.5) to

$$\mu^{**} = \frac{\Sigma \, \alpha_i^2 \cdot \Sigma \, X_{(i)} - \Sigma \, \alpha_i \cdot \Sigma \, \alpha_i X_{(i)}}{\Delta}$$

or

$$\mu^{**} = \sum_{i=1}^{n} b_i X_{(i)},$$

where

$$b_i = \frac{\Sigma \, \alpha_i^2 - \alpha_i \Sigma \, \alpha_i}{\Delta}$$

$$= \frac{\Sigma \, (\alpha_i - \bar{\alpha})^2 + \bar{\alpha} \Sigma \, \alpha_i - \alpha_i \cdot \Sigma \, \alpha_i}{\Delta}$$

$$= \frac{1}{n} - \frac{\bar{\alpha}(\alpha_i - \bar{\alpha})}{\Sigma \, (\alpha_i - \alpha)^2}. \qquad (6.2.13)$$

For a symmetrical parent this gives $\mu^{**} = \bar{X}$. Likewise

$$\sigma^{**} = \Sigma \, c_i X_{(i)},$$

where

$$c_i = \frac{\alpha_i - \bar{\alpha}}{\Sigma \, (\alpha_i - \bar{\alpha})^2}. \qquad (6.2.14)$$

Unpromising as this crude approach may seem, it gives surprisingly good results, at least in the normal case. This is discussed further in the next section. Since the necessity of inverting the $n \times n$ covariance matrix is completely avoided in the above method, one might consider it for occasional use even when this matrix is known. Ali and Chan (1964) show that in the normal case σ^{**} is asymptotically normal and fully efficient; what is more, the reduction in the efficiency of σ^{**} compared to that of σ^* is negligible even in small samples. This is clear from Table 6.2 (Chernoff and Lieberman,

1954; Sarhan and Greenberg, 1956; Ali and Chan, 1964), which gives the variances for $n = 2(1)10$ of σ^*, σ^{**}, and also of the unbiased maximum-likelihood estimator

$$\hat{\sigma} = \frac{\Gamma[\frac{1}{2}(n-1)]}{\sqrt{2}\,\Gamma(\frac{1}{2}n)}[\Sigma(X_i - \bar{X})^2]^{\frac{1}{2}}.$$

For $n \leq 10$ the efficiency of σ^{**} (relative to $\hat{\sigma}$) is lowest at $n = 6$, when it is 98.7%.

Table 6.2. *Comparison of three unbiased estimators of σ for a normal parent*

n	var $(\hat{\sigma}/\sigma)$	var (σ^*/σ)	var (σ^{**}/σ)
2	.57080	.57080	.57080
3	.27324	.27548	.27548
4	.17810	.18005	.18013
5	.13177	.13332	.13342
6	.10447	.10571	.10580
7	.08650	.08750	.08759
8	.07379	.07461	.07469
9	.06432	.06502	.06509
10	.05701	.05760	.05766

Blom's Estimates

An interesting and rather general approach to the estimation of location and scale parameters was advanced by Blom (1958) and later summarized by him (1962). His "unbiased nearly best linear" estimates require, as do Gupta's, the exact expectations of the order statistics $Y_{(r)}$ for the reduced variate $Y = (X - \mu)/\sigma$ with cdf $P(y)$, but use asymptotic approximations to the covariance matrix. If exact unbiasedness is given up, one may also approximate asymptotically to the expectations and obtain "nearly unbiased, nearly best" estimates.

The starting point of Blom's procedure is to approximate the covariance of $Y_{(r)}$ and $Y_{(s)}$ ($r \leq s$) by the first order term in (4.5.5), namely,

$$\text{cov}(Y_{(r)}, Y_{(s)}) \sim \frac{p_r q_s}{(n+2)p(Q_r)p(Q_s)},$$

where $p_r = r/(n+1)$, $q_s = 1 - p_s$, $Q_r = P^{-1}(p_r)$, and $p(Q_r)$ is the pdf of Y evaluated at Q_r. Now write $f_i = p(Q_i)$ ($i = 1, 2, \ldots, n$). Then

$$\text{cov}(f_i Y_{(i)}, f_j Y_{(j)}) \sim \frac{p_i q_j}{n+2} \qquad i \leq j;$$

and, setting

$$Z_{(i)} = f_{i+1} Y_{(i+1)} - f_i Y_{(i)} \qquad f_0 = f_{n+1} = 0, \qquad (6.2.15)$$

we have *independently of P* for $0 \le i \le j \le n$

$$\text{var } Z_{(i)} \sim \frac{n}{(n+1)^2(n+2)}, \quad \text{cov}(Z_{(i)}, Z_{(j)}) \sim \frac{-1}{(n+1)^2(n+2)}. \quad (6.2.16)$$

To estimate the general linear combination $\eta = k_1\mu + k_2\sigma$ of μ and σ, write the linear estimator $\tilde{\eta}$ of η as

$$\tilde{\eta} = \sum_{i=1}^{n} g_i X_{(i)} = \Sigma g_i(\mu + \sigma Y_{(i)}).$$

Noting from (6.2.15) that

$$f_i Y_{(i)} = \sum_{j=0}^{i-1} Z_{(j)},$$

and replacing the g_i by new coefficients $h_0, h_1, \ldots, h_n$ defined (except for an additive constant) by

$$g_i = f_i(h_i - h_{i-1}),$$

we can rewrite $\tilde{\eta}$ as

$$\tilde{\eta} = \mu \sum_{i=1}^{n} f_i(h_i - h_{i-1}) + \sigma \sum_{i=1}^{n} (h_i - h_{i-1}) \sum_{j=0}^{i-1} Z_{(j)}$$

$$= \sum_{i=0}^{n} h_i[\mu(f_i - f_{i+1}) - \sigma Z_{(i)}]. \quad (6.2.17)$$

Then

$$\mathscr{E}\tilde{\eta} = \mu \sum_{i=0}^{n} C_{1i}h_i + \sigma \sum_{i=0}^{n} C_{2i}h_i,$$

where

$$C_{1i} = f_i - f_{i+1}, \quad C_{2i} = f_i\alpha_i - f_{i+1}\alpha_{i+1}.$$

Also

$$\text{var } \tilde{\eta} = \sigma^2\left[\sum_{i=0}^{n} h_i^2 \text{ var } Z_{(i)} + \sum_{i \ne j}\sum h_i h_j \text{ cov}(Z_{(i)}, Z_{(j)})\right]$$

$$\sim \frac{\sigma^2}{(n+1)(n+2)}\sum_{i=0}^{n}(h_i - \bar{h})^2 \quad (6.2.18)$$

by (6.2.16), where $\bar{h} = \Sigma h_i/(n+1)$. Thus, by means of the approximations used, the problem of estimating η has been reduced to minimizing (6.2.18) subject to the unbiasedness conditions

$$\sum_{i=0}^{n} C_{1i}h_i = k_1, \quad \sum_{i=0}^{n} C_{2i}h_i = k_2.$$

This leads by standard methods to the solutions

$$h_i = \bar{h} + a_1 C_{1i} + a_2 C_{2i},$$

where a_1, a_2 are Lagrange multipliers given by

$$a_1 = d^{11}k_1 + d^{12}k_2, \quad a_2 = d^{21}k_1 + d^{22}k_2,$$

the $d^{\alpha\beta}$ ($\alpha, \beta = 1, 2$) being the elements of the matrix inverse to the 2×2 matrix $\mathbf{D} = (d_{\alpha\beta})$ with

$$d_{\alpha\beta} = \sum_{i=0}^{n} C_{\alpha i} C_{\beta i} \quad \alpha, \beta = 1, 2. \tag{6.2.19}$$

With the h_i so defined, $\tilde{\eta}$ of (6.2.17) is the unbiased nearly best linear estimator of η. Returning to the original order statistics $X_{(i)}$, we have in particular, taking in turn $k_1 = 1$, $k_2 = 0$ and $k_1 = 0$, $k_2 = 1$,

$$\tilde{\mu} = \sum_{i=1}^{n} g_{1i} X_{(i)}, \quad \tilde{\sigma} = \sum_{i=1}^{n} g_{2i} X_{(i)}, \tag{6.2.20}$$

where

$$g_{\alpha i} = f_i [d^{\alpha 1}(C_{1i} - C_{1, i-1}) + d^{\alpha 2}(C_{2i} - C_{2, i-1})] \quad \alpha = 1, 2. \tag{6.2.21}$$

Also from (6.2.18) we have

$$\begin{pmatrix} \text{var } \tilde{\mu} & \text{cov} (\tilde{\mu}, \tilde{\sigma}) \\ & \text{var } \tilde{\sigma} \end{pmatrix} \sim \frac{\sigma^2}{(n+1)(n+2)} \begin{pmatrix} d^{11} & d^{12} \\ & d^{22} \end{pmatrix}. \tag{6.2.22}$$

Thus, whatever n, matrix inversion is reduced in this method to that of a 2×2 matrix. Although the asymptotic approximations made may not be particularly good in small samples, the estimates are likely to be highly efficient in a wide set of circumstances, since considerable accumulated evidence indicates that the efficiencies of linear systematic statistics are not very sensitive to changes in the coefficients. How to carry out the various methods of estimation described in this section is illustrated in Section 6.3 on a small censored sample from a normal parent.

Other Methods of Estimation

There have been yet other attempts to derive linear estimators of μ and σ without knowledge of the **B** matrix and in some instances also of the **α** vector. Since these quantities are becoming available for more and more parent distributions and for ever larger sample sizes, the need for methods alternative to Lloyd's is less strong than it once was. The value of such methods remains, however, for fresh parents, for large samples, and for some theoretical purposes.

"Asymptotically best linear estimates," which are systematic statistics defined by continuous weight functions, have been studied by Bennett (1952), Jung (1955, 1962), Chernoff et al. (1967), and Chan (1967a).

Downton (1966a) has proposed "linear estimates with polynomial coefficients," in which "the general structure of the coefficients is chosen for mathematical tractability, both in the determination of these coefficients and also in the computation of the standard error of the estimates so obtained."

A somewhat different idea (McCool, 1965; Chu and Ya'coub, 1968) is to base the estimation in large samples on the average of μ^*'s and σ^*'s obtained for subsamples small enough to allow the use of tables. The drawback to this approach is the arbitrariness in the choice of subsamples.

6.3. ESTIMATION OF LOCATION AND SCALE PARAMETERS FOR CENSORED DATA

By censored data we shall mean that, in a potential sample of n, a known number of observations is missing at either end (single censoring) or at both ends (double censoring). An important example occurs in life testing when it is decided to stop experimentation as soon as N ($< n$) items under test have failed. Here, by censoring at the right, we may be able to obtain reasonably good estimates of the parameters much sooner than by waiting for all items to fail. A particularly simple way of proceeding is to stop as soon as the sample median m can be calculated (i.e., after $[\frac{1}{2}n] + 1$ observations) and to use m as the estimate of mean life μ. When lifetimes are normally distributed (possibly after transformation of the data), μ may in large samples be estimated with the same accuracy from m in samples of $n(\frac{1}{2}\pi) \doteq 1.57n$ as from the mean of complete samples of n. On the other hand, the expected times to completion of the experiments are respectively μ and $\mu + \sigma\Phi^{-1}[n/(n+1)]$.

The type of censoring just described is often called *Type II censoring* (Gupta, 1952) to distinguish it from the situation in which the sample is curtailed below and/or above a fixed point. In such *Type I censoring* the number of censored observations is a random variable. Both forms of censoring are different from *truncation*, where the population rather than the sample is curtailed and the number of lost observations is unknown.

The methods of Section 6.2 developed for complete samples are immediately applicable to Type II censored observations. All that needs to be done is to interpret the vector **α** and the matrix **B** as the vector of means and the covariance matrix of the uncensored ordered $Y_{(r)}$ variates. (In fact, observations missing in the body of the sample would introduce no difficulties either.) Of course, each pattern of censoring requires separate calculations. For a normal parent extensive tables of the coefficients of the order statistics giving the estimators μ^* and σ^* have been prepared by Sarhan and Greenberg (1962, pp. 218–51). These tables cover all cases of single and double censoring in samples of $n \leq 20$. The variances and covariances of these estimates and their efficiencies relative to the best linear estimators in *uncensored* samples are also given. Not surprisingly, the loss in efficiency due to censoring is much more pronounced for σ^* than for μ^*. For example, for $n = 10$ and one observation censored at each end, the relative efficiencies are 95.85% for μ^* and 69.88% for σ^*. It should be noted that we can take advantage of the simplification for symmetric populations only if the censoring is also symmetric.

Gupta's alternative estimators are particularly simple for Type II censoring: the sums in (6.2.13) and (6.2.14) now extend over the surviving $n - r_1 - r_2$ observations, r_1 and r_2 being the number censored on the left and on the right, respectively. The efficiencies of these estimators relative to the corresponding best linear estimators have been tabulated by Sarhan and Greenberg (1962, pp. 266–8) in all cases of single and double censoring for $n = 10, 12, 15$. In most cases the values are over 90%, the lowest ones obtained being 84.66% for μ^{**} ($n = 15$, r_1 or $r_2 = 10$) and 86.75% for σ^{**} ($n = 15$, r_1 or $r_2 = 9$). Incidentally, for complete samples $\mu^{**} = \mu^*$, and the relative efficiency of σ^{**} is 99.9% for $n \leq 15$.

For a few simple parent distributions it is possible to invert **B** algebraically, and so to obtain general expressions for μ^* and σ^*. Sarhan (1955) deals with uniform and exponential parents, and for $n \leq 5$ also with several other distributions (see Ex. 6.3.1).

Type I censoring obviously does not lend itself equally well to analysis by order statistics. In fact, van Zwet (1966) points out that in many practical cases unbiased estimation of any kind is impossible. Here the method of maximum likelihood, although complicated and leading to estimates with largely unknown small-sample properties, offers a general approach covering truncation as well.

Method of Maximum Likelihood

Following Cohen (1959, 1961), we present a unified treatment, for a normal parent, of one-sided censoring of both types and of truncation. We begin with truncation (on the left) and suppose that N observations are available from the distribution

$$\frac{(2\pi)^{-1/2}\sigma \exp\left[-\dfrac{1}{2\sigma^2}(x - \mu)^2\right]}{\displaystyle\int_{x_0}^{\infty}(2\pi)^{-1/2}\sigma \exp\left[-\dfrac{1}{2\sigma^2}(x - \mu)^2\right]dx} \qquad x \geq x_0.$$

With $\xi = (x_0 - \mu)/\sigma$, the denominator is just $1 - \Phi(\xi)$, and the likelihood function may be written

$$L = [1 - \Phi(\xi)]^{-N}(2\pi\sigma^2)^{-\frac{1}{2}N} \exp\left[-\sum_1^N \frac{(x_i - \mu)^2}{2\sigma^2}\right],$$

giving

$$\frac{\partial \log L}{\partial \mu} = \frac{-N\phi(\xi)}{\sigma[1 - \Phi(\xi)]} + \frac{1}{\sigma^2} \sum_1^N (x_i - \mu), \tag{6.3.1}$$

$$\frac{\partial \log L}{\partial \sigma} = \frac{-N\xi\phi(\xi)}{\sigma[1 - \Phi(\xi)]} - \frac{N}{\sigma} + \frac{1}{\sigma^3} \sum_1^N (x_i - \mu)^2 \tag{6.3.2}$$

Setting

$$\bar{x} = \sum_1^N \frac{x_i}{N}, \quad s'^2 = \sum_1^N \frac{(x_i - \bar{x})^2}{N}, \quad \text{and} \quad A(y) = \frac{\phi(y)}{1 - \Phi(y)},$$

we have the corresponding likelihood equations

$$\bar{x} - \hat{\mu} = \hat{\sigma} A(\hat{\xi}), \tag{6.3.3}$$

$$s'^2 + (\bar{x} - \hat{\mu})^2 = \hat{\sigma}^2 [1 + \hat{\xi} A(\hat{\xi})]. \tag{6.3.4}$$

Eliminating $\bar{x} - \hat{\mu}$ and writing $\hat{A}$ for $A(\hat{\xi})$, we obtain

$$\hat{\sigma}^2 = s'^2 + \hat{\sigma}^2 \hat{A}(\hat{A} - \hat{\xi}). \tag{6.3.5}$$

But

$$\hat{\sigma}\hat{\xi} = x_0 - \hat{\mu} = x_0 - \bar{x} + \hat{\sigma}\hat{A}$$

so that

$$\hat{\sigma} = \frac{\bar{x} - x_0}{\hat{A} - \hat{\xi}}. \tag{6.3.6}$$

Together with (6.3.5) this gives

$$\hat{\sigma}^2 = s'^2 + \frac{\hat{A}(\bar{x} - x_0)^2}{\hat{A} - \hat{\xi}}$$

$$= s'^2 + \theta(\bar{x} - x_0)^2, \tag{6.3.7}$$

where

$$\theta = \frac{\hat{A}}{\hat{A} - \hat{\xi}}. \tag{6.3.8}$$

Then by (6.3.3), (6.3.6), and (6.3.8)

$$\hat{\mu} = \bar{x} - \hat{\sigma}\hat{A} = \bar{x} - \theta(\bar{x} - x_0), \tag{6.3.9}$$

and by (6.3.7) and (6.3.6)

$$\frac{s'^2}{(\bar{x} - x_0)^2} = \frac{\hat{\sigma}^2}{(\bar{x} - x_0)^2} - \frac{\hat{A}}{\hat{A} - \hat{\xi}}$$

$$= \frac{1 - \hat{A}(\hat{A} - \hat{\xi})}{(\hat{A} - \hat{\xi})^2}. \tag{6.3.10}$$

Now from (6.3.7) and (6.3.9) we could determine $\hat{\mu}$ and $\hat{\sigma}$ if we knew the auxiliary function θ. But θ is a function of $\hat{\xi}$ and hence of the RHS of (6.3.10).

Thus from $s'^2/(\bar{x} - x_0)^2$ ($= \hat{\gamma}$ in Cohen's notation) we can find $\hat{\theta}$—Cohen's (1961) Table 1—and hence $\hat{\mu}$ and $\hat{\sigma}$ (cf. Example 6.3.1).

For two other methods, in the first instance approximate solutions of the likelihood equations but of more general interest, see Plackett (1958) and Tiku (1967a).

Type I Censoring. N now denotes the random number of observations $\geq x_0$, the number censored ($< x_0$) being r out of a total of n. Thus $N + r = n$. The corresponding likelihood function is

$$L = \frac{n!}{r!N!}[\Phi(\xi)]^r (2\pi\sigma^2)^{-\frac{1}{2}N} \exp\left[-\sum_1^N \frac{(x_i - \mu)^2}{2\sigma^2}\right]. \qquad (6.3.11)$$

Then

$$\frac{\partial \log L}{\partial \mu} = \frac{-r\phi(\xi)}{\sigma\Phi(\xi)} + \frac{1}{\sigma^2}\sum_1^N (x_i - \mu),$$

giving, with $h = r/n$,

$$\bar{x} - \hat{\mu} = \frac{r}{N}\hat{\sigma}A(-\hat{\xi}) = \frac{h}{1-h}\hat{\sigma}A(-\hat{\xi})$$

$$= \hat{\sigma}\hat{B},$$

parallel to (6.3.3), where

$$\hat{B}(h, \xi) = \frac{h}{1-h}A(-\hat{\xi}).$$

Likewise

$$s'^2 + (\bar{x} - \hat{\mu})^2 = \hat{\sigma}^2(1 + \hat{\xi}\hat{B}),$$

so that analogously to (6.3.7), (6.3.9), and (6.3.10) we now have

$$\hat{\sigma}^2 = s'^2 + \hat{\lambda}(\bar{x} - x_0)^2, \qquad (6.3.12)$$

$$\hat{\mu} = \bar{x} - \hat{\lambda}(\bar{x} - x_0), \qquad (6.3.13)$$

$$\hat{\gamma} = \frac{s'^2}{(\bar{x} - x_0)^2} = \frac{1 - \hat{B}(\hat{B} - \hat{\xi})}{(\hat{B} - \hat{\xi})^2}, \qquad (6.3.14)$$

where $\hat{\lambda} = \hat{B}/(\hat{B} - \hat{\xi})$. The auxiliary function $\hat{\lambda}$ depends on two variables h and $\hat{\xi}$, or equivalently on h and $\hat{\gamma}$; but, subject to interpolation, $\hat{\lambda}$ can be obtained at once from Cohen's (1961) Table 2, reproduced as our Table 6.3. The estimation procedure is otherwise exactly as for truncation.

Type II Censoring. Let us denote the $r = N - n$ observations censored on the left by $x_1' \leq x_2' \leq \cdots \leq x_r'$, and, for the moment, let $\underline{x}$ be the smallest

Table 6.3. *Auxiliary estimation function* $\lambda(h, \hat{\gamma})$ *for singly censored samples from a normal population*

$\hat{\gamma}$ \ h	.01	.02	.03	.04	.05	.06	.07	.08	.09	.10	.15	.20	h \ $\hat{\gamma}$
.00	.010100	.020400	.030902	.041583	.052507	.063627	.074953	.086488	.09824	.11020	.17342	.24268	.00
.05	.010551	.021294	.032225	.043350	.054670	.066189	.077909	.089834	.10197	.11431	.17935	.25033	.05
.10	.010950	.022082	.033398	.044902	.056596	.068483	.080568	.092852	.10534	.11804	.18479	.25741	.10
.15	.011310	.022798	.034466	.046318	.058356	.070586	.083009	.095629	.10845	.12148	.18985	.26405	.15
.20	.011642	.023459	.035453	.047629	.059990	.072539	.085280	.098216	.11135	.12469	.19460	.27031	.20
.25	.011952	.024076	.036377	.048858	.061522	.074372	.087413	.10065	.11408	.12772	.19910	.27626	.25
.30	.012243	.024658	.037249	.050018	.062969	.076106	.089433	.10295	.11667	.13059	.20338	.28193	.30
.35	.012520	.025211	.038077	.051120	.064345	.077756	.091355	.10515	.11914	.13333	.20747	.28737	.35
.40	.012784	.025738	.038866	.052173	.065660	.079332	.093193	.10725	.12150	.13595	.21139	.29260	.40
.45	.013036	.026243	.039624	.053182	.066921	.080845	.094958	.10926	.12377	.13847	.21517	.29765	.45
.50	.013279	.026728	.040352	.054153	.068135	.082301	.096657	.11121	.12595	.14090	.21882	.30253	.50
.55	.013513	0.27196	.041054	.055089	.069306	.083708	.098298	.11308	.12806	.14325	.22235	.30725	.55
.60	.013739	.027649	.041733	.055995	.070439	.085068	.099887	.11490	.13011	.14552	.22578	.31184	.60
.65	.013958	.028087	.042391	.056874	.071538	.086388	.10143	.11666	.13209	.14773	.22910	.31630	.65
.70	.014171	.028513	.043030	.057726	.072605	.087670	.10292	.11837	.13402	.14987	.23234	.32065	.70
.75	.014378	.028927	.043652	.058556	.073643	.088917	.10438	.12004	.13590	.15196	.23550	.32489	.75
.80	.014579	.029330	.044258	.059364	.074655	.090133	.10580	.12167	.13773	.15400	.23858	.32903	.80
.85	.014775	.029723	.044848	.060153	.075642	.091319	.10719	.12325	.13952	.15599	.24158	.33307	.85
.90	.014967	.030107	.045425	.060923	.076606	.092477	.10854	.12480	.14126	.15793	.24452	.33703	.90
.95	.015154	.030483	.045989	.061676	.077549	.093611	.10987	.12632	.14297	.15983	.24740	.34091	.95
1.00	.015338	.030850	.046540	.062413	.078471	.094720	.11116	.12780	.14465	.16170	.25022	.34471	1.00

$\hat{\gamma}$ \ h	.25	.30	.35	.40	.45	.50	.55	.60	.65	.70	.80	.90	h \ $\hat{\gamma}$
.00	.31862	.4021	.4941	.5961	.7096	.8368	.9808	1.145	1.336	1.561	2.176	3.283	.00
.05	.32793	.4130	.5066	.6101	.7252	.8540	.9994	1.166	1.358	1.585	2.203	3.314	.05
.10	.33662	.4233	.5184	.6234	.7400	.8703	1.017	1.185	1.379	1.608	2.229	3.345	.10
.15	.34480	.4330	.5296	.6361	.7542	.8860	1.035	1.204	1.400	1.630	2.255	3.376	.15
.20	.35255	.4422	.5403	.6483	.7678	.9012	1.051	1.222	1.419	1.651	2.280	3.405	.20
.25	.35993	.4510	.5506	.6600	.7810	.9158	1.067	1.240	1.439	1.672	2.305	3.435	.25
.30	.36700	.4595	.5604	.6713	.7937	.9300	1.083	1.257	1.457	1.693	2.329	3.464	.30
.35	.37379	.4676	.5699	.6821	.8060	.9437	1.098	1.274	1.476	1.713	2.353	3.492	.35
.40	.38033	.4755	.5791	.6927	.8179	.9570	1.113	1.290	1.494	1.732	2.376	3.520	.40
.45	.38665	.4831	.5880	.7029	.8295	.9700	1.127	1.306	1.511	1.751	2.399	3.547	.45
.50	.39276	.4904	.5967	.7129	.8408	.9826	1.141	1.321	1.528	1.770	2.421	3.575	.50
.55	.39870	.4976	.6051	.7225	.8617	.9950	1.155	1.337	1.545	1.788	2.443	3.601	.55
.60	.40447	.5045	.6133	.7320	.8625	1.007	1.169	1.351	1.561	1.806	2.465	3.628	.60
.65	.41008	.5114	.6213	.7412	.8729	1.019	1.182	1.366	1.577	1.824	2.486	3.654	.65
.70	.41555	.5180	.6291	.7502	.8832	1.030	1.195	1.380	1.593	1.841	2.507	3.679	.70
'75	.42090	.5245	.6367	.7590	.8932	1.042	1.207	1.394	1.608	1.858	2.528	3.705	.75
.80	.42612	.5308	.6441	.7676	.9031	1.053	1.220	1.408	1.624	1.875	2.548	3.730	.80
.85	.43122	.5370	.6515	.7761	.9127	1.064	1.232	1.422	1.639	1.892	2.568	3.754	.85
.90	.43622	.5430	.6586	.7844	.9222	1.074	1.244	1.435	1.653	1.908	2.588	3.779	.90
.95	.44112	.5490	.6656	.7925	.9314	1.085	1.255	1.448	1.668	1.924	2.607	3.803	.95
1.00	.44592	.5548	.6724	.8005	.9406	1.095	1.267	1.461	1.682	1.940	2.626	3.827	1.00

For all values $0 \leq \hat{\gamma} \leq 1$, $\lambda(0, \hat{\gamma}) = 0$.
(From Cohen, 1961, with permission of the author and the Editor of *Technometrics*.)

observed value. Then the likelihood function is given by

$$L = \left[\prod_{i=1}^{N} p(x_i) \right] \cdot \int_{-\infty}^{x} \int_{-\infty}^{x_r{}'} \cdots \int_{-\infty}^{x_2{}'} n! \, p(x_1{}') \cdots p(x_{r-1}{}') p(x_r{}') \, dx_1{}' \cdots dx_{r-1}{}' \, dx_r{}'$$

$$= \frac{n!}{r!} P^r(\underline{x}) \prod_{i=1}^{N} p(x_i)$$

$$= \frac{n!}{r!} \left[\Phi\left(\frac{x - \mu}{\sigma} \right) \right]^r (2\pi\sigma^2)^{-\frac{1}{2}N} \exp\left[-\sum_{1}^{N} \frac{(x_i - \mu)^2}{2\sigma^2} \right].$$

Comparing with (6.3.11), we see that ξ need merely be replaced by $y = (x - \mu)/\sigma$. In this case the bias of the estimates has been examined by Saw (1961), who finds that for severe censoring the bias may be considerable, for example, as high as 13% for $n = 19$, $N = 7$.

It is clear that for truncation on the right $1 - \Phi(\xi)$ has to be replaced by $\Phi(\xi)$, that is, ξ by $-\xi$. This makes no difference in the estimation procedure. The same remark applies to censoring except that for Type II censoring x has to be replaced by the largest observed value.

Asymptotic Variances and Covariances of the ML estimators. We consider Type II censoring. The first derivatives of $\log L$ may be written as

$$\frac{\partial \log L}{\partial \mu} = \frac{-rA(-y)}{\sigma} + \sum_1^N \frac{x_i - \mu}{\sigma^2},$$

$$\frac{\partial \log L}{\partial \sigma} = \frac{-ryA(-y)}{\sigma} - \frac{N}{\sigma} - \sum_1^N \frac{(x_i - \mu)^2}{\sigma^3}.$$

Since

$$A(-y) = \frac{\phi(y)}{\Phi(y)},$$

$$\frac{\partial A(-y)}{\partial y} = \frac{\Phi(y)[-y\phi(y)] - \phi^2(y)}{[\Phi(y)]^2}$$

$$= -A^-(A^- + y) \quad \text{where} \quad A^- = A(-y).$$

Hence

$$-\frac{\partial^2 \log L}{\partial \mu^2} = \frac{r}{\sigma^2} A^-(A^- + y) + \frac{N}{\sigma^2}$$

$$= \frac{N}{\sigma^2}\left[\frac{h}{1 - h} A^-(A^- + y) + 1\right],$$

or

$$\omega_{11} = \frac{-\sigma^2}{N} \frac{\partial^2 \log L}{\partial \mu^2} = B(A^- + y) + 1.$$

Also

$$\omega_{12} = \frac{-\sigma^2}{N} \frac{\partial^2 \log L}{\partial \mu\, \partial \sigma} = B[1 + y(A^- + y)],$$

$$\omega_{22} = \frac{-\sigma^2}{N} \frac{\partial^2 \log L}{\partial \sigma^2} = 2 + y\omega_{12}.$$

Asymptotically, as $N \to \infty$, $y \to y_0 = \Phi^{-1}(h)$. With this substitution the asymptotic covariance matrix may be obtained on inversion of $\begin{pmatrix} \omega_{11} & \omega_{12} \\ \omega_{12} & \omega_{22} \end{pmatrix}$, for example,

$$\text{var } \hat{\mu} \sim \frac{\sigma^2}{N} \frac{\omega_{22}}{\omega_{11}\omega_{22} - \omega_{12}^{\,2}} = \frac{\sigma^2}{n} \mu_{11}, \text{ etc.,}$$

thereby defining μ_{11}, and similarly μ_{12}, μ_{22}. Cohen (1961) tables these and $\rho(\hat{\mu}, \hat{\sigma})$ as functions of y_0. To obtain estimates of these quantities one may enter the tables with $\hat{y} = (x_{\min} - \hat{\mu})/\hat{\sigma}$. The same tables apply also to Type I censoring if entered with $\hat{\xi} = (x_0 - \hat{\mu})/\hat{\sigma}$. Separate similar tables are given for truncation.

The large-sample properties of ML estimators for singly-censored (but called "truncated") samples have been studied by Halperin (1952).

Example 6.3.1.   Gupta (1952) presents the following data, showing the number x' of days to death of the first 7 in a sample of 10 mice after inoculation with a uniform culture of human tuberculosis:

x'	41	44	46	54	55	58	60
$x = \log_{10} x'$	1.613	1.644	1.663	1.732	1.740	1.763	1.778

Gupta takes $\log x'$ to be normally distributed. We shall estimate the mean and s.d. of $x = \log_{10} x'$ by the various methods of this chapter.

(i) *Maximum Likelihood.*   This is a case of Type II censoring on the right. We have

$$r = 3, \quad n = 10, \quad h = 0.3, \quad \bar{x} = 1.70471, \quad s'^2 = \tfrac{1}{7}\Sigma(x_i - \bar{x})^2 = 0.003514.$$

Then from (6.3.14)

$$\hat{\gamma} = s'^2/(\bar{x} - 1.778)^2 = 0.654.$$

Entering Table 6.3, we find $\hat{\lambda} = 0.512$, and hence from (6.3.13) and (6.3.12)

$$\hat{\mu} = 1.742 \quad \text{and} \quad \hat{\sigma} = 0.079.$$

Also Cohen's (1961) Table 3 gives approximately

$$\mu_{11} = 1.14, \quad \mu_{22} = 0.82, \quad \rho = 0.21,$$

leading to the following estimates of error:

$$\text{s.e. of } \hat{\mu} = 0.079 \times (1.14/10)^{\frac{1}{2}} = 0.027,$$
$$\text{s.e. of } \hat{\sigma} = 0.079 \times (0.82/10)^{\frac{1}{2}} = 0.023.$$

(ii) *Best Linear Estimates.*   From Sarhan and Greenberg (1962, p. 222) we have, on applying (6.2.5′) to censored data,

$$\mu^* = (0.0244)(1.613) + (0.0636)(1.644) + \cdots$$
$$+ (0.5045)(1.778)$$
$$= 1.746,$$
$$\sigma^* = (-0.3252)(1.613) + (-0.1758)(1.644) + \cdots$$
$$+ (0.6107)(1.778)$$
$$= 0.091.$$

Also from p. 253 of the same reference

$$\text{var } \mu^* = 0.1167\sigma^2, \quad \text{var } \sigma^* = 0.0989\sigma^2, \quad \text{cov } (\mu^*, \sigma^*) = 0.0260\sigma^2,$$

giving as estimates of error

$$\text{s.e. of } \mu^* = 0.091 \times (0.1167)^{\frac{1}{2}} = 0.031,$$
$$\text{s.e. of } \sigma^* = 0.091 \times (0.0989)^{\frac{1}{2}} = 0.029.$$

(iii) *Simplified Linear Estimates.* Here the coefficients (6.2.13) and (6.2.14) are applicable (on adaptation to the censored case). Gupta (1952) gives

$$\mu^{**} = (-0.0433)(1.613) + (0.0491)(1.644) + \cdots$$
$$+ (0.2861)(1.778)$$
$$= 1.748,$$
$$\sigma^{**} = (-0.4077)(1.613) + (-0.2053)(1.644) + \cdots$$
$$+ (0.3136)(1.778)$$
$$= 0.094,$$
$$\text{s.e. of } \mu^{**} = 0.033, \quad \text{s.e. of } \sigma^{**} = 0.031.$$

The coefficients have not been tabulated in this case, so that (iii) is actually more laborious than (ii). The point of (iii) is, of course, that the coefficients can be calculated whenever the expected values of the order statistics are available.

It will be noted that the ML estimators have the lowest standard errors. However, this is due largely to the fact that $\hat{\sigma}$ has come out lower than the (unbiased) estimates σ^* and σ^{**}. The efficiencies of μ^{**} relative to μ^* and of σ^{**} relative to σ^* are, respectively, 0.960 and 0.920 (SG, p. 266).

Most of the results of this numerical example were given by Gupta, who was the first to consider estimation of the parameters of a normal population from Type II censored samples. Due to an error in his computations and also because of the availability of more accurate tables our numerical results in (i) and (ii) differ somewhat from his.

Finally we illustrate Blom's method on these same data.

(iv) **Blom's Unbiased Nearly Best Estimates.** The estimates of (6.2.20) continue to hold in the censored case provided the C_{1i}, C_{2i} in (6.2.21) are replaced by C_{1i}^*, C_{2i}^*, defined as follows for r_1 censored observations on the left and r_2 on the right:

C_{1i}^*	C_{2i}^*	
$-f_{r_1+1}/(r_1 + 1)$	$-f_{r_1+1}\alpha_{r_1+1}/(r_1 + 1)$	$0 \leq i \leq r_1$
C_{1i}	C_{2i}	$r_1 + 1 \leq i \leq n - r_2 - 1$
$f_{n-r_2}/(r_2 + 1)$	$f_{n-r_2}\alpha_{n-r_2}/(r_2 + 1)$	$n - r_2 \leq i \leq n$

Since the numerical work is laborious, the following auxiliary table is formed for our example:

i	f_i	C_{1i}	$f_i\alpha_i$	C_{2i}
0	0	−.1636	0	.2517
1	.1636	−.1004	−.2517	.0127
2	.2640	−.0683	−.2644	−.0464
3	.3323	−.0431	−.2180	−.0769
4	.3754	−.0209	−.1411	−.0925
5	.3963	0	−.0486	−.0972
6	.3963	.0209	.0486	−.0925
7, 8, 9, 10	...	.0938	.1411	.0353

The f_i are most conveniently obtained from Table 5 of Pearson and Hartley (1966). We now have

$$d_{11} = 0.0794 \quad \text{and} \quad d^{11} = 13.44,$$
$$d_{12} = -0.0227 \quad d^{12} = 2.96,$$
$$d_{22} = 0.1031 \quad d^{22} = 10.35,$$

giving

$$\tilde{\mu} = 1.746 \quad \text{and} \quad \tilde{\sigma} = 0.090.$$

Also from (6.2.22) we find the estimates of standard error:

$$\text{s.e. of } \tilde{\mu} = 0.029, \quad \text{s.e. of } \tilde{\sigma} = 0.025.$$

Grouped Data

Since grouping represents a partial ordering, it seems particularly natural to ask to what extent the various preceding methods, suitably modified, can be carried over to grouped observations. As far back as 1942, Hartley studied the distribution of the range in grouped samples from a normal parent and found the mean range very little affected even by quite coarse grouping for $n \leq 20$. Similar results were obtained by David and Mishriky (1968) for the means of all order statistics for $n \leq 100$, although the effect of grouping (a) is more important for the central order statistics than for the extremes (which are less crowded) and (b) increases with n. Subject to similar remarks, the variances of order statistics in grouped samples are well approximated by those in ungrouped samples after the application of Sheppard's correction $h^2/12$, where h is the grouping interval of the original, and hence also of the ordered, observations. From the general theory of Sheppard's correction no such adjustment is needed for the covariances. These results taken together strongly suggest that any of the methods appropriate for ungrouped normal samples can also be used in the grouped case, that is, the same weights multiplying the order statistics now multiply the midpoints of the corresponding class intervals.

Example 6.3.2. The first 20 random normal $N(0, 1)$ numbers given in Beyer (1968) are:

$$0.464, \quad 0.060, \quad 1.486, \quad 1.022, \quad 1.394, \quad 0.906, 1.179, -1.501,$$
$$-0.690, \quad 1.372, \quad -0.482, \quad -1.376, \quad -1.010, \quad -0.005, \quad 1.393, \quad -1.787,$$
$$-0.104, \quad -1.339, \quad 1.041, \quad 0.279.$$

Grouped in intervals of width $h = 0.5$ proceeding from the origin, the group midpoints and the associated frequencies are:

$$-1.75 \ (2) \quad -1.25 \ (3) \quad -0.75 \ (1) \quad -0.25 \ (3)$$
$$0.25 \ (3) \quad 0.75 \ (1) \quad 1.25 \ (7)$$

Denoting the grouped case by a subscript g, we have for the mean and standard deviation of the sample

$$\bar{x} = 0.115 \qquad s = 1.105,$$
$$\bar{x}_g = 0.075 \qquad s_g = 1.066 \text{ (with Sheppard's correction)}.$$

Of course, $\bar{x}$ and $\bar{x}_g$ are also the estimates of $\mu \ (= 0)$ obtained by the use of order statistics. The best linear estimate of $\sigma \ (= 1)$, with coefficients from SG, p. 248, is

$$\sigma^* = (-1.787)(-0.1128) + (-1.501)(-0.0765) + \cdots + (1.486)(0.1128)$$
$$= 1.096.$$

Correspondingly the grouped estimate is

$$\sigma_g^* = (-1.75)(-0.1128 - 0.0765) + \cdots + (1.25)$$
$$\qquad \times (0.0241 + 0.0318 + 0.0402 + 0.0497 + 0.0611 + 0.0765 + 0.1128)$$
$$= 1.067.$$

Of course, no optimal properties are claimed for this procedure. In any case, a price for grouping must be paid in the form of increased variance of the estimators. Nevertheless, the same method should often prove useful for parents other than normal, and for censored as well as complete samples.

A different approach, also based on Lloyd's work, is given by Hammersley and Morton (1954). See also Grundy (1952) and Swamy (1962).

Censoring of the Multivariate Normal

A number of new situations arise when multivariate data are curtailed in some manner. For definiteness we consider Type II censoring and take the bivariate case for illustration. Before any censoring is done, and corresponding to the usual ordering of one set of variates, say the x's, the associated y's may be denoted by $y_{[i]}$ ($i = 1, 2, \ldots, n$), where $y_{[i]}$ is the y value corresponding to $x_{(i)}$. The $y_{[i]}$ are not necessarily in ascending order. For some properties under bivariate normality $N(\mu_x, \mu_y, \sigma_x^2, \sigma_y^2, \rho)$, see Ex. 3.2.3.

Three kinds of censoring may usefully be distinguished (Watterson, 1959): (A) censoring of certain $x_{(i)}$'s and of the associated $y_{[i]}$'s; (B) censoring of $y_{[i]}$'s only; and (C) censoring of $x_{(i)}$'s only. For example, (B) (or, more fully, Type IIB) occurs when the $x_{(i)}$ ($i = 1, 2, \ldots, n$) are entrance scores and the $y_{[i]}$ ($i = r + 1, r + 2, \ldots, n$) later scores of the successful candidates. On the other hand, Type IIC applies in a life test terminated after $n - r$ failures when measurements on some associated variable are available for all n items. Watterson obtains estimators based on the coefficients of the best linear and simplified estimates in the univariate case. The estimators are unbiased, but their variances depend on ρ. Use of the simplified coefficients turns out to give estimators both simple to compute and generally low in variance. See also Cohen (1955a, 1957) and Singh (1960).

Censoring in Non-normal Distributions

We now list a number of papers dealing with the estimation of location and scale parameters from censored data for various univariate populations. The literature is too large to allow a more detailed treatment here, and the methods used are in any case mostly straightforward modifications of those for a normal parent. As is clear from Cohen's approach, the estimation of parameters from truncated data proceeds similarly, but no attempt is made to cover that subject systematically. The exponential case is handled separately in Section 6.4 in connection with life testing. For $n \leq 5$, best linear estimates are given for several parents by Sarhan and Greenberg (1962, pp. 391–7). Federer's (1963) bibliography on screening in selection has many relevant entries.

Truncated normal: Cohen (1955c).
Log normal: Harter and Moore (1966), Tiku (1968a).
Gamma: Cohen (1955b), Harter (1967), Harter and Moore (1967b), Wilk *et al.* (1962b, 1963b, 1966).

Chi (1 DF): Govindarajulu and Eisenstat (1965).
Beta: Gnanadesikan *et al.* (1967).
Double exponential: Govindarajulu (1966).
Extreme value: Harter and Moore (1967a, 1968a, b), Lieblein (1954a), Lieblein and Zelen (1956), Mann (1967b), White (1964), Winer (1963).
Weibull: Bain and Antle (1967), Cohen (1965), Gumbel (1958), Harter and Moore (1965, 1967b), Mann (1967a), Menon (1963).
Logistic: Gupta *et al.* (1967), Harter and Moore (1967c), Tiku (1968b).
Poisson: Cohen (1954), Doss (1963).
Power function: Likeš (1967).

Note that, if X has the Weibull distribution with cdf

$$P(x) = 1 - \exp\left[-(x/\theta)\right]^\kappa \qquad x \geq 0,$$

then $\log X$ has the extreme-value distribution (of the *smallest* extreme) with

$$P(x) = 1 - \exp\left\{-\exp\left[(x - \mu)/\sigma\right]\right\},$$

where $\mu = \log \theta$ and $\sigma = 1/\kappa$. Thus the ML (but not the linear) estimation for the two distributions is essentially the same. Likewise, if X has the power-function distribution with pdf

$$p(x) = k\theta^{-k}x^{k-1} \qquad k > 0; \qquad 0 \leq x \leq \theta,$$

then $\log X$ is exponential with start $-\log \theta$ and scale parameter $1/k$.

6.4. LIFE TESTING, WITH SPECIAL EMPHASIS ON THE EXPONENTIAL DISTRIBUTION

If n items such as radio tubes, wire fuses, or light bulbs are put through a life test, the weakest item will fail first, followed by the second weakest, and so on until all have failed. Thus, if the lifetime X of a randomly chosen item has pdf $p(x)$, the life test generates in turn ordered observations $x_{(1)}, x_{(2)}, \ldots, x_{(n)}$ from this distribution. Switching from the physical to the biological sciences, we may also interpret X as, for example, the time to death after n animals are subjected to a common dose of radiation. The practical importance of such experiments is evident. They also afford an ideal application of order statistics, since by the nature of the experiment the observations arrive in ascending order of magnitude and do not have to be ordered after collection of the data. Moreover, as was already alluded to in Section 6.3, the possibility is now open of terminating the experiment before its conclusion, by stopping after a given time (Type I censoring) or after a given number of failures (Type II censoring). Provided the form of $p(x)$ is well known from similar experiments, the estimation of parameters may often proceed with a loss in efficiency small compared to the gain in time.

There are many plausible candidates for the distribution of X, including the Weibull, gamma, log normal, and even the normal.[4] By far the most attention in the literature has been devoted to the exponential distribution (which is, of course, a special case of both the Weibull and the gamma). The exponential occupies as commanding a position in life testing as does the normal elsewhere in parametric theory. It must be confessed that this is (in both cases) partly a matter of convenience, since simple, elegant results are possible. As Zelen and Dannemiller (1961) have shown, departures from the exponential distribution may seriously upset procedures valid under exponentiality. However, the exponential holds exactly when failures follow a Poisson process or, putting it differently, when the failure rate[5] (= conditional pdf of X, given $X > x$) $p(x)/[1 - P(x)]$ of a given item remains constant so that the item is as good as new over its lifetime. In practice this means *inter alia* that wear must play a negligible role compared to accidental causes of failure.

For further discussion the reader may consult Barlow and Proschan (1965), although the main aim of their book is to replace specific distributional assumptions by the requirement that failure rates vary monotonically with time. The monograph by Cox and Lewis (1966) is also pertinent.

We turn now to a more detailed discussion of the exponential case but for greater generality take the density in the two-parameter form:

$$p(x) = \sigma^{-1} e^{-(x-\theta)/\sigma} \qquad x \geq \theta. \qquad (6.4.1)$$

Here X has mean $\theta + \sigma$ and s.d. σ. In the context of life testing θ may be interpreted as an unknown point at which "life" begins or as a "guarantee time" during which failure cannot occur (Epstein and Sobel, 1954). Another interpretation arises in what is sometimes termed *interval analysis*; θ may represent the "dead time" of a Geiger counter, X being the interval between successive counts. We will leave it to the reader to verify results obtained by Sukhatme as early as 1937, namely, that for a (complete) sample of n from the pdf (6.4.1) ML estimators of θ and σ are

$$\hat{\theta} = X_{(1)}, \quad \hat{\sigma} = \sum_{i=2}^{n} \frac{(X_{(i)} - X_{(1)})}{n}, \qquad (6.4.2)$$

that these are jointly sufficient statistics, and that the best unbiased estimators are, respectively,

$$\theta^* = X_{(1)} - \frac{\sigma^*}{n}, \quad \sigma^* = \frac{n\hat{\sigma}}{n - 1}.$$

---

[4] Since X is non-negative, the coefficient of variation of the normal distribution must be small enough to ensure a negligible probability for $X < 0$.

[5] Aliases: hazard rate, intensity function, force of mortality.

Moreover (cf. Section 2.7) the quantities

$$Y_i = \frac{(n - i + 1)(X_{(i)} - X_{(i-1)})}{\sigma} \qquad i = 1, 2, \ldots, n \quad (X_{(0)} = \theta) \quad (6.4.3)$$

are independent variates from $p(y) = e^{-y}$ $(y \geq 0)$. Since

$$\sum_{i=2}^{n} Y_i = \sum_{i=2}^{n} \frac{(X_{(i)} - X_{(1)})}{\sigma},$$

it follows that $2(n - 1)\sigma^*/\sigma$ is distributed as χ^2 with $2(n - 1)$ DF.[6] This result can be used immediately to construct confidence intervals and tests of significance for σ, in striking analogy to the normal case. In the same spirit, confidence intervals and tests on θ may be based on the ratio

$$T = \frac{n(X_{(1)} - \theta)}{\sigma^*}$$

which from (6.4.3) has an F-ratio distribution with 2 and $2(n - 1)$ DF. Extensions to two and several sample tests, also given by Sukhatme, are straightforward continuations of the analogy with normal theory (e.g., Ex. 6.4.1).

Type II censoring on the right can be handled almost without change in view of (6.4.3). We simply work with the successive differences of the available first N failure times and estimate σ as

$$\sigma^* = \sum_{i=2}^{N} \frac{(n - i + 1)(X_{(i)} - X_{(i-1)})}{N - 1}$$

$$= \sum_{i=2}^{N} \frac{(X_{(i)} - X_{(1)}) + (n - N)(X_{(N)} - X_{(1)})}{N - 1},$$

where now $2(N - 1)\sigma^*/\sigma$ is a χ^2 with $2(N - 1)$ DF. Correspondingly $\theta^* = X_{(1)} - \sigma^*/n$. The reader may again wish to verify that $\hat{\sigma} = (N - 1)\sigma^*/N$ and $X_{(1)}$ are the ML estimators and that they are jointly sufficient. They are also complete (Epstein and Sobel, 1954), which formally establishes that σ^* and θ^* are, respectively, unique UMVU estimators of σ and θ. Various tests of significance are treated by Epstein and Tsao (1953).

The important case $\theta = 0$ is discussed separately in a number of papers, notably that of Epstein and Sobel (1953). Type I censoring and a generalization permitting termination after time x_0 *or* after the Nth failure, whichever comes first, are considered by Epstein (1954); see Ex. 6.4.3.[7] The idea here is

---

[6] The general distribution of a linear function, $\sum_{i=1}^{n} c_i X_{(i)}$, can also be obtained (cf. Likeš, 1967).

[7] Epstein somewhat misleadingly calls such a procedure a *truncated life test*.

that to test $H\colon \sigma \leq \sigma_0$ against $K\colon \sigma > \sigma_0$ we can decide in favor of H without waiting until time x_0 if the Nth failure occurs early enough. See also Bartholomew (1963) for the estimation of σ in Type I censoring.

A review of these and related life-testing procedures is given by Epstein (1960a, b).

Closely related to life testing are problems of *reliability*. The reliability of an item required to perform satisfactorily for at least time x (fixed) is defined as

$$R(x) = 1 - P(x) = \Pr\{X > x\},$$

which for a (one-parameter) exponential is just $R(x) = e^{-x/\sigma}$ $(x \geq 0)$. Hypotheses about R are therefore immediately reducible to hypotheses about σ. The UMVU estimation of R for the truncated exponential distribution is treated by Sathe and Varde (1969). Note also that a series system of n items (or "components") with individual reliabilities $R_i(x)$ has reliability

$$\Pr\{\text{system life} > x\} = \Pr\{\text{each component life} > x\} = \prod_{i=1}^{n} R_i,$$

whereas for a parallel system

$$\Pr\{\text{system life} > x\} = \Pr\{\text{at least one component life} > x\}$$

$$= 1 - \prod_{i=1}^{n}(1 - R_i).$$

These general formulae assume that failures occur independently, an assumption worth checking in individual applications. We see incidentally that for the exponential the lifetime X_S of the series system has pdf

$$-\frac{d}{dx}\exp\left[-\Sigma\left(\frac{x}{\sigma_i}\right)\right] = \Sigma\left(\frac{1}{\sigma_i}\right)\exp\left[-x\,\Sigma\left(\frac{1}{\sigma_i}\right)\right];$$

that is, X_S has a one-parameter exponential distribution with mean life $1/\Sigma(1/\sigma_i)$, which for identical components reduces to σ/n.

Censoring on the Left and Two-Sided Censoring in the Exponential Case

It will have been noted that in the preceding discussion any censoring has been on the right. Censoring on the left—fortunately less important—does not permit equally elegant results. However, one can fall back on Lloyd's general approach (Sarhan, 1955) or use simplified estimates (Epstein, 1956), a summary being provided in Sarhan and Greenberg (1962). See also Tiku (1967b). The general estimates for two-sided censoring are given in Ex. 6.4.5. It should be noted that these include the results of censoring on the right, but the previous approach is necessary to establish uniformly minimum variance unbiasedness in the class of all estimators, not merely among linear estimators. Extensive tables for $n \leq 10$ are given by Sarhan and Greenberg (1957).

Some References

The literature on life testing and reliability is huge. For papers of statistical interest see the bibliographic guide by Buckland (1964) and the bibliography by Mendenhall (1958) with a supplement by Govindarajulu (1964). Below is a short classified list of references (in addition to those already cited in this section) in which order statistics play a role.

1. *Life testing (parametric):* Bain and Weeks (1965), Barlow and Proschan (1967), Barlow *et al.* (1968), Bartholomew (1957), Basu (1965, 1968), Churchman and Epstein (1946), Cohen (1963, 1966), Cox (1959, 1964), David (1957), Doksum (1967), Gani and Yeo (1962), Garner (1958), Goodman and Madansky (1962), Gupta (1962), Gupta and Groll (1961), Gupta and Sobel (1958), Hogg and Tanis (1963), Jacobson (1947), Likeš (1962, 1967), Madansky (1962), Mantel and Pasternak (1966), Miller (1960), Proschan and Pyke (1967), Rao (1962), Tanis (1964), Tiku (1968c), Zelen (1959).

2. *Life testing (distribution-free):* Barlow and Gupta (1966), Basu (1967), Eilbott and Nadler (1965), Shorack (1967), Walsh (1956).

3. *Reliability:* Babik (1968), Birnbaum and Saunders (1958), Birnbaum *et al.* (1961), Esary and Proschan (1963), Johns and Lieberman (1966), Lentner and Buehler (1963), Morrison and David (1960), Rutemiller (1966), Saunders (1968), Zacks and Even (1966).

4. *Interval analysis:* Barnard (1953), Maguire *et al.* (1952, 1953).

6.5. ROBUST ESTIMATION

So far in this chapter, apart from fleeting references to nonparametric methods, we have considered the use of order statistics when the form of the parent distribution is known. In practice, we can seldom be certain of such distributional assumptions, and two kinds of questions arise:

1. How will estimates constructed—and perhaps optimal in some sense— for one kind of population behave when another parental form in fact holds?

2. Can we *construct* estimators which perform well (i.e., are *robust*) for a variety of distributions and/or in the presence of "contamination" leading to outlying observations?

These questions are of very general interest and are by no means restricted to estimators which are linear functions of the order statistics. Nevertheless it is clear that the lowest and highest few observations in a sample are the most likely to be the results of failure of distributional assumptions or of contamination. One approach (to be pursued in Section 8.5) is to remove such extreme observations by means of some test of significance and to base estimates only on the remaining observations. Here, however, we are interested in robustness without such preliminary screening of data.

Let us now take a closer look at questions 1 and 2. Tukey, who may fairly be called the father of the subject of robust estimation, points out forcefully (1960) that, whereas for a sample from $N(\mu, \sigma^2)$ the mean deviation has asymptotic efficiency 0.88 relative to the standard deviation in estimating σ, the situation is changed drastically if some contamination by a wider normal, say $N(\mu, 9\sigma^2)$, is present: as little as 0.008 of the wider population will render the mean deviation asymptotically superior. Results for the estimation of μ are not quite so spectacular but are even more important. For a complete random sample from an unknown parent no estimator of μ is more widely used than the sample mean $\bar{X}$ or has a more impressive set of credentials: unbiasedness for all populations possessing a mean; sufficiency, completeness, and hence full efficiency for, for example, normal, gamma, Poisson parents; and under wide conditions a convenient normal large-sample distribution which in many cases holds approximately even for moderate sample sizes. Nevertheless there are flaws: the efficiency of the mean is zero for a uniform parent, and for any parent a single sufficiently wild observation may render $\bar{X}$ useless. It has long been known that the midpoint is optimal in the former case but much worse than $\bar{X}$ in the latter, and that the median is preferable in the latter case but worse in the former. The obvious moral: we must not expect an estimator to be good under too wide a set of circumstances.

Crow and Siddiqui (1967) accordingly consider the robust estimation of location relative to a class $\mathscr{F}$ consisting of at least two pencils of the following symmetrical distributions: rectangular (R), parabolic (P), triangular (T), normal (N), double exponential (DE), and Cauchy (C). The problem is to investigate various classes of estimators with the aim of finding those which although perhaps not optimal for *any* of the foregoing distributions, perform reasonably well for all or for a chosen subset. Note that Crow and Siddiqui do not specifically allow for contamination, but estimators which perform well for long-tailed distributions (such as DE and C) are likely to be robust in the presence of outliers. Since their study is confined to symmetrical distributions, the authors consider only estimators of the form

$$Z_n(\mathbf{a}) = \sum_{i=1}^{n} a_i X_{(i)}, \quad \text{with } a_{n-i+1} = a_i, \quad \sum_{i=1}^{n} a_i = 1, \qquad (6.5.1)$$

and in particular, with $p = \frac{1}{2} - r/n$,

(a) Winsorized means:

$$W_n(p) = \frac{(r+1)(X_{(r+1)} + X_{(n-r)}) + \sum\limits_{i=r+2}^{n-r-1} X_{(i)}}{n} \qquad 0 < r < \tfrac{1}{2}(n-1),$$

$$W_n\left(\frac{1}{2n}\right) = X_{((n+1)/2)} \quad \text{if } n \text{ is odd}; \qquad r = \tfrac{1}{2}(n-1);$$

(b) Trimmed means:

$$T_n(p) = \sum_{i=r+1}^{n-r} \frac{X_{(i)}}{n-2r};$$

(c) Linearly weighted means:

$$L_n(p) = \sum_{j=1}^{\frac{1}{2}n-r} \frac{(2j-1)(X_{(r+j)} + X_{(n-r+1-j)})}{2(\tfrac{1}{2}n - r)^2} \qquad n \text{ even,}$$

$$= \sum_{j=1}^{\frac{1}{2}(n-1)-r} \frac{(2j-1)(X_{(r+j)} + X_{(n-r+1-j)}) + (n-2r)X_{((n+1)/2)}}{[\tfrac{1}{2}(n-1) - r]^2 + [\tfrac{1}{2}(n+1) - r]^2} \qquad n \text{ odd;}$$

(d) Median and two other symmetric order statistics:

$$Y_n(p,a) = a(X_{(r+1)} + X_{(n-r)}) + (\tfrac{1}{2} - a)(X_{(\frac{1}{2}n)} + X_{(\frac{1}{2}n+1)}) \qquad n \text{ even,}$$

$$= a(X_{(r+1)} + X_{(n-r)}) + (1 - 2a)X_{((n+1)/2)} \qquad n \text{ odd.}$$

For $n = 4$, only $r = 1$ is possible, so that (a)–(c) coincide and are just the special case $a = \tfrac{1}{2}$ of (d), which is the most general linear systematic statistic,

$$Z_4(a) = a(X_{(2)} + X_{(3)}) + (\tfrac{1}{2} - a)(X_{(1)} + X_{(4)}).$$

It is easy to show that var $[Z_4(a)]$ is a minimum for

$$a = a_0 = \frac{\sigma_{11} + \sigma_{14} - \sigma_{12} - \sigma_{13}}{2(\sigma_{22} + \sigma_{23} + \sigma_{11} + \sigma_{14} - 2\sigma_{12} - 2\sigma_{13})},$$

the variances and covariances for R, P, T, and N being known (Chapter 3).[8]

Figure 6.5 gives the efficiencies of $Z_4(a)$ relative to $Z_4(a_0)$ as a function of a. The figure shows that the sample mean ($a = \tfrac{1}{4}$) is actually the estimator most robust over the four distributions, and that it has a guaranteed efficiency of 0.80. On the other hand, if we are concerned with the possibility of outlying observations and confine consideration to N and DE, then $a = 0.36$ is best and $Z_4(0.36)$ has guaranteed efficiency 0.95 over this smaller class of distributions. See Crow and Siddiqui (1967) for further details and also for results when $n = 8$, 16, and ∞. Additional asymptotic results are given by Siddiqui

[8] For the Cauchy distribution all second order moments are infinite for $n = 4$.

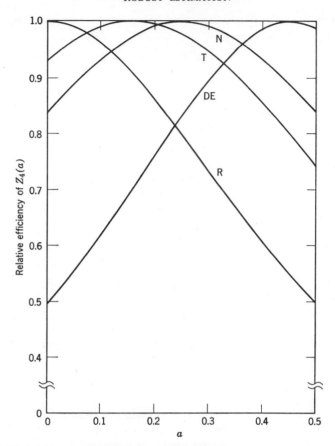

Fig. 6.5. Efficiencies of linear systematic statistics for $n = 4$. N = normal pencil of distributions; T = triangular; DE = double exponential; R = rectangular (from Crow and Siddiqui, 1967, with permission of the authors and the Editor of the *Journal of the American Statistical Association*).

and Raghunandanan (1967). Both small-sample and asymptotic results in the same spirit are obtained by Gastwirth and Cohen (1968), who pay special attention to scale-contaminated normal distributions (see the end of A3.1). Filliben (1969) studies robustness with the help of Tukey's λ-distribution. A slightly different approach to the same problem of robust estimation of location for symmetrical distributions has been developed by Birnbaum and Laska (1967); see Ex. 6.5.1.

Hodges and Lehmann (1963) have pointed out that for a symmetric population any rank test of location can be converted into an estimator of μ.

When the rank test is Wilcoxon's (one-sample) test, the estimator is, with $M_{ij} = \frac{1}{2}(X_{(i)} + X_{(j)})$,

$$T = \underset{i \leq j}{\text{med}} \, M_{ij},$$

that is, T is the median of the $\frac{1}{2}n(n + 1)$ pairwise means, including the observations themselves. This estimator has desirable large-sample properties and may intuitively be expected to have considerable robustness against outliers.

Related statistics studied in an interesting paper by Hodges (1967) are

$$U = \underset{i < j}{\text{med}} \, M_{ij}$$

and

$$D = \underset{i=1,2,\ldots,[\frac{1}{2}n]}{\text{med}} \, M_{i,n-i+1}.$$

The D statistic is much simpler to compute than the other two. Since the means of symmetrically placed order statistics are clearly the most relevant among all pairwise means for the estimation of μ in a symmetric parent, one may hope that D is not much less efficient than T. Hodges shows this to be so for $n = 18$ by an ingenious sampling experiment, finding the efficiencies and associated standard errors 0.949 ± 0.007 for T, 0.956 ± 0.006 for U, and 0.954 ± 0.007 for D. How does D compare with the trimmed and Winsorized means? Table 6.5 reproduces Hodges's results for normal samples of 18,

Table 6.5.   Efficiencies of trimmed and Winsorized means in normal samples of 18

r	Trimmed Mean	Winsorized Mean
0	1.00000	1.00000
1	.97462	.98116
2	.94084	.95581
3	.90367	.92501
4	.86429	.88896
5	.82314	.84749
6	.78030	.80021
7	.73535	.74649
8	.68563	.68563

(From Hodges, 1967)

the efficiencies being readily obtainable from tables of the covariances of normal order statistics. $r = 0$ gives the sample mean, and $r = 8$ the sample median. We see that D is about as efficient as the Winsorized mean W for $r = 2$. But note that W becomes useless when there are more than 2 (r in general) out and outliers—to use Ferguson's happy phrase—on one side, whereas D can tolerate 4 such. In fact, Hodges formally defines *tolerance* along these lines, showing it to be $[\frac{1}{4}(n - 2)]$ for D_n. Thus, from the combined standpoint of efficiency in normal samples and of tolerance, D comes out on top, at least for $n = 18$. Nor does any other linear estimator of the same tolerance as W provide higher efficiency (to three decimals) than does W (Dixon, 1960; cf. Section 7.2). For asymptotic results on D see Bickel and Hodges (1967).

Another interesting approach to robust estimation has recently been put forward by Hogg (1967). An example of his class of statistics is the following estimator H of the center of a symmetric distribution:[9]

$$H = \begin{cases} \bar{X}^c(\frac{1}{4}) & b_2 < 2.0, \\ \bar{X} & 2.0 \le b_2 \le 4.0, \\ \bar{X}(\frac{1}{4}) & 4.0 < b_2 \le 5.5, \\ M & 5.5 < b_2, \end{cases}$$

where $\bar{X}^c(\frac{1}{4})$ is the mean of the $[\frac{1}{4}n]$ smallest and $[\frac{1}{4}n]$ largest observations, $\bar{X}(\frac{1}{4})$ is the mean[10] of the remaining interior observations, $\bar{X}$ and M are sample mean and median, and b_2 is the sample coefficient of kurtosis:

$$b_2 = \frac{n \Sigma (x_i - \bar{x})^4}{[\Sigma (x_i - \bar{x})^2]^2}.$$

Thus the choice of estimator is made to depend on a preliminary calculation (which need not be confined to b_2). In so far as $\bar{X}^c(\frac{1}{4})$ has good properties over the region $\beta_2 < 2.0$, etc., where β_2 is the population kurtosis, H may be expected to perform well, and Hogg finds it on the whole slightly superior to the Hodges-Lehmann T in a sampling experiment of 200 samples of $n = 7$ and 25 from each of four symmetric distributions with approximate β_2 values of 1.9, 2.7, 3.9, and 9.9.

Other work on robust estimation rests largely on asymptotic theory. Notable are papers by Bickel (1965), Gastwirth (1966), Gastwirth and Rubin (1969), and Huber (1964). For samples of 20 from a target normal distribution contaminated by another normal distribution Leone *et al.*

---

[9] H is our symbol: Hogg calls it T.
[10] $\bar{X}(\frac{1}{4}) = T_n(\frac{1}{4})$ when $\frac{1}{4}n$ is integral. Also termed the midmean, this statistic has gained considerable support (Tukey, Crow and Siddiqui, Gastwirth and Cohen) as a general robust estimator when little is known about the class of parent distributions.

(1967) in a Monte Carlo study compare Huber's estimators with the sample mean and Hodges-Lehmann estimators. Attention is also paid to the distribution of the estimators. We may mention here a proposal by Tukey and McLaughlin (1963) for *testing* the mean of a symmetrical population by means of a suitably trimmed version (trimming in both numerator and denominator) of Student's *t*. They advocate such a statistic for general use (in symmetric parents) since it guards against outliers and appears to be robust in validity and power for long-tailed distributions such as frequently occur in practice. A corresponding Winsorized *t* is examined by Dixon and Tukey (1968). See also the review paper by Huber (1968).

EXERCISES

6.1.1. If $u(x)$ is continuous, show by differentiation with respect to θ that (6.1.1) implies $u(x) = 0$ for all $x \geq 0$.

When $u(x)$ is not necessarily continuous, write $u(x) = u^+(x) - u^-(x)$, where u^+ and u^- denote the positive and negative parts of u, respectively. Hence show that in this case (6.1.1) implies $u(x) = 0$ a.e. for $x \geq 0$.

(Lehmann, 1959, p. 131)

6.1.2. For the uniform distribution of Example 6.1.2 find expressions for the efficiency of the median as an estimator of μ for (*a*) *n* odd and (*b*) *n* even. Show that in both cases the asymptotic efficiency is zero.

6.1.3. Discuss the estimation of θ in the following case:

$$p(x, \theta) = 1/\theta \qquad k\theta \leq x \leq (k + 1)\theta; \theta > 0, k > 0.$$

(cf. Kendall and Stuart, 1961, p. 30)

6.1.4. Show that the shortest confidence interval for θ in (6.1.5) for confidence coefficient $1 - \alpha$ is

$$C^{-1}[\alpha^{-1/n}C(z)] \leq \theta \leq z.$$

(Huzurbazar, 1955)

6.1.5. Let $Y_1, Y_2, \ldots, Y_k$ be the maxima of, respectively, $n_1, n_2, \ldots, n_k$ $(\sum_{i=1}^{k} n_i = n)$ variates, all mutually independently drawn from a $R(0, 1)$ parent. Then the pdf of Y_i is

$$f(y_i) = n_i y_i^{n_i-1} \qquad 0 \leq y_i \leq 1.$$

Also let

$$z = \max_i y_i, \quad y = \prod_{i=1}^{k} y_i{}^{n_i}, \quad \text{and } v = \frac{u}{z^n}.$$

Prove that

$$(a) \quad -2 \log U \frown \chi_{2k}{}^2,$$

$$(b) \quad -2 \log V \frown \chi_{2(k-1)}^2.$$

[For (*b*) show first that V and Z are statistically independent.]

(Hogg, 1956)

6.1.6. Let X have the pdf

$$p(x; \theta) = C(\theta)g(x) \qquad a \le x \le b(\theta)$$
$$= 0 \qquad \text{elsewhere,}$$

where $g(x)$ is a single-valued positive continuous function of x, and $b(\theta)$ is strictly increasing in θ. Let Y_i ($i = 1, 2, \ldots, k$) and Z be defined as in Ex. 6.1.5 except that Y_i is now the maximum of n_i variates with pdf $p(y_i; \theta_i)$.

Show that the likelihood ratio λ for testing

$$H_0 : \theta_1 = \theta_2 = \cdots = \theta_k \qquad k > 1$$

against a general alternative is

$$\lambda = \frac{[C(t)]^n}{\displaystyle\prod_{i=1}^{k} C(t_i)^{n_i}},$$

where $t_i = b^{-1}(y_i)$ and $t = b^{-1}(z)$, and that, on H_0, $-2 \log \lambda$ has a $\chi^2_{2(k-1)}$ distribution.

(Hogg, 1956)

6.2.1. Verify that for X uniform in $(\mu - \tfrac{1}{2}\omega, \mu + \tfrac{1}{2}\omega)$ Lloyd's method gives the optimal estimators M and W' of Example 6.1.2.

(Lloyd, 1952)

6.2.2. For the right-triangular distribution

$$p(x) = \frac{(x - \mu)/\sigma + 2\sqrt{2}}{9\sigma} \qquad \mu - 2\sqrt{2}\sigma \le x \le \mu + \sqrt{2}\sigma,$$

show that

$$\alpha_r = \frac{\dfrac{6n(n - 1) \cdots (r + 1)r \cdot 2^{n-r+1}}{(2n + 1)(2n - 1) \cdots (2r + 3)(2r + 1)} - 4}{\sqrt{2}}$$

$$= \frac{6\alpha_r' - 4}{\sqrt{2}} \text{ (say)},$$

$$\beta_{rr} = 18\left(\frac{r}{n + 1} - \alpha_r'^2\right),$$

$$\beta_{rs} = \frac{(s - 1)(s - 2) \cdots (r + 1)r \cdot 2^{s-r}\beta_{ss}}{(2s - 1)(2s - 3) \cdots (2r + 3)(2r + 1)} \qquad r < s;$$

and that the estimators of μ and σ are

$$\mu^* = \frac{\dfrac{n - 1}{3}\left[\sum \dfrac{X_{(i)}}{i} + 2X_{(1)} + \dfrac{2n + 1}{n - 1}X_{(n)}\sum_{i=2}^{n}\dfrac{1}{i}\right]}{n\sum\dfrac{1}{i} - 1},$$

$$\sigma^* = \frac{\dfrac{2n - 1}{6\sqrt{2}}\left[\left(\sum \dfrac{1}{i} + 2\right)X_{(n)} - 2X_{(1)} - \sum \dfrac{X_{(i)}}{i}\right]}{n\sum\dfrac{1}{i} - 1}.$$

(Downton, 1954)

6.2.3. Show that, in a random sample from any population depending only on location and scale parameters, the mean and the range are uncorrelated if

$$\sum_{i=1}^{n} \beta_{i1} = \sum_{i=1}^{n} \beta_{in}.$$

(Aiyar, 1963)

6.2.4. If in (6.2.1) σ is known, show that the BLUE $*\mu$ of μ and its variance are given by

$$*\mu = \mu^* - (\sigma^* - \sigma) \, \text{cov} \, (\mu^*, \sigma^*)/\text{var} \, \sigma^*,$$

$$\text{var} \, *\mu = \text{var} \, \mu^* - [\text{cov} \, (\mu^*, \sigma^*)]^2/\text{var} \, \sigma^*,$$

and that corresponding results when μ is the known parameter follow from interchanging μ and σ.

(Hudson, 1968)

6.2.5. Show that, when one of σ and μ is known, Blom's unbiased nearly best estimator of the other is, respectively,

$$\tilde{\mu} = \frac{1}{d_{11}} \left[\sum_{i=1}^{n} f_i (C_{1i} - C_{1,i-1}) X_{(i)} - \sigma d_{12} \right],$$

$$\tilde{\sigma} = \frac{1}{d_{22}} \left[\sum_{i=1}^{n} f_i (C_{2i} - C_{2,i-1}) X_{(i)} - \mu d_{12} \right].$$

(Blom, 1958, p. 121)

6.3.1. For a doubly-censored (r_1 to the left, r_2 to the right) sample from the uniform distribution

$$p(x) = \frac{1}{\omega} \qquad \mu - \tfrac{1}{2}\omega \le x \le \mu + \tfrac{1}{2}\omega$$

obtain the following results in the notation of Section 6.2:

$$\Omega = (n+1)(n+2) \begin{bmatrix} \dfrac{r_1+2}{r_1+1} & -1 & 0 & \cdots & & & 0 \\ & 2 & -1 & \cdots & & & 0 \\ & & 2 & \cdots & & & 0 \\ & & & \cdots & & & \\ & & & & & \dfrac{n+1}{(n-r_2)(r_2+1)} & + \dfrac{n-r_2-1}{n-r_2} \end{bmatrix}$$

$$\mu^* = \frac{1}{2(n - r_1 - r_2 - 1)} \, [(n - 2r_2 - 1)X_{(r_1+1)} + (n - 2r_1 - 1)X_{(n-r_2)}],$$

$$\omega^* = \frac{n+1}{n - r_1 - r_2 - 1} \, (X_{(n-r_2)} - X_{(r_1+1)}),$$

$$\text{var} \, \mu^* = \frac{(r_1 + 1)(n - 2r_2 - 1) + (r_2 + 1)(n - 2r_1 - 1)}{4(n+1)(n+2)(n - r_1 - r_2 - 1)} \, \omega^2,$$

$$\text{var} \, \omega^* = \frac{r_1 + r_2 + 2}{(n+2)(n - r_1 - r_2 - 1)} \, \omega^2.$$

(Sarhan, 1955; Sarhan and Greenberg, 1959)

6.3.2. For a Type II censored sample $x_{(1)} \leq x_{(2)} \leq \cdots \leq x_{(N)}$ from a distribution with mean μ and variance σ^2, let

$$\bar{x}_{N-1} = \frac{\sum\limits_{i=1}^{N-1} x_{(i)}}{N-1} \quad \zeta_{10} = \mathcal{E}\left[\frac{\bar{X}_{(N-1)} - \mu}{\sigma}\right], \quad \zeta_{01} = \mathcal{E}\left[\frac{\bar{X}_{(N)} - \mu}{\sigma}\right],$$

$$\eta_{ij} = \sigma^{-2(i+j)} \mathcal{E}\left\{\left[\sum_{t=1}^{N-1} (X_{(t)} - X_{(N)})^2\right]^i \left[\sum_{t=1}^{N-1} (X_{(t)} - X_{(N)})\right]^{2j}\right\} \quad i, j = 0, 1, 2.$$

Show that the following estimators, symmetrical in $X_{(1)}, X_{(2)}, \ldots, X_{(N-1)}$, are unbiased for μ and σ^2, respectively:

$$\varepsilon \bar{X}_{N-1} + (1 - \varepsilon) X_{(N)} \quad \text{with} \quad \varepsilon = \zeta_{01}/(\zeta_{01} - \zeta_{10}),$$

$$\alpha \sum_{i=1}^{N-1} (X_{(i)} - X_{(N)})^2 + \beta\left[\sum_{i=1}^{N-1} (X_{(i)} - X_{(N)})\right]^2$$

with

$$\alpha = (\eta_{10}\eta_{02} - \eta_{01}\eta_{11})/(\eta_{10}{}^2\eta_{02} + \eta_{01}{}^2\eta_{20} - 2\eta_{10}\eta_{01}\eta_{11})$$
$$\beta = (\eta_{01}\eta_{20} - \eta_{10}\eta_{11})/(\eta_{10}{}^2\eta_{02} + \eta_{01}{}^2\eta_{20} - 2\eta_{10}\eta_{01}\eta_{11})$$

and that the second estimator is of minimum variance over (α, β).

(Saw, 1959)

6.4.1. Let x_{ij} $(i = 1, 2, \ldots, k; j = 1, 2, \ldots, n_i; \sum_{i=1}^{k} n_i = N)$ be k independent samples, with x_{ij} drawn from the pdf $\sigma^{-1} e^{-(x-\theta_i)/\sigma}$ $(x \geq \theta_i)$. If

$$x_{i(1)} = \min_j x_{ij} \quad \text{and} \quad x_{(1)(1)} = \min_{ij} x_{ij},$$

show that equality of all θ_i may be tested by referring

$$\frac{\sum\limits_{i} n_i(x_{i(1)} - x_{(1)(1)})/(k-1)}{\sum\limits_{i}\sum\limits_{j} (x_{ij} - x_{i(1)})/(N-k)}$$

to tables of the F ratio with $2(k-1)$ and $2(N-k)$ DF.

(Sukhatme, 1937)

6.4.2. Let x_i $(i = 1, 2, \ldots, m)$, y_j $(j = 1, 2, \ldots, n)$ be independent samples from

$$p(x) = \sigma_x^{-1} e^{-(x-\theta)/\sigma_x} (x \geq \theta) \quad \text{and} \quad p(y) = \sigma_y^{-1} e^{-(y-\theta)/\sigma_y} (y \geq \theta),$$

respectively. Also let $u = 2m(x_{(1)} - y_{(1)})/\sigma_x$ for $x_{(1)} > y_{(1)}$, $v = 2n(y_{(1)} - x_{(1)})/\sigma_y$ for $y_{(1)} > x_{(1)}$, and $w = u$ for $x_{(1)} > y_{(1)}$, $w = v$ for $y_{(1)} > x_{(1)}$. Show that

(a) $\Pr\{Y_{(1)} > X_{(1)}\} = \dfrac{m/\sigma_x}{m/\sigma_x + n/\sigma_y}$.

(b) U, V, W are all distributed as χ^2 with 2 DF.

(Epstein and Tsao, 1953)

6.4.3. A life test on n items with independent lifetimes X_i having pdf $\sigma^{-1}e^{-x/\sigma}$ $(x \geq 0)$ is terminated as soon as N items have failed or after time x_0, whichever comes first. Show that under this procedure

(a) the expected number of failures is

$$np \sum_{i=0}^{N-2} \binom{n-1}{i} p^i (1-p)^{n-1-i} + N \sum_{i=N}^{n} \binom{n}{i} p^i (1-p)^{n-i},$$

where $p = 1 - e^{-x_0/\sigma}$; and

(b) the expected duration of the test is

$$\sum_{i=1}^{N-1} \binom{n}{i} p^i (1-p)^{n-i} \mathscr{E} X_{(i)} + \sum_{i=N}^{n} \binom{n}{i} p^i (1-p)^{n-i} \mathscr{E} X_{(N)},$$

where

$$\mathscr{E} X_{(r)} = \sigma \left(\frac{1}{n} + \frac{1}{n-1} + \cdots + \frac{1}{n-r+1} \right) \qquad r = 1, 2, \ldots, n.$$

Derive also the corresponding results when items are replaced immediately upon failure so that n items are always under test.

(Epstein, 1954)

6.4.4. Show that for a Type II censored sample from the uniform $R(0, 1)$ distribution the statistic

$$Y = \sum_{i=1}^{N} X_{(i)} + (m - N) X_{(N)} \qquad m > N - 1$$

has pdf

$$f(y) = \frac{n}{(N-1)!} \left\{ \binom{N-1}{0} \frac{(m-y)^{n-1}}{m^{n-N+1}} - \binom{N-1}{1} \frac{(m-1-y)^{n-1}}{(m-1)^{n-N+1}} + \cdots \right\}$$

for $0 \leq y \leq m$, where m is not necessarily an integer and the summation continues as long as $m - y, m - 1 - y, \ldots$ are positive.

[Hint: Show first that the joint pdf of $V = X_{(N)}$ and $W = \sum_{i=1}^{N-1} X_{(i)}/X_{(N)}$ is

$$f(v, w) = f(v) f(w \mid v) \qquad 0 \leq v \leq 1, 0 \leq w \leq N - 1$$

$$= \frac{n!}{(N-1)!(n-N)!} v^{N-1} (1-v)^{n-N}$$

$$\cdot \frac{1}{(N-2)!} \left\{ \binom{N-1}{0} w^{N-2} - \binom{N-1}{1} (w-1)^{N-2} + \cdots \right\}.]$$

(Gupta and Sobel, 1958)

6.4.5. Show that for a sample of n from $p(x) = \sigma^{-1} e^{-(x-\theta)/\sigma}$ ($x \geq \theta$), with r_1 missing to the left and r_2 to the right, the BLUE's are

$$\theta^* = C \left\{ \left[\frac{1}{C} + (n - r_1) \sum_{i=1}^{r_1+1} \frac{1}{n-i+1} \right] x_{(r_1+1)} \right.$$

$$\left. - \left(r_2 \sum_{i=1}^{r_1+1} \frac{1}{n-i+1} \right) x_{(n-r_2)} - \sum_{i=1}^{r_1+1} \frac{1}{n-i+1} \sum_{i=r_1+1}^{n-r_2} x_{(i)} \right\}$$

and

$$\sigma^* = C\left[\sum_{i=r_1+1}^{n-r_2} x_{(i)} - (n - r_1)x_{(r_1+1)} + r_2 x_{(n-r_2)}\right],$$

where $1/C = n - r_1 - r_2 - 1$.

Also obtain

$$\text{var } \theta^* = \left[C\left(\sum_{i=1}^{r_1+1} \frac{1}{n-i+1}\right)^2 + \sum_{i=1}^{r_1+1} \frac{1}{(n-i+1)^2}\right]\sigma^2$$

and

$$\text{var } \sigma^* = C\sigma^2.$$

(Sarhan, 1955)

6.4.6. For the exponential distribution

$$p(x) = \sigma^{-1}e^{-x/\sigma} \qquad x \geq 0$$

let C_L be the random variable $\int_{L(X)}^{\infty} p(x)\,dx$, where $L(X)$ is a function of the sample $X_1, X_2, \ldots, X_n$. It is desired to find a tolerance interval $(L(X), \infty)$ such that

$$\Pr\{C_L \geq \gamma\} = \beta$$

and that for $\gamma' > \gamma$

$$\Pr\{C_L \geq \gamma'\} \leq \delta.$$

Show that

$$L(X) = \frac{2\left[\sum_{i=1}^{r} X_{i:n} + (n - r)X_{r:n}\right](-\log \gamma)}{\chi_{1-\beta}^2(2r)}$$

satisfies both requirements provided r is chosen so large that

$$\left(\frac{\log \gamma'}{\log \gamma}\right)\chi_{1-\beta}^2(2r) \leq \chi_{1-\delta}^2(2r).$$

(n may then be chosen $\geq r$.)

(Faulkenberry and Weeks, 1968)

6.5.1. Let X have pdf $(1/\sigma)g((x - \mu)/\sigma, \lambda)$, where λ is an unknown shape parameter, $\lambda \in \Lambda$. Suppose that $g(y, \lambda)$ is symmetric in $y = (x - \mu)/\sigma$ for given λ. Write

$$\mathscr{E}(Y_{(r)} \mid \lambda) = \alpha_r\lambda, \quad \text{cov}(Y_{(r)}, Y_{(s)} \mid \lambda) = \beta_{rs}\lambda.$$

Next, suppose that λ has cdf $H(\lambda)$ over Λ and define

$$\alpha_r H \equiv \mathscr{E}(Y_{(r)} \mid H) = \int_\Lambda \alpha_r^\lambda \, dH(\lambda)$$

and let $\beta_{rs}H$ be the covariance of $Y_{(r)}$, $Y_{(s)}$ under this mixture model.
Show that

(a) $B_{rs}H = \int \beta_{rs}^\lambda \, dH(\lambda) + \int \alpha_r^\lambda \alpha_s^\lambda \, dH(\lambda) - \int \alpha_r^\lambda \, dH(\lambda) \int \alpha_s^\lambda \, dH(\lambda)$;
(b) if the $\alpha_r H$, $\beta_{rs}H$ are given, μ can be estimated (under the mixture model) by the Gauss-Markov theorem as

$$\mu_H^* = \frac{\mathbf{1}'\mathbf{\Omega}^H\mathbf{X}}{\mathbf{1}'\mathbf{\Omega}^H\mathbf{1}}, \quad \text{where} \quad \mathbf{\Omega}^H = (\mathbf{B}^H)^{-1} = (\beta_{rs}H)^{-1}$$

with

$$\text{var } \mu_H{}^* = \frac{\sigma^2}{1'\Omega^H 1} \; ;$$

(c) under shape λ

$$\mathscr{E}(\mu_H{}^* \mid \lambda) = \mu,$$

$$\text{var } (\mu_H{}^* \mid \lambda) = \frac{\sigma^2 1'\Omega^H B^\lambda \Omega^H 1}{(1'\Omega^H 1)^2}.$$

By taking $H(\lambda)$ to be a two-point distribution with $h = \Pr \{\lambda = 1\}$ and $1 - h = \Pr \{\lambda = 0\}$, indicate how the above approach can be used to determine the "most robust mixture" of two distributions, that is, the mixture maximizing the minimum efficiency relative to the BLUE under either distribution.

(Birnbaum and Laska, 1967)

6.5.2. Show that, for samples of $n = 2m + 1$ $(m = 0, 1, 2, \ldots)$ from the Cauchy distribution

$$p(x) = \frac{1}{\pi[1 + (x - \mu)^2]} \qquad -\infty < x < \infty,$$

the trimmed mean

$$\frac{1}{n - 2[nk]} \sum_{i=m-[nk]}^{m+[nk]} X_{(i)} \qquad 0 \le k \le \tfrac{1}{2}$$

is an unbiased estimator of μ with asymptotic variance

$$\frac{1}{nk} \left[\frac{1 - k}{k} \tan^2 \left(\tfrac{1}{2}\pi k \right) + \frac{2}{\pi k} \tan \left(\tfrac{1}{2}\pi k \right) - 1 \right].$$

(This is a minimum for $k = 0.24$, when the trimmed mean is almost the midmean.)

(Rothenberg et al., 1964)

CHAPTER 7

Short-Cut Procedures

7.1. INTRODUCTION

Whereas Chapter 6 has dealt mainly with estimation and testing procedures which are optimal in some sense, we now turn to methods, mostly designed for normal samples, whose principal merit is their simplicity. In some instances there are additional benefits such as robustness (see, e.g., the early paper by Benson, 1949). Censored data, for which optimal methods can be very laborious, afford a particularly good opportunity for a drastic reduction in labor. However, we will illustrate briefly in this introduction a number of general features of short-cut methods by means of the (sample) range w, probably the most widely used of all "quick" statistics. True, the computational advantage of w over the sample standard deviation $s = [\Sigma(x_i - \bar{x})^2/ (n - 1)]^{1/2}$ has become less important since the advent of high-speed computers, but there remain many instances when the simplicity of w and its possible use by the unskilled represent major advantages. Thus w has almost completely ousted s from the quality control field, where small samples are taken at frequent intervals and their means and ranges plotted (Section 7.9).

Although we are here concerned with range as a quick measure of variability, it should be noted that there are situations in which the range is the only or the most appropriate measure. Thus the range of the times of occurrence of n events provides the most relevant assessment of the closeness to simultaneity of the events; the range of length or thickness of n objects best gauges departures from requirements for balance (e.g., table legs) or for uniformity of appearance (Eisenhart et al., 1947, Chapter 5).

Tables of percentage points of the range and the studentized range are referred to in A2.3 and A5.2, respectively. That W, like S, is independent of $\bar{X}$ in normal samples has been pointed out in Section 2.7. Can we, with the help of these basic tools, use W widely in estimation and testing? The answer is yes, although not as widely as S and not without additional tables and occasional approximations. But suitable tables are available for many

137

purposes, rendering the range techniques extremely easy to use—see Section 7.3 for estimation, Section 7.7 for testing. In these sections questions of efficiency, power, and robustness are also considered. All three properties, although of the utmost importance, are concerned with the long-run individual performances of range methods compared to standard methods. One may also ask: To what extent, in the long run, do quick methods give the same verdict as standard methods when both are applied to the same sets of data? This question was examined experimentally as early as 1935 by Pearson and Haines, who plotted w against s for a series of small samples of real data. With the help of the known standard deviation σ, they were able to insert on their diagrams a grid of several upper and lower percentage points of both W and S, so that the corresponding degrees of significance attained by any sample, and the agreement in this respect, could be seen on inspection. More theoretical approaches are taken by Cox (1956) and David and Perez (1960). Although W and S are not normally distributed, a good idea of the agreement, at least in normal samples, may be gained from the correlation coefficient $\rho(W, S)$. This is easily calculated from the relation

$$\rho(W, S) = (\text{eff } \hat{\sigma}_w)^{1/2},$$

where eff $\hat{\sigma}_w$, the efficiency of the range estimator of σ, is given in Table 7.3.1. See Ex. 7.1.1.

Quick measures of dispersion other than the range and the mean range are discussed in Section 7.4, quick measures of location in Section 7.2, and quick estimates in bivariate samples in Section 7.5.

More dubiously placed in the present chapter are methods involving "optimal spacing" of order statistics (Section 7.6). These are short-cut procedures in that the estimation of parameters in medium and large samples is based on a set of k ($\ll n$) order statistics chosen so as to provide asymptotically optimal estimators in the class of linear functions of k order statistics.

Probability plotting methods (Section 7.8), although not new, have attained new flexibility in recent years and are playing an increasingly important role in informal methods of data analysis. Statistical quality control, heavily dependent on the use of the range, is treated briefly in Section 7.9.

7.2. QUICK MEASURES OF LOCATION

In view of the computational simplicity of the sample mean, other measures of location may hardly seem to fall under our chapter heading of short-cut procedures. This is indeed so for complete samples but not if we are faced with censored data. Even in complete samples other measures may have advantages in robustness, as already discussed in Section 6.5. Here we confine ourselves to simple measures, the oldest, most robust, and simplest being the *sample median* M. For a normal parent, investigations of the distribution and moments of M go back to Hojo (1931, 1933) and K. and M. V. Pearson (1931). More recently, Cadwell (1952) developed an approximation to the pdf of M which gives results very close to the true values of $\sigma^2(M)$ and $\beta_2(M)$ even for small samples (for n odd see Ex. 7.2.1). Since $\beta_2(M) = 3.0347$ for $n = 3$ and is closer to 3 for larger n, the tendency to normality appears remarkably swift. Chu (1955) confirms this result theoretically. However, the simple asymptotic formula

$$\sigma^2(M) = \{4n[p(\mu)]^2\}^{-1}$$

does not give equally good approximations to the exact variance. Chu and Hotelling (1955) therefore investigate a variety of approximation methods applicable also to other parents. One of their methods is further studied by Siddiqui (1962), who compares approximate and exact values of $\sigma^2(M)$ in odd samples for the following distributions:

(i) $p(x) = \pi^{-1}(1 - x^2)^{-\frac{1}{2}}$, $x^2 \leq 1$;
(ii) uniform;
(iii) $p(x) = \frac{3}{4}(1 - x^2)$ $x^2 \leq 1$;
(iv) normal;
(v) $p(x) = \frac{1}{2}e^{-|x|}$ $x^2 \leq \infty$;
(vi) Cauchy;
(vii) exponential.

Hodges and Lehmann (1967) table the efficiency of M in normal samples for $n \leq 20$, both exactly and by the usual asymptotic formula taken to order $1/n$ (see also Ex. 4.5.1). Although this efficiency is higher than $2/\pi \doteq 0.637$, the asymptotic value, it is not good, being 0.743 for $n = 3$, 0.838 for $n = 4$, and lower thereafter. This is, of course, the basic reason for seeking other measures of location, preferably still reasonably robust.

Dixon (1957) gives the efficiencies in normal samples of $n \leq 20$ of the trimmed mean

$$T = \frac{\sum_{i=2}^{n-1} X_{(i)}}{n - 2}$$

and of the two-point mean

$$\frac{1}{2}(X_{(i)} + X_{(j)}),$$

where i and j are chosen for maximum efficiency. It is interesting to note that the efficiency of T *relative to the BLUE* is never less than 0.99, while that of $\frac{1}{2}(X_{(i)} + X_{(j)})$ is somewhat higher than its asymptotic value of 0.81. For $n > 5$ the optimal i and j are not far away from the 27th and 73rd percentiles given by asymptotic theory (Section 7.6). For censored data Dixon (1960) suggests consideration of the Winsorized mean

$$W = \frac{1}{n} [(i + 1)x_{(i+1)} + x_{(i+2)} + \cdots + x_{(n-1-i)} + (i + 1)x_{(n-i)}]$$

when i observations are censored at one end and $j \leq i$ at the other. Although this estimate ignores $i - j$ observations, the efficiency of W relative to the BLUE, based on all available $n - i - j$ observations, is never less than 0.956 for $n \leq 20$ and $i \leq 6$ ($n \geq 2i + 1$).

Another estimate of location is the midpoint or midrange $\frac{1}{2}(x_{(1)} + x_{(n)})$. This is, of course, completely lacking in robustness to outliers. However, it is optimal for uniform populations (Section 6.1) and retains good properties for other symmetric finite-range distributions with small β_2 values (Rider, 1957).

Harter (1961b, 1964a) discusses the estimation of the parameters of a negative exponential distribution by one or two order statistics. For the one-parameter exponential he gives both point and confidence interval estimates based on the best single order statistic. Since the sample mean is optimal, the main purpose is again to obtain an estimate which can be used even when some of the largest observations are censored or unreliable.

7.3. RANGE AND MEAN RANGE AS MEASURES OF DISPERSION

The exact distribution of the range W in continuous populations has been derived in Section 2.3. Results for discrete parents are given in Ex. 2.4.2. In the normal $N(\mu, \sigma^2)$ case, with which we shall be mainly concerned here, extensive tables of percentage points, cdf, and moments of W are available (see A2.3 and A3.2). A very simple unbiased estimator $\hat{\sigma}_w$ of σ is obtained on multiplying W by $1/d_n$, where $d_n = \mathscr{E}(W/\sigma)$ in normal samples of n. Table 7.3.1 gives $1/d_n$ together with

$$\text{eff}(\hat{\sigma}_w) = \frac{\text{var } S'}{\text{var } \hat{\sigma}_w},$$

where S' is the unbiased rms estimator of σ, well known to be UMVU, namely,

$$S' = \frac{\Gamma[\frac{1}{2}(n - 1)]}{\sqrt{2}\ \Gamma(\frac{1}{2}n)} [\Sigma(X_i - \bar{X})^2] \ . \tag{7.3.1}$$

From the table the efficiency of $\hat{\sigma}_w$ is seen to be adequate for $n \leq 12$ and very good for the small sample sizes (typically $n = 5$) generally used in quality control work. For $n > 12$ the efficiency can be increased by random division of the sample of n into smaller subsamples, for which purpose a subsample

Table 7.3.1.   Multipliers and efficiency for the range estimator $\hat{\sigma}_w = W/d_n$ *of* σ

n	$1/d_n$	eff $\hat{\sigma}_w$	$d_n^2/V_n{}^*$	n	$1/d_n$	eff $\hat{\sigma}_w$	d_n^2/V_n
				11	.315	.831	16.2
2	.886	1	1.75	12	.307	.814	17.5
3	.591	.992	3.63	13	.300	.797	18.8
4	.486	.975	5.48	14	.294	.781	19.9
5	.430	.955	7.25	15	.288	.766	21.1
6	.395	.933	8.93	16	.283	.751	22.2
7	.370	.911	10.53	17	.279	.738	23.3
8	.351	.890	12.06	18	.275	.725	24.3
9	.337	.869	13.52	19	.271	.712	25.3
10	.325	.850	14.91	20	.268	.700	26.3

* See (7.3.2) for use of this ratio.

size of 8 is optimal (Grubbs and Weaver, 1947). However, in view of the arbitrariness of the subdivision one of the methods of the next section is to be preferred. The *mean range* $\bar{w}_{n,k}$, the average of k ranges each of size n, does play a useful role in the estimation of σ in a one-way classification of k groups of n ($\hat{\sigma}_{\bar{w}} = \bar{w}_{n,k}/d_n$) and the associated analysis of variance (see Section 7.7).

Approximations to the Mean Range

While the efficiency of $\hat{\sigma}_{\bar{w}}$ in normal samples can be readily calculated from tables, the exact distribution of $\overline{W}_{n,k}$ (or $\overline{W}$ for short) is awkward to handle for $k > 1$ (see, e.g., Bland *et al.*, 1966). There are, however, several useful approximations available, namely,

(i) $\overline{W}/\sigma = c\chi_v/v^{1/2}$ (Parnaik, 1950)
(ii) $\overline{W}/\sigma = c\chi_v^2/v$ (Cox, 1949)
(iii) $\overline{W}/\sigma = (\chi_v^2/c)^\alpha$ (Cadwell, 1953b)

The constants v (the fractional degrees of freedom) and c in (i) and (ii) are determined by equating the first two moments of left- and right-hand sides. For the three-parameter approximation (iii) third moments are also equated.

In general, the approximations become more accurate with increasing k (for given n), since all of them and also $\overline{W}$ tend to normality; taking $\overline{W}$ to be normally distributed may indeed give sufficient accuracy for some purposes. For the case of a single range ($k = 1$), when the approximations are most severely tested, Pearson (1952) has made a detailed comparison of (i) and (ii) for $n = 4, 6, 10$, and 15. He concludes that for $n < 10$ the χ approximation is the more accurate, that there is little difference for $n = 10$, and that for $n > 10$ the χ^2 approximation becomes the better. Approximation (iii) is appreciably more accurate and seems adequate whenever a specially good representation of the range is required. See also Pillai (1950).

With such a variety of reasonably accurate approximations available, considerable flexibility is possible. The χ approximation has been the most popular, since, in addition to its high accuracy for small n, it takes $\overline{W}$ as proportional to $S_\nu = \chi_\nu \sigma / \nu^{1/2}$. It therefore permits at once the imitation of tests such as Student's t through the replacement of the usual rms estimate of σ by $\overline{W}/c$, the only change being a small reduction in the degrees of freedom (see Section 7.6). All approximations allow variance ratios to be replaced by range ratios or powers thereof.

The usefulness of the three approximations rests, of course, on the availability of suitable tables. These are particularly simply obtained in case (ii), for which clearly

$$
\left.
\begin{aligned}
d_n &= c, \quad \frac{V_n}{k} = \frac{2c^2}{\nu}, \\
c &= d_n, \quad \nu = \frac{2kd_n^2}{V_n},
\end{aligned}
\right\}
$$

or   (7.3.2)

where $V_n = \text{var}\,(W/\sigma)$ in normal samples of n. The ratio d_n^2/V_n is tabulated in Table 7.3.1. Correspondingly, (i) gives

$$
\left.
\begin{aligned}
d_n &= \frac{c\sqrt{2}}{\nu^{1/2}} \frac{\Gamma[(\nu + 1)/2]}{\Gamma(\frac{1}{2}\nu)} \\
&= c\left(1 - \frac{1}{4\nu} + \frac{1}{32\nu^2} + \frac{5}{128\nu^3} - \frac{21}{2048\nu^4} - \cdots\right), \\
\frac{V_n}{k} &= \mathscr{E}\left(\frac{c\chi_\nu}{\nu^{1/2}}\right)^2 - \left[\mathscr{E}\left(\frac{c\chi_\nu}{\nu^{1/2}}\right)\right]^2 \\
&= c^2 - d_n^2.
\end{aligned}
\right\}
$$

(7.3.3)

Thus c is found easily from (7.3.3), whereas ν is obtained most conveniently by inversion of the series expansion (in powers of $1/\nu$) of $A = 2V_n/kd_n^2$. This yields

$$
\nu = A^{-1} + \frac{1}{4} - \frac{3}{16}A + \frac{3}{64}A^2 + \frac{33}{256}A^3 - \frac{105}{1024}A^4 \cdots. \quad (7.3.4)
$$

Table 7.3.2 gives ν and c for $n \le 10$ and all k.[1] Note that the table provides an immediate (approximate) measure of the efficiency of $\hat{\sigma}_{\overline{w}}$ through the

---

[1] The coefficient of A^4 in (7.3.4) is given incorrectly in David (1962, p. 98) but correctly in Ghosh (1963). Note that Patnaik's (1950) earlier less convenient approach and less accurate table for ν and c continue to be frequently cited.

Table 7.3.2. *Scale factor* c *and equivalent degrees of freedom* v *appropriate to a one-way classification into* k *groups of* n *observations*

k \ n	2 v	2 c	3 v	3 c	4 v	4 c	5 v	5 c	6 v	6 c	7 v	7 c	8 v	8 c	9 v	9 c	10 v	10 c
1	1.00	1.41	1.98	1.91	2.93	2.24	3.83	2.48	4.68	2.67	5.48	2.83	6.25	2.96	6.98	3.08	7.68	3.18
2	1.92	1.28	3.83	1.81	5.69	2.15	7.47	2.40	9.16	2.60	10.8	2.77	12.3	2.91	13.8	3.02	15.1	3.13
3	2.82	1.23	5.66	1.77	8.44	2.12	11.1	2.38	13.6	2.58	16.0	2.75	18.3	2.89	20.5	3.01	22.6	3.11
4	3.71	1.21	7.49	1.75	11.2	2.11	14.7	2.37	18.1	2.57	21.3	2.74	24.4	2.88	27.3	3.00	30.1	3.10
5	4.59	1.19	9.30	1.74	13.9	2.10	18.4	2.36	22.6	2.56	26.6	2.73	30.4	2.87	34.0	2.99	37.5	3.10
6	5.47	1.18	11.1	1.73	16.7	2.09	22.0	2.35	27.0	2.56	31.8	2.73	36.4	2.87	40.8	2.99	45.0	3.09
7	6.35	1.17	12.9	1.73	19.4	2.09	25.6	2.35	31.5	2.55	37.1	2.72	42.5	2.86	47.6	2.99	52.4	3.09
8	7.23	1.17	14.8	1.72	22.1	2.08	29.2	2.35	36.0	2.55	42.4	2.72	48.5	2.86	54.3	2.98	59.9	3.09
9	8.11	1.16	16.6	1.72	24.9	2.08	32.9	2.34	40.4	2.55	47.6	2.72	54.5	2.86	61.1	2.98	67.3	3.09
10	8.99	1.16	18.4	1.72	27.6	2.08	36.5	2.34	44.9	2.55	52.9	2.72	60.6	2.86	67.8	2.98	74.8	3.09
d_n		1.13		1.69		2.06		2.33		2.53		2.70		2.85		2.97		3.08
CD*	0.88		1.82		2.74		3.62		4.47		5.27		6.03		6.76		7.45	

* CD = constant difference; for example, $n = 5$, $k = 12$ gives $v = 36.5 + 2(3.62) = 43.7$. [This table is reproduced (with extension) from David (1951), with permission of the Managing Editor of *Biometrika*.]

143

associated equivalent DF. Thus for $n = 6$, $k = 5$ we have $\nu = 22.6$ as against 25 DF for the mean square estimator.

In the case of a single range Grubbs et al. (1966) have considered the evaluation of c and ν for Patnaik's approximation by other methods, such as equating means and variances of W^2/σ^2 in (i). Their two candidates appear to do slightly better than Patnaik's in approximating upper percentage points of W.

See Tukey (1955) for approximations (essentially interpolation formulae) applicable also for large n to a variety of quantities related to range, for example, d_n, V_n, $w_{0.05}$.

Effect of Parent Non-normality

It is stated in several elementary textbooks that the range, since it involves only the extreme observations, is bound to be inefficient and oversensitive to the shape of the underlying distribution. We have seen that the first claim is misleading: the loss in efficiency is of no practical importance in the routine applications for which range estimates of σ are usually recommended. The second claim is harder to dispose of but even less justified: the indications are that $\hat{\sigma}_w$ and $\hat{\sigma}_{\bar{w}}$ stand up under non-normality as well as the rms estimate s, and perhaps better, at least for $n \leq 6$ (this despite the fact that $\mathscr{E}S^2 = \sigma^2$ for all distributions possessing a variance). Note that we maintain, not that the range is very robust, but only that it compares favorably with s in small samples.

Perhaps the result of greatest interest is the remarkable stability of the ratio $\mathscr{E}(W_n/\sigma)$, the quantity whose normal-theory value d_n determines the width of control limits in quality control charts for the mean (see Section 7.9). Early work, mainly empirical, by Pearson and Adyanthāya (1928)—see also Pearson (1950)—foreshadowed this finding. As was shown in Section 4.2, $\mathscr{E}(W_n/\sigma)$ is in fact bounded above, the upper bound being twice the "upper bound (symmetrical parent)" given in Table 4.2. That table also shows the upper bound to be little in excess of d_n for $n \leq 12$; even for a uniform parent, $\mathscr{E}(W_n/\sigma)$ is not too different from d_n. The same holds for a variety of other "reasonable" distributions (David, 1962), although, as pointed out at the end of Section 4.2, $\mathscr{E}(W_n/\sigma)$ can be made arbitrarily close to zero for pathological parent populations.

Actually the ratio $\mathscr{E}(W_n/d_n\sigma)$ was found by Cox (1954) to be generally slightly less than unity in results for $n \leq 5$ obtained both theoretically and empirically for a large number of distributions. Cox suggests that $\mathscr{E}(W_n/d_n\sigma)$ does not depend on the parent β_1 and tabulates "average" values of this quantity as a function of β_2. Thus w_n/d_n will tend to underestimate σ very slightly for most non-normal parents, but with approximate knowledge of β_2

use of Cox's tables will largely correct even this small bias.[2] See also Tsukibayashi (1958), who considers the behavior, for various non-normal distributions, of range estimators of σ as well as σ^2 which are unbiased under normality.

The dependence of the coefficient of variation of range on β_2 has been similarly examined by Cox. Here the changes are much more pronounced, and this is also true of upper percentage points (see also Belz and Hooke, 1954).

Some of the foregoing results have been challenged by Bhattacharjee (1965), who claims, for example, that on the basis of tail probabilities (probabilities of exceeding upper normal-theory values) W appears to be more affected than S by departures from normality, especially when n is large. (Actually in his table for $\beta_2 = 4$ and 5 the effect begins at $n = 4$.) It is doubtful, however, that Bhattacharjee's use of the first four terms of the Edgeworth series to represent non-normality is adequate for his purposes; at any rate, his results are at variance in a number of places with more direct calculations. The same approach (with some duplication) is used by Singh (1967) to study the effect of non-normality on the extremes as well as on the range.

We conclude this section by pointing out that the range may be used to spot gross errors in the calculation of s in samples from *any* parent (Thomson, 1955). This follows at once from the boundedness of the ratio w/s. The upper bound results from a sample configuration with $n - 2$ observations at the sample mean and the other two observations at equal distances from the mean. The lower bound corresponds to half the observations at one extreme and the other half (plus 1 if n is odd) at the other extreme. The respective bounds are

$$w/s \leq [2(n-1)]^{1/2}, \tag{i}$$

$$w/s \geq \begin{cases} 2[(n-1)/n]^{1/2} & n \text{ even,} \tag{ii} \\ 2[n/(n+1)]^{1/2} & n \text{ odd.} \tag{iii} \end{cases}$$

These values are, incidentally, also the upper and lower 0% points of W/S and as such are tabulated in Table 29c of Pearson and Hartley (1966). Only (iii) needs proof. Let $\bar{x}(i)$ $(i = 1, 2, \ldots, n)$ denote the mean of the first i x's observed. Then

$$\sum_{i=1}^{n} [x_i - \bar{x}(n)]^2 = \sum_{i=1}^{n-1} [x_i - \bar{x}(n-1)]^2 + (n-1)[\bar{x}(n-1) - \bar{x}(n)]^2 + [x_n - (\bar{x}n)]^2.$$

[2] Since empirically d_n is approximately equal to $n^{1/2}$ $(n \leq 10)$, the ratio w_n/n provides an immediate general, if rough, estimate of the standard error of the mean, namely, of $\sigma/n^{1/2}$ (Mantel, 1951).

Clearly, each term on the right is maximized if the first $n - 1$ x's are chosen as for (ii) and if x_n is then taken as $\bar{x}(n - 1) \pm \frac{1}{2}w$.

7.4. OTHER QUICK MEASURES OF DISPERSION

Since the efficiency of W as an estimator of σ in normal samples falls off quickly with increasing n, one may ask at what point a quasi-range

$$W_{(i)} = X_{(n+1-i)} - X_{(i)} \qquad 2 \leq i \leq [\tfrac{1}{2}n]$$

will do better.[3,4] Cadwell (1953a) has shown that $W_{(1)} = W$ is more efficient than any quasi-range for $n \leq 17$, but that thereafter $W_{(2)}$ is more efficient, to be in turn replaced by $W_{(3)}$ for $n \geq 32$, and so on (cf. Section 7.6). He has tabulated moment constants and percentage points of $W_{(2)}$ and given a series expansion for $f(w_{(2)})$. Quasi-ranges are useful in censored samples and obviously have some robustness against outliers. In complete samples their efficiency is not very high, but suitable linear combinations of W and $W_{(i)}$ can provide very efficient estimators. A simple way of doing this is to use the "thickened range"

$$J_i = W + W_{(2)} + \cdots + W_{(i)},$$

which was introduced by Jones (1946). However, Dixon (1957) has shown that still better results can be obtained by summing over suitably selected, rather than consecutive, quasi-ranges (including the range). For example, for $n = 16$, $W + W_{(2)} + W_{(4)}$ has efficiency 97.5%. More extensive results are given by Harter (1959).

In censored samples various simplified estimates of σ (as of μ) have been proposed by Dixon (1960). As has been pointed out toward the end of Section 6.3, censoring produces a marked loss in efficiency; the additional loss for suitably chosen Dixon statistics tends to be slight in comparison. When fewer than one fourth of the observations are censored at each end, that old standby, the interquartile range (or rather its small-sample version $W_{(i)}$ with $i = [\tfrac{1}{4}n] + 1$) provides a general, very simple, but usually rather inefficient estimator.

From the foregoing considerations it is clear that robustness against even a very few extreme outliers by the exclusion of extreme observations can be secured only at the expense of a major drop in efficiency. Barnett *et al.* (1967) study Downton's (1966b) unbiased estimator

$$``\sigma" = \frac{2\sqrt{\pi}}{n(n - 1)} \sum_{i=1}^{n} [i - \tfrac{1}{2}(n + 1)]X_{(i)}, \qquad (7.4.1)$$

[3] Here $[\tfrac{1}{2}n]$ denotes the integral part of $\tfrac{1}{2}n$.

[4] Divisors rendering $W_{(i)}$ and other estimators unbiased are readily available from tables of expected values of order statistics. Harter (1959) gives $\mathscr{E}(W_{(i)}/\sigma)$ for $n \leq 100$ and $i \leq 9$.

pointing out that it is highly efficient (> 97.79%) and "not so influenced by outliers as is either the range or the root-mean-square deviation." It might be added that "σ" places less weight on the extremes than does the best linear estimator σ^* and therefore, like J_i or Dixon's (1957) estimators, provides some modest protection against outliers with little loss in efficiency. Actually, "σ" is just another form (except for a multiplicative constant) of a statistic of respectable vintage, namely, Gini's (1912) mean difference

$$G = \frac{1}{n(n-1)} \sum_{i,j=1}^{n} |X_i - X_j|, \qquad (7.4.2)$$

already studied by Helmert in 1876 and not brand new then! See Ex. 7.4.1. Probably G is most conveniently computed as (von Andrae, 1872)

$$G = \sum_{i=1}^{[\frac{1}{2}n]} \frac{(n - 2i + 1)W_{(i)}}{n(n-1)}.$$

For $n \leq 10$ Nair (1950) compares the efficiencies of J_i, G, the mean deviation (MD), and the best linear estimator σ^*. G is almost as efficient as σ^*, with MD far behind. From $n = 6$ onwards, J_2 is best among the J statistics. For example, for $n = 10$ the percentage efficiencies are as follows:

J_1	J_2	J_3	J_4	J_5	MD	G	σ^*
85.0	96.4	95.9	92.2	89.4	91.0	98.1	99.0

Barnett et al. (1967) find a Pearson Type III approximation satisfactory for the distribution of "σ". It seems likely that the distributions of the various estimators considered in this section can be approximated adequately by the methods used for the range (see Cadwell, 1953b; F. N. David and Johnson, 1956).

The estimation of σ by the measurement (e.g., weighing) of *groups* of ranked observations has been considered by Mead (1966) in cases where individual measurement is much more difficult than ranking.

Simplified estimators of Weibull parameters, based on two order statistics, are obtained by Dubey (1967).

Standard Error of $X_{(r)}$

On the basis of F. N. David and Johnson's (1954) approach Walsh (1958) suggests the following estimator of $\sigma(X_{(r)})$:

$$a(X_{(r+i)} - X_{(r-i)}),$$

where

$$i \doteq (n+1)^{2/5}, \quad \text{and} \quad a = \tfrac{1}{2}(n+1)^{-3/10}\left[\left(\frac{r}{n+1}\right)\left(1 - \frac{r}{n+1}\right)\right]^{1/2}.$$

7.5. QUICK ESTIMATES IN BIVARIATE SAMPLES

Dispersion in the Circular Normal Distribution

The distribution of impact points (X, Y) arising from the firing of a rifle at a vertical target or from guns or missiles aimed at a ground target is often well described by the circular normal distribution

$$p(x, y) = \frac{1}{2\pi\sigma^2} \exp\left\{ -\frac{1}{2\sigma^2}\left[(x - \mu_x)^2 + (y - \mu_y)^2\right]\right\},$$

where $\mu_x = \mathscr{E}X$, $\mu_y = \mathscr{E}Y$, and σ^2 is the common variance. It is well known that the UMVU estimator of σ^2 from a sample of n points (X_i, Y_i) is just

$$S^2 = \frac{1}{2(n-1)}\left[\sum_{i=1}^{n}(X_i - \bar{X})^2 + \sum_{i=1}^{n}(Y_i - \bar{Y})^2\right],$$

and that $2(n-1)S^2/\sigma^2$ has a χ^2 distribution with $2(n-1)$ DF. The UMVU estimator of σ is therefore

$$S' = \frac{\Gamma(n-1)}{\sqrt{2}\,\Gamma(n - \frac{1}{2})}[\Sigma(X_i - \bar{X})^2 + \Sigma(Y_i - \bar{Y})^2]^{\frac{1}{2}}.$$

Quick estimators of σ include (i) the radius (RC) of the covering circle (i.e., the smallest circle containing all the points of impact), first studied by Daniels (1952); (ii) the extreme spread or bivariate range

$$R = \max_{i,j} \left[(X_i - X_j)^2 + (Y_i - Y_j)^2\right]^{\frac{1}{2}} \qquad i, j = 1, 2, \ldots, n;$$

and (iii) the diagonal $D = (W_x^2 + W_y^2)^{\frac{1}{2}}$, where $W_x = X_{(n)} - X_{(1)}$, $W_y = Y_{(n)} - Y_{(1)}$. These and other measures are conveniently summarized and discussed by Grubbs (1964). See also Moranda (1959) and Cacoullos and DeCicco (1967). RC and R are intriguing, simple to measure, but less efficient than D. Knowledge on R is confined largely to Monte Carlo sampling. The distribution of D can be handled with the help of Patnaik's approximation (Section 7.3), according to which D is seen to be distributed approximately as $c\sigma\chi_{2\nu}/\nu^{\frac{1}{2}}$, where c and ν are the scale factor and the equivalent degrees of freedom for either W_x or W_y. For $n \leq 20$ Grubbs tabulates $\mathscr{E}(D/\sigma)$, $1/\mathscr{E}(D/\sigma)$, s.e. (D/σ), and corresponding quantities for the other measures.

Regression Coefficient

If the regression of Y on the nonstochastic variable x is linear, namely,

$$\mathscr{E}(Y \mid x) = \alpha + \beta x, \tag{7.5.1}$$

then β may be estimated by the ratio statistic

$$b' = \frac{\overline{Y}'_{[k]} - \overline{Y}_{[k]}}{\overline{x}'_{(k)} - \overline{x}_{(k)}}, \tag{7.5.2}$$

where

$$\overline{x}'_{(k)} = \frac{1}{k} \sum_{i=1}^{k} x_{(n+1-i)}, \qquad \overline{x}_{(k)} = \frac{1}{k} \sum_{i=1}^{k} x_{(i)},$$

and

$$\overline{Y}'_{[k]} = \frac{1}{k} \sum_{i=1}^{k} Y_{[n+1-i]}, \qquad \overline{Y}_{[k]} = \frac{1}{k} \sum_{i=1}^{k} Y_{[i]},$$

$Y_{[i]}$ being the value of Y (not necessarily the ith largest) corresponding to $x_{(i)}$ (cf. Ex. 3.2.3).

If X is itself a chance variable, we may interpret (7.5.1) as conditional on $X = x$ and have from (7.5.2)

$$\mathscr{E}(b' \mid x_1, x_2, \ldots, x_n) = \beta. \tag{7.5.3}$$

Since (7.5.3) holds whatever the x_i, it also holds unconditionally; that is,

$$B' = \frac{\overline{Y}'_{[k]} - \overline{Y}_{[k]}}{\overline{X}'_{(k)} - \overline{X}_{(k)}} \tag{7.5.4}$$

is also an unbiased estimator of β. Note that this result does not require either the X's or the Y's to be identically distributed. Barton and Casley (1958) show that B' has an efficiency of 75–80% when X, Y are bivariate normal, provided k is chosen as about $0.27n$.

Correlation Coefficient

Since $\rho = \beta \sigma_x / \sigma_y$, (7.5.4) suggests

$$\hat{\rho}' = B' \cdot \frac{(\overline{X}'_{(k)} - \overline{X}_{(k)})/c_{n,x}}{(\overline{Y}'_{(k)} - \overline{Y}_{(k)})/c_{n,y}} = \frac{(\overline{Y}'_{[k]} - \overline{Y}_{[k]})/c_{n,x}}{(\overline{Y}'_{(k)} - \overline{Y}_{(k)})/c_{n,y}}$$

as an estimator of ρ, where $c_{n,x} = \mathscr{E}(\overline{X}'_{(k)} - \overline{X}_{(k)})/\sigma_x$, etc. If X and Y have the same marginal distributional form (e.g., both normal), $\hat{\rho}'$ simplifies to

$$\hat{\rho}' = \frac{\overline{Y}'_{[k]} - \overline{Y}_{[k]}}{\overline{Y}'_{(k)} - \overline{Y}_{(k)}}. \tag{7.5.5}$$

This estimator has been suggested by Tsukibayashi (1962) for $k = 1$ when the denominator is just W_y, the range of the Y's, and also for a mean range denominator. He points out that (7.5.5) can be calculated even if only the ranks of the X's are known. If both X and Y are measured, their asymmetrical treatment in (7.5.5) is displeasing, and $\hat{\rho}'$ may be replaced by

$$\frac{1}{2} \left(\frac{\overline{Y}'_{[k]} - \overline{Y}_{[k]}}{\overline{Y}'_{(k)} - \overline{Y}_{(k)}} + \frac{\overline{X}'_{[k]} - \overline{X}_{[k]}}{\overline{X}'_{(k)} - \overline{X}_{(k)}} \right).$$

However, not much seems to be known about the properties of this estimator or, for that matter, of $\hat{\rho}'$. For an estimate of cov (X, Y) see Ex. 7.5.1.

7.6. OPTIMAL SPACING OF ORDER STATISTICS IN LARGE SAMPLES

Although a more detailed account of the large-sample theory of order statistics is deferred to Chapter 9, we will consider here the following problem: Given a large sample from a population with pdf

$$\frac{1}{\sigma} p \left(\frac{x - \mu}{\sigma} \right)$$

we wish to estimate μ or σ or both from a fixed small number k of order statistics. How shall we choose or *space* the order statistics to obtain good estimates? It is clear that in small samples the problem can always be solved numerically provided means, variances, and covariances of $Y_{(i)} = (X_{(i)} - \mu)/\sigma$ $(i = 1, 2, \ldots, n)$ are available: we need only find the variances of the $\binom{n}{k}$ possible BLUE's of μ, say, corresponding to all selections of k order statistics, and then pick the one with the smallest variance; likewise for σ. If a single set of k order statistics is to provide good estimators of *both* μ and σ, we can similarly minimize var $\mu^* + c$ var σ^* after having decided on a suitable constant c, where μ^* and σ^* denote BLUE's of μ and σ based on the same set of k order statistics. Of course, some short cuts to this process are usually possible. However, the problem is most important for moderate and large sample sizes, where the saving in effort, particularly if the variables are already ordered, may more than compensate for the loss in efficiency. There are also interesting possibilities for data compression (Eisenberger and Posner, 1965), since a large sample (e.g., of particle counts taken on a space craft) may be replaced by enough order statistics to allow (on the ground) satisfactory estimation of parameters as well as a test of the assumed underlying distributional form.

Our starting point is the joint distribution as $n \to \infty$ of k order statistics $X_{(n_j)}$ $(j = 1, 2, \ldots, k)$, where $n_j = [n\lambda_j] + 1$ and $0 < \lambda_1 < \lambda_2 < \cdots < \lambda_k < 1$. Under mild restrictions this limiting distribution is k-variate normal (see Section 9.2) and may be written as

$$h = (2\pi\sigma^2)^{-\frac{1}{2}k} f_1 f_2 \cdots f_k [\lambda_1(\lambda_2 - \lambda_1) \cdots (\lambda_k - \lambda_{k-1})(1 - \lambda_k)]^{-\frac{1}{2}} n^{\frac{1}{2}k} e^{-\frac{1}{2}(n/\sigma^2)S},$$

$$(7.6.1)$$

where f_j is the pdf of Y evaluated at ξ_{λ_j}, the quantile of order λ_j of Y, and (with $\lambda_0 = 0$, $\lambda_{k+1} = 1$)

$$S = \sum_{j=1}^{k} \frac{\lambda_{j+1} - \lambda_{j-1}}{(\lambda_{j+1} - \lambda_j)(\lambda_j - \lambda_{j-1})} f_j^2 (x_{(n_j)} - \mu - \sigma\xi_{\lambda_j})^2$$

$$- 2\sum_{j=2}^{k} \frac{f_j f_{j-1}}{\lambda_j - \lambda_{j-1}} (x_{(n_j)} - \mu - \sigma\xi_{\lambda_j})(x_{(n_{j-1})} - \mu - \sigma\xi_{\lambda_{j-1}}). \quad (7.6.2)$$

Following Ogawa (1951 or 1962), who should be consulted for further details, we consider first the case of σ known. The BLUE μ_0^* of μ corresponding to the order statistics $X_{(n_j)}$ is given by $(\partial S/\partial\mu)_{\mu = \mu_0^*} = 0$, that is, by

$$\left[\sum_{j=1}^{k} \frac{\lambda_{j+1} - \lambda_{j-1}}{(\lambda_{j+1} - \lambda_j)(\lambda_j - \lambda_{j-1})} f_j^2 - 2\sum_{j=2}^{k} \frac{f_j f_{j-1}}{\lambda_j - \lambda_{j-1}} \right] \mu_0^*$$

$$= -\sum_{j=1}^{k} \left(\frac{f_{j+1} - f_j}{\lambda_{j+1} - \lambda_j} - \frac{f_j - f_{j-1}}{\lambda_j - \lambda_{j-1}} \right) f_j(x_{(n_j)} - \sigma\xi_{\lambda_j}) \quad (7.6.3)$$

(with $f_0 = f_{k+1} = 0$), or

$$\mu_0^* = \frac{Z_1 - \sigma K_3}{K_1}, \quad (7.6.3')$$

where

$$K_1 = \sum_{j=1}^{k+1} \frac{(f_j - f_{j-1})^2}{\lambda_j - \lambda_{j-1}}, \quad (7.6.4)$$

$$K_3 = \sum_{j=1}^{k+1} \frac{(f_j - f_{j-1})(f_j\xi_{\lambda_j} - f_{j-1}\xi_{\lambda_{j-1}})}{\lambda_j - \lambda_{j-1}}, \quad (7.6.5)$$

and

$$Z_1 = \sum_{j=1}^{k+1} \frac{(f_j - f_{j-1})(f_j X_{(n_j)} - f_{j-1}X_{(n_{j-1})})}{\lambda_j - \lambda_{j-1}}. \quad (7.6.6)$$

Since by (4.5.5) we have asymptotically

$$\operatorname{cov}(X_{(n_j)}, X_{(n_{j'})}) \sim \frac{\lambda_j(1 - \lambda_{j'})}{nf_jf_{j'}},$$

it follows that

$$\operatorname{var} \mu_0^* \sim \frac{\sigma^2}{n} \frac{1}{K_1}, \quad (7.6.7)$$

showing that μ_0^* has (asymptotic) efficiency K_1 relative to $\bar{X}$, the mean of the original sample.

Similar results hold for the estimation of σ for μ known, and also for the combined estimation of μ and σ. Thus in the former case one finds

$$\sigma_0^* = \frac{Z_2 - \mu K_3}{K_2}, \tag{7.6.8}$$

where

$$K_2 = \sum_{j=1}^{k+1} \frac{(f_j \xi_{\lambda_j} - f_{j-1} \xi_{\lambda_{j-1}})^2}{\lambda_j - \lambda_{j-1}}, \tag{7.6.9}$$

$$Z_2 = \sum_{j=1}^{k+1} \frac{(f_j \xi_{\lambda_j} - f_{j-1} \xi_{\lambda_{j-1}})(f_j X_{(n_j)} - f_{j-1} X_{(n_{j-1})})}{\lambda_j - \lambda_{j-1}}, \tag{7.6.10}$$

and

$$\operatorname{var} \sigma_0^* \sim \frac{\sigma^2}{n} \frac{1}{K_2}; \tag{7.6.11}$$

and in the latter case

$$\mu_1^* = \frac{1}{\Delta}(K_2 Z_1 - K_3 Z_2), \quad \sigma_1^* = \frac{1}{\Delta}(-K_3 Z_1 + K_1 Z_2), \tag{7.6.12}$$

where

$$\Delta = K_1 K_2 - K_3^2,$$

and

$$\left. \begin{array}{c} \operatorname{var} \mu_1^* \sim \dfrac{\sigma^2}{n} \dfrac{K_2}{\Delta}, \quad \operatorname{var} \sigma_1^* \sim \dfrac{\sigma^2}{n} \dfrac{K_1}{\Delta}, \\[2mm] \operatorname{cov}(\mu_1^*, \sigma_1^*) \sim -\dfrac{\sigma^2}{n} \dfrac{K_3}{\Delta}. \end{array} \right\} \tag{7.6.13}$$

If $p(y)$ is symmetric and the spacing is also symmetric, then for all j

$$\left. \begin{array}{c} \lambda_j + \lambda_{k+1-j} = 1, \quad n_j + n_{k+1-j} = n, \\[2mm] \xi_{\lambda_j} + \xi_{\lambda_{k+1-j}} = 0, \quad f_j = f_{k+1-j}, \end{array} \right\} \tag{7.6.14}$$

so that

from which it follows that $K_3 = 0$. Thus in this important case

$$\mu_0^* = \mu_1^* = \frac{Z_1}{K_1}, \quad \sigma_0^* = \sigma_1^* = \frac{Z_2}{K_2}.$$

Also μ_1^* and σ_1^* are uncorrelated and asymptotically independent.

Since by (7.6.1)

$$\log h = -k \log \sigma - \frac{\frac{1}{2}nS}{\sigma^2} + \text{terms free of } \mu \text{ and } \sigma,$$

we have

$$\frac{-\partial^2 \log h}{\partial \mu^2} = \frac{n}{\sigma^2} K_1, \qquad \frac{-\partial^2 \log h}{\partial \sigma^2} = \frac{2k}{\sigma^2} + \frac{n}{\sigma^2} K_2.$$

Thus the asymptotic $(n \to \infty, k/n \to 0)$ variances of the most efficient estimators based on the same $X_{(n_j)}$ as $\mu_0{}^*$, $\sigma_0{}^*$ are, respectively,

$$\frac{\sigma^2}{nK_1} \quad \text{and} \quad \frac{\sigma^2}{nK_2},$$

showing that $\mu_0{}^*$ and $\sigma_0{}^*$ are fully efficient in the sense of extracting all the information available from the chosen order statistics. Clearly these results continue to hold, for both μ and σ unknown, in the situation of (7.6.14).

So far the n_j have been taken as given. It is now evident that for the optimal estimation of $\mu(\sigma)$ for known $\sigma(\mu)$ the n_j must be chosen so as to maximize $K_1(K_2)$. Moreover, if for a symmetric parent this leads to (7.6.14), then the optimality will also apply for both μ and σ unknown. It should be stressed that the optimization refers to the separate estimation of μ and σ, requiring two different spacings. If a single spacing is to serve, we may again proceed by attempting to minimize var $\mu_0{}^* + c$ var $\sigma_0{}^*$ for a suitable value of c.

Although the maximization is straightforward in principle, the numerical problems involved may be substantial in practice. We confine ourselves to the normal case and begin with $k = 2$. Clearly, for any symmetrical parent the optimal spacing is symmetric, so that from (7.6.4)

$$K_1 = \frac{2f_1{}^2}{\lambda_1} = \frac{2p^2(x_{\lambda_1})}{P(x_{\lambda_1})}.$$

Thus we have to find x minimizing $P(x)/p^2(x)$, that is, satisfying (for any symmetric parent)

$$2P(x)p'(x) = p^2(x).$$

For $p(x) = \phi(x)$ this reduces to $xP(x) = -\tfrac{1}{2}p(x)$, which has a minimum at $x = -0.6121$. Thus

and
$$\left.\begin{array}{ll} x_{\lambda_1} = -0.6121, & x_{\lambda_2} = 0.6121, \\ \lambda_1 = 0.2702, & \lambda_2 = 0.7298, \end{array}\right\} \qquad (7.6.15)$$

this being the result cited in Section 7.2. Since $\bar{X}$ is efficient the (asymptotic) efficiency of $\mu_0{}^*$ (for all k) is K_1 by (7.6.7), giving in particular 81% for the efficiency of $\tfrac{1}{2}(X_{(n\lambda_1)} + X_{(n\lambda_2)})$. As usual, $n\lambda_1$ is to be interpreted as the integral part of $n\lambda_1 + 1$, etc., so that for $n = 100$ we use

$$\mu_0{}^* = \tfrac{1}{2}(X_{(28)} + X_{(73)}).$$

Similarly, since var $S \sim \sigma^2/2n$, the efficiency of σ_0^* is by (7.6.11) $\frac{1}{2}K_2$ for all k. For $k = 2$ this is maximized when $\lambda_1 = 0.0694$ and $\lambda_2 = 1 - \lambda_1$.[5] Also $\sigma_0^* (= Z_2/K_2)$ reduces to $0.337(X_{(n\lambda_2)} - X_{(n\lambda_1)})$ and has efficiency 65%.

For $k > 2$ it can be formally shown that there is a unique optimal spacing for the estimation of μ and that this spacing is symmetric. Similar results seem to hold for the estimation of σ, although no general proof is available. From a practical point of view the problem is solved, since explicit expressions giving the estimators μ_0^* and σ_0^* corresponding to the respective optimal spacings have been tabulated for most values of $k \leq 20$ (Ogawa, 1962; Eisenberger and Posner, 1965). The latter authors also give the estimators minimizing var $\mu_0^* + c$ var σ_0^* for $c = 1, 2, 3$.

Example 7.6. If $k = 4$ the estimators minimizing var μ_0^*, var σ_0^*, var $\mu_0^* + c$ var σ_0^* ($c = 1, 3$), together with their efficiencies, are, respectively, as follows:

Estimators	Efficiency
$.1918(X_{(.1068n)} + X_{(.8932n)}) + .3082(X_{(.3512n)} + X_{(.6488n)})$	.920
$.116(X_{(.9770n)} - X_{(.0230n)}) + .236(X_{(.8729n)} - X_{(.1271n)})$	.824
$\lbrace.1414(X_{(.0668n)} + X_{(.9332n)}) + .3586(X_{(.2912n)} + X_{(.7088n)})$	.908
$\lbrace.2581(X_{(.9332n)} - X_{(.0668n)}) + .2051(X_{(.7088n)} - X_{(.2912n)})$	.735
$\lbrace.0971(X_{(.0389n)} + X_{(.9611n)}) + .4029(X_{(.2160n)} + X_{(.7840n)})$	.857
$\lbrace.1787(X_{(.9611n)} - X_{(.0389n)}) + .2353(X_{(.7840n)} - X_{(.2160n)})$	.792

Here the bracketed estimators are based on a common spacing, the estimators in, for example, lines 3 and 4 being, respectively, the best linear four-point estimators of μ and σ when var $\mu_0^* + $ var σ_0^* is minimized. Similar results for $k = 2$ are given in Ex. 7.6.2.

Ogawa (1962) deals also with the one-parameter exponential, and Saleh and Ali (1966) consider the two-parameter exponential. See also Gupta and Gnanadesikan (1966) for the logistic, Bloch (1966) for the Cauchy, and Hassanein (1968) for the extreme-value distribution.

There has been a series of papers on small-sample results for the exponential; these are listed in the most recent addition by Saleh (1967).

A simplified approximate approach leading to "nearly optimal" spacings, somewhat in the manner of Blom's nearly best estimators, has been developed by Särndal (1962, Chapter 4) and is applied by him to several parent distributions. Särndal (1964) conveniently summarizes his methods and deals in detail with the (separate) estimation of μ and σ in the gamma distribution for known $p (= 1(1)5)$:

$$\frac{1}{\sigma\Gamma(p)} \left(\frac{x - \mu}{\sigma}\right)^{p-1} e^{-(x-\mu)/\sigma} \qquad \mu \leq x < \infty.$$

[5] This result as well as (7.6.15) was obtained by Karl Pearson in 1920.

Särndal (1962, Chapter 7) points out, as have others, that the problem of optimum spacing is closely connected with the following two problems:

(a) optimum grouping of observations;

(b) optimum strata limits in proportionate sampling.

Tests of significance using optimally spaced order statistics have also been studied by Ogawa (1962) and Eisenberger (1968).

7.7. QUICK TESTS

It will be clear to the reader that many of the estimation procedures discussed so far are readily converted into quick tests of significance, or provide a basis for the construction of such tests. From the user's point of view the availability of suitable tables is crucial: without them the tests will cease to be quick! We will therefore emphasize this practical aspect. However, there may also be incidental side benefits, such as superior robustness, for certain quick tests which must therefore not be dismissed out of hand as second best even by those looking for theoretically optimal methods. In the normal case we consider in turn tests on variances, substitute t tests, and the use of range in the analysis of variance. See also F. N. David and Johnson (1956).

Tests on Variances

(i) Single-sample test, $H_0 : \sigma = \sigma_0$. Harter's (1964b) tables giving confidence intervals for σ based on suitably chosen quasi-ranges provide a convenient way of testing, from a normal sample of n, the null hypothesis $H_0 : \sigma = \sigma_0$ against either one- or two-sided alternatives.[6] For example, for $n = 40$ the tables give two-sided 95% confidence intervals for σ as $(0.267153w_{(3)}, 0.419858w_{(3)})$. If the interval covers σ_0, we accept H_0 at the 5% level against $\sigma \neq \sigma_0$, etc.

(ii) Two-sample tests, $H_0 : \sigma_1 = \sigma_2$. Extending work of Link in 1950, Harter (1963) has prepared, by numerical integration, tables of the upper percentage points of $_1W/_2W$, the range ratio in normal samples of n_1 and n_2, for n_1, $n_2 \leq 15$.

(iii) k-sample tests, $H_0 : \sigma_1 = \sigma_2 = \cdots = \sigma_k$. Let $_tS^2$ $(t = 1, 2, \ldots, k)$ be the usual unbiased mean square estimator of σ_t^2 based on ν DF. Then H_0 may be tested very simply (Hartley, 1950b) by referring $s^2_{\max}/s^2_{\min}$ to Table 31 of Pearson and Hartley (1966). Note that this test ratio and other "extremal

---

[6] More fully we are interested in three sets of hypotheses: $\sigma \leq \sigma_0$ versus $\sigma > \sigma_0$, $\sigma \geq \sigma_0$ versus $\sigma < \sigma_0$, and $\sigma = \sigma_0$ versus $\sigma \neq \sigma_0$. The statement in the main text should be interpreted as an equivalent short form. Similar remarks apply to other tests.

quotients" (Gumbel and Herbach, 1951) are closely related to the range, since

$$\log (s^2_{\max}/s^2_{\min}) = \text{range} (\log {}_t s^2).[7]$$

In small samples ($\nu \leq 10$, say) we may wish to simplify further and to use the ratio $W_{\max}/W_{\min}$ (Cadwell, 1953b), whose upper percentage points have been computed in detail by Leslie and Brown (1966). Two other test statistics which might be used in testing homogeneity of variance from samples of equal size have been mentioned in Section 5.4, namely

$$S^2_{\max} \Big/ \sum_{t=1}^{k} {}_t S^2 \quad \text{and} \quad W_{\max} \Big/ \sum_{t=1}^{k} {}_t W.$$

The power, or some other measure of performance, of the various tests cited has been considered in the references. There is an inevitable loss in power due to the replacement of mean squares by ranges, but this is small for $\nu \leq 10$. On the other hand, since Bartlett's well-known M test has no strong optimality properties, Hartley's test (and presumably even Cadwell's) *need* not be inferior but may be considerably so, depending on the particular configuration of the σ_t^2. Intuitively, the two short-cut tests may be expected to be near their best for the alternatives $\sigma_1^2 < \sigma_2^2 = \cdots = \sigma_{k-1}^2 < \sigma_k^2$ with $\sigma_1 \sigma_k = \sigma_2^2$. This is borne out by some experimental sampling results of Pearson (1966), who, however, rather surprisingly finds M never inferior. Likewise $S^2_{\max}/\sum_t S^2$ and $W_{\max}/\sum_t W$ may be expected to do best against the slippage alternative $\sigma_1^2 = \cdots = \sigma_{k-1}^2 < \sigma_k^2$ (cf. Section 8.3, comment 7). Unlike all the short-cut tests, the M test is, of course, applicable also for samples of unequal sizes.

As has been forcefully pointed out by Box (1953), all the foregoing tests are very sensitive to the assumption of normality. This does not deprive them of all value, but there is little to recommend them as mere preliminary tests preceding a test for homogeneity of means. See Miller (1968) for more robust procedures.

Sequential range tests of $H_0 : \sigma = \sigma_0$ versus $H_1 : \sigma = \delta\sigma$ (δ specified) and of $H_0 : \sigma_1 = \sigma_2$ versus $H_1 : \sigma_1 = \delta\sigma_2$ were first considered by Cox (1949). The two-sample situation is treated further, with different approaches, by Rushton (1952) and Ghosh (1963). Although the range is convenient because of its simplicity, the estimation of σ and of σ_1, σ_2 must in each case be performed in stages (subgroups of, say, 4 or 8 observations) rather than after each individual observation. This is readily done from the sum of ranges over the respective successive subgroups.

---

[7] For the use of $s^2_{\max}/s^2_{\min}$ when the observations within samples are equicorrelated and for testing the homogeneity of column variances in a two-way classification see Han (1968, 1969).

Substitute t-tests

The idea of using the range in place of the sample standard deviation in a single-sample t test ($H_0: \mu = \mu_0$) was put forward first by Daly (1946) and in more detail by Lord (1947), who also considered the two-sample situation ($H_0: \mu_1 = \mu_2$). Lord gives upper percentage points of

$$R_1 = \frac{|\bar{X} - \mu_0|}{W} \quad \text{and} \quad R_2 = \frac{|\bar{X}_1 - \bar{X}_2|}{\frac{1}{2}(_1W + _2W)},$$

for $n \le 20$, the ranges $_1W$ and $_2W$ in the two-sample case both being over n observations. Extensions to cover unequal sample sizes $n_1, n_2 \le 20$ are given by Moore (1957). For larger sample sizes one may, at the price of some arbitrariness, use mean ranges in place of single ranges. This was also considered by Lord, who used heroic quadrature methods. More convenient (although very slightly less powerful) tables are due to Jackson and Ross (1955), who give upper percentage points of

$$G_1 = \frac{|\bar{X} - \mu_0|}{W_{n',k}} \quad \text{and} \quad G_2 = \frac{|\bar{X}_1 - \bar{X}_2|}{W_{n',k_1+k_2}},$$

where n' is the subgroup size (preferably 6–10) and k, k_1, k_2 the number of subgroups. A few observations may have to be thrown out so that n' can be appropriately chosen.

The loss in power by the use of the above methods instead of the optimal t test is quite small (see, e.g., Lord, 1950). Note that by means of Patnaik's approximation R_1, R_2, G_1, G_2 all become approximate t statistics with the somewhat reduced degress of freedom given by Table 7.3.2 entered with $n = n'$, and $k = k$ (for G_1) or $k = k_1 + k_2$ (for G_2).

Example 7.7. To test whether the means of the following two samples differ significantly.

First sample: 35.5, 23.4, 45.0, 20.4, 74.4, 46.7, 27.6, 47.6, 35.4, 38.9
($n_1 = 10$, $\bar{x}_1 = 39.49$; mean of first 9 observations: 39.56).
Second sample: 46.5, 63.9, 48.6, 43.6, 33.3, 38.7, 49.6, 56.1, 43.7, 51.3, 69.1, 51.8, 78.1, 57.2, 72.5, 74.2, 53.4, 66.9 ($n_2 = 18$, $\bar{x}_2 = 55.47$).

To achieve a common subgroup size we drop the last observation of the first sample and find the three ranges of 9 to be 54.0, 30.6, 26.4, giving $\bar{w} = 37.1$ and $G_2 = 15.91/37.1 = 0.43$, which exceeds the 1% point, 0.38. For Patnaik's approximation we may retain all observations in the *numerator* and find from Table 7.3.2 $c = 3.01$, $v = 20.5$, giving

$$\frac{15.98 \times 3.01}{37.1 \times (\frac{1}{10} + \frac{1}{18})^{1/2}} = 3.29,$$

which again is significant at the 1% level (1% point = 2.84).

Both single- and two-sample tests are, of course, readily converted into confidence statements (cf. Noether, 1955). For example, if $R_2(\alpha)$ is the upper α significance point of R_2, then the interval

$$\bar{X}_1 - \bar{X}_2 + \tfrac{1}{2}R_2(\alpha)(_1W + _2W)$$

covers $\mu_1 - \mu_2$ with probability $1 - \alpha$.

Corresponding sequential tests based on the mean range are considered by Gilchrist (1961). For a range version of Stein's test see Knight (1963).

Another possible substitute for the single-sample t test uses the range-midrange statistic

$$\frac{\tfrac{1}{2}(X_{(1)} + X_{(n)}) - \mu_0}{X_{(n)} - X_{(1)}}.$$

This test, first proposed by E. S. Pearson (1929), is found to be reasonably efficient and rather robust in very small samples by Walsh (1949c), who gives upper percentage points for $n \leq 10$.

Short-Cut Analysis of Variance

For the one-way classification of nk observations x_{ij} ($i = 1, 2, \ldots, k$; $j = 1, 2, \ldots, n$) into k groups of n, the usual F-ratio test statistic may be replaced by

$$Q_{k,\nu} = \frac{n^{1/2} \operatorname*{range}_i \bar{X}_{i\cdot}}{S_\nu}, \qquad (7.7.1)$$

where S_ν is the usual rms estimator of σ based on $\nu = k(n-1)$ DF. In view of the mutual independence of S_ν and all the $\bar{X}_{i\cdot}$, it follows that $Q_{k,\nu}$ is the studentized range of Section 5.2. With the help of tables of percentage points of Q the above substitute test is easily carried out but represents trivial computational saving over the F test. The use of (7.7.1), however, is of considerable value as the first step in multiple decision procedures, due to Tukey and Duncan, having the aim of helping us arrive at a ranking or partial ranking of the groups (or "treatments"). Such procedures do not really come under the heading of short-cut tests, and the reader is referred to Miller (1966, Chapter 2). Note, however, that (7.7.1) applies equally to other orthogonal classifications (e.g., randomized blocks and Latin squares) if ν is adjusted to the appropriate error degrees of freedom.

For a one-way classification a computationally simpler test statistic is obtained if S_ν in (7.7.1) is replaced by $\bar{W}/c$, where $\bar{W}$ is the mean of the k within-group ranges and c is the constant of Table 7.3.2. By virtue of Patnaik's

(1950) approximation, the resulting ratio,

$$\frac{cn^{1/2}\ \text{range}\ \bar{X}_{i.}}{\bar{W}},$$

is distributed approximately as $Q_{k,v}$, but with v now denoting the equivalent degrees of freedom also given in Table 7.3.2. More convenient still is the use of the equivalent statistic

$$Q' = \frac{\text{range}\ \left(\sum_j X_{ij}\right)}{\sum_i W}, \tag{7.7.2}$$

the range of group totals divided by the sum of the within-group ranges, whose upper 5% and 1% points are tabulated in Beyer (1968, p. 386).[8] For extensions to two-way and certain other classifications see Hartley (1950a), Staude (1959), Mardia (1967),[9] and David (1951). Corresponding multiple comparison procedures (in the manner of Tukey) for balanced one- and two-way classifications have been studied in some detail by Kurtz et al. (1965a, b). Sequential range tests for components of variance are proposed by Ghosh (1965).

When we turn to the power of the above tests, we must distinguish between different probability models. For the components-of-variance (or random) model

$$X_{ij} = \mu + A_i + Z_{ij} \qquad i = 1, 2, \ldots, k; j = 1, 2, \ldots, n, \tag{7.7.3}$$

where μ is a constant and the A_i, Z_{ij} are all mutually independent normal deviates with respective variances σ'^2, σ^2, it is known that the standard F test is the UMP test of $H_0: \sigma'^2 = 0$ against $H_1: \sigma'^2 > 0$. Range tests, however, are little inferior (David, 1953; cf. Ex. 7.7.1). In the fixed-effects (or systematic) model

$$X_{ij} = \mu + \alpha_i + Z_{ij},$$

where the α_i are constants subject to $\Sigma\ \alpha_i = 0$, the situation is not so simple. The power of the range test, unlike that of the F test, is not expressible as a function of a single parameter ($\Sigma\ \alpha_i^2$) but depends on $k - 1$ parameters (Ex. 7.7.2). Let $\alpha_{(1)} \leq \alpha_{(2)} \leq \cdots \leq \alpha_{(k)}$ be the ordered α_i. Then, corresponding to a fixed value of $\Sigma\ \alpha_i^2$ (and hence of the power of the F test), the maximum power of the studentized range test (7.7.1) occurs for

$$\alpha_{(1)} = -\alpha_{(k)}, \quad \alpha_{(2)} = \cdots = \alpha_{(k-1)} = 0,$$

[8] This variant of Patnaik's test was described by the author in Sarhan and Greenberg (1962), but a wrong table slipped in (p. 121).

[9] This paper should be read in conjunction with that of Smith and Hartley (1968).

and the minimum power (with k even) for

$$-\alpha_{(1)} = \cdots = -\alpha_{(\frac{1}{2}k)} = \alpha_{(\frac{1}{2}k+1)} = \cdots = \alpha_{(k)}.$$

In the former case the range test is superior, in the latter distinctly inferior (David, 1953; Lachenbruch and David, 1968). The use of (7.7.2) will, of course, lead to slightly lower power than (7.7.1), but no numerical results are available.

Some Quick Tests for Discrete Variates

The range of k independent binomial $b(p, n)$ variates r_i $(i = 1, 2, \ldots, k)$ has been proposed by Siotani (1957) to test the hypothesis that in k binomial $b(p_i, n)$ trials $p_1 = p_2 = \cdots = p_k = p$. Tables are given by Siotani and Ozawa (1958), who suggest that for unknown p their tables (which are confined to $n \geq 10$) be entered with $p = \hat{p} = \Sigma r_i / kn$. The exact distribution of the range for unknown p, based on the hypergeometric distribution corresponding to fixed marginal totals (namely, n and Σr_i), is tabled by Ishii and Yamasaki (1961) for $n \leq 10$.

Again, for a multinomial distribution with observed frequency y_i and probability parameter p_i in the ith class $(i = 1, 2, \ldots, k; \Sigma y_i = N,$ $\Sigma p_i = 1)$ a test of homogeneity $(p_i = 1/k,$ all $i)$ may be based on max y_i / N, on min $y_i / \max y_i$, or on range y_i / N. Johnson and Young (1960) have considered various approximate methods for obtaining upper percentage points of these statistics (see also Ex. 5.3.6). One of their approximations consists in noting that the standardized variates

$$Z_i = \frac{Y_i - N/k}{\left[N \cdot \dfrac{1}{k} \cdot \dfrac{k-1}{k} \right]^{\frac{1}{2}}} \qquad i = 1, 2, \ldots, k$$

have (asymptotically as $N \to \infty$) the same degenerate k-variate normal distribution with correlation coefficient $-1/(k-1)$ as do the variates, $[k/(k-1)]^{\frac{1}{2}}(Z_i - \bar{Z})$, where the Z_i are independent unit normal variates. Thus approximately

$$\text{range } \frac{Y_i}{N} \doteq \frac{1}{(Nk)^{\frac{1}{2}}} \text{ range } Z_i = \frac{1}{(Nk)^{\frac{1}{2}}} W.$$

Some exact tables of percentage points have been constructed by Bennett and Nakamura (1968). Since the conditional distribution of k iid Poisson variates X_i, given $\Sigma X_i = N$, is just the joint distribution of the Y_i above, it follows that the distribution of range X_i, given $\bar{X} = \bar{x}(= N/k)$, is approximately the same as that of $W(\bar{x})^{\frac{1}{2}}$. This result has been used by Pettigrew and Mohler (1967) as a quick alternative to the Poisson index of dispersion test.

7.8. PROBABILITY PLOTTING

A graphical method of estimating the parameters of continuous distributions which are completely specified except for location and scale is as follows: Plot the ordered observations $x_{(i)}$ $(i = 1, 2, \ldots, n)$ against $v_i = P^{*-1}(p_i)$, where P^* is the cdf of the standardized variate $Y = (X - \mu)/\sigma$ and the p_i are intuitively plausible and simple probability levels, such as

$$p_i = \frac{i}{n+1} \quad \text{or} \quad p_i = \frac{i - \frac{1}{2}}{n}.$$

The plotting is facilitated if probability paper corresponding to P^* is available, but such paper is by no means essential. Now fit a line by eye through the points $(v_i, x_{(i)})$. (If such a straight-line fit seems unreasonable, the original distribution assumption needs to be re-examined, an important use of probability paper to which we will turn shortly.) The line may be regarded as an approximation to the (unweighted) regression line of $x_{(i)}$ on v_i, namely,

$$x = \mathring{\mu} + \mathring{\sigma} v, \tag{7.8.1}$$

where the estimates of μ and σ are

$$\mathring{\mu} = \bar{x} - \mathring{\sigma}\bar{v} \quad \text{and} \quad \mathring{\sigma} = \frac{\sum x_{(i)}(v_i - \bar{v})}{\sum (v_i - \bar{v})^2}. \tag{7.8.2}$$

Graphically, $\mathring{\sigma}$ is just the slope of the regression line; $\mathring{\mu}$ may then be found from (7.8.2). If Y has a symmetrical distribution, then $\bar{v} = 0$ and $\mathring{\mu} = \bar{x}$, which is the ordinate on (7.8.1) corresponding to $v = 0$ or $p = \frac{1}{2}$; in this case, $\mathring{\sigma}$ is conveniently obtained as the difference in ordinates on (7.8.1) for $v = 0$ and $v = 1$, where in the normal case the latter corresponds to $p = 0.8413$.

Chernoff and Lieberman have studied the method theoretically both for normal (1954) and generalized (1956) probability paper.[10] In the normal case they show that the choice of the frequently recommended plotting positions $p_i = i/(n + 1)$ leads to much poorer estimates of σ, as measured by mean square error, than those obtained with $p_i = (i - \frac{1}{2})/n$. Of course, both methods—and all other reasonable methods—give $\mathring{\mu} = \bar{x}$. The authors then raise the interesting question: What choice of the p_i leads to the best (= smallest mean square error) estimates of σ (a) among unbiased estimators and (b) allowing bias? For $n \leq 10$ they tabulate the p_i corresponding to (a) and (b) and compare mean square deviations of six estimators (Table 7.8). It will be noted in particular that the simple choice $p_i = (i - \frac{1}{2})/n$, which results in some bias, performs quite well. See also Blom (1958, p. 143), who

[10] Note that Chernoff and Lieberman take the $x_{(i)}$ to be the abscissae and the v_i the ordinates, the reverse of the above.

Table 7.8.   Comparison of the mean square deviations from σ of various estimates of σ

n	(1)	(2)	(3)	(4)	(5)	(6)
2	.57080	.57084	.36338	.36340	1.07533	.42611
3	.27324	.27549	.21460	.21599	.49856	.22649
4	.17810	.18006	.15117	.15259	.31559	.15558
5	.13177	.13332	.11643	.11764	.22751	.11872
6	.10447	.10571	.09459	.09560	.17630	.09605
7	.08650	.08714	.07961	.08015	.14306	.08067
8	.07379	.07469	.06872	.06950	.11987	.06954
9	.06432	.06501	.06044	.06105	.10283	.06111
10	.05701	.05759	.05393	.05445	.08981	.05449

[Reproduced from Chernoff and Lieberman (1954), with permission of the authors and the Editor of the *Journal of the American Statistical Association*.]

(1) Variance of the minimum variance nonlinear unbiased estimate S' of (7.3.1).

(2) Variance of the minimum variance unbiased estimate that is linear in the ordered observations.

(3) Mean square deviation from σ of the nonlinear biased estimate that has minimum mean square deviation.

(4) Mean square deviation from σ of the biased estimate that is linear in the ordered observations and has minimum mean square deviation.

(5) Mean square deviation from σ of the biased estimate based upon the ordinates $i/(n + 1)$.

(6) Mean square deviation from σ of the biased estimate based upon the ordinates $(i - \frac{1}{2})/n$.

points out that this choice goes back to Bliss and Stevens (1937) and also that $p_i = (i - \frac{3}{8})/(n + \frac{1}{4})$ leads to a practically unbiased estimator of σ with a mean square error about the same as that of the best linear unbiased estimator. A further discussion of plotting positions, with special reference to the extreme-value distribution, is given by Kimball (1960).

The use of probability paper as a quick means of checking on an assumed distributional form provides a most valuable and versatile aid to the applied statistician. In addition to the normal and extreme-value distributions the gamma distribution (with three parameters) has been studied from this point of view (Wilk *et al.*, 1962a).

For supposedly normal data the informal visual approach of probability plotting may—but often need not be—followed up by a test of normality (see, e.g., Ex. 7.8.1) or a test for outliers (Section 8.2). To test for departures from the exponential distribution $p(x) = \sigma^{-1}e^{-x/\sigma}$ ($x \geq 0$) Jackson (1967) has suggested the statistic $\Sigma \mu_{i:n} x_{(i)} / \Sigma x_{(i)}$, that is, the cross product of the ordered observations and the expectations $\mu_{i:n} = \mathscr{E}(X_{(i)}/\sigma)$, suitably standardized.

A neat extension of the graphical approach is due to Daniel (1959), who suggests the probability plotting of the $2^n - 1$ ordered absolute contrasts in a 2^n factorial experiment. With the standard assumptions for such experiments, the contrasts are, on the null hypothesis of no treatment effects, independent half-normal variates with common variance. Marked departures of the largest contrasts from a straight line through the origin on half-normal probability paper indicate the existence of the corresponding main effects or interactions. A partial formalization of this approach as a multiple decision procedure has been attempted by Birnbaum (1959, 1961). Estimation of the error variance from the m (say) smallest absolute contrasts is considered by Wilk et al. (1963a). Cox and Lauh (1967) adapt Daniel's methods to the graphical analysis of multidimensional contingency tables. See also Wilk and Gnanadesikan (1968) for a more general discussion of probability plotting and related methods, and Cox (1968, p. 276) and Healy (1968) for methods of detecting outliers in multivariate data.

7.9. QUALITY CONTROL

In statistical quality control, small samples, commonly of size $n = 5$, are taken at intervals from a production process. For each sample the mean and often also the range are plotted on separate control charts, giving a running picture of such means and ranges. In the case of the mean $\bar{x}$ the control chart consists of three horizontal lines: a central line set at $\bar{\bar{x}}$, the grand mean of a large previous number N of samples of n, and the upper and lower control limits at

$$\bar{\bar{x}} \pm 3\hat{\sigma}_w/n^{1/2} = \bar{\bar{x}} \pm A_2 \bar{w}.$$

Here $\hat{\sigma}_{\bar{w}}$ is the mean range estimate of σ discussed in Section 7.3, namely, $\bar{w}/d_n$, $\bar{w}$ being the grand mean range corresponding to $\bar{\bar{x}}$. Note that d_n is often called d_2 in the quality control literature. Correspondingly, $A_2 = 3/(d_n n^{1/2})$ is a widely tabulated convenient constant (see, e.g., Duncan, 1965). From a statistical point of view the idea of the control chart is simply this: for sufficiently large N, when $\bar{\bar{x}}$ and $\hat{\sigma}_{\bar{w}}$ may be taken equal to their respective expected values μ and σ, the probability that a particular mean will fall outside the control limits is 0.0027 on the usual normal theory. An occurrence with such low probability of happening by pure chance may be reasonably interpreted to mean that the process has got out of control, thus indicating the need for remedial action. Similar remarks apply to the range chart, for which upper and lower control lines are commonly set at

$$\bar{w} \pm 3 \cdot (\text{range estimate of s.d. of range } W_n) = \begin{cases} \bar{w} + 3(V_n)^{1/2}\bar{w}/d_n = D_4\bar{w}, \\ \bar{w} - 3(V_n)^{1/2}\bar{w}/d_n = D_3\bar{w}, \end{cases}$$

where again $D_4 = 1 + 3(V_n)^{1/2}/d_n$ and D_3 are widely tabulated. Since the range is not normally distributed, the 3σ limits are even more arbitrary than in the case of the mean chart, but there is still a very small probability on normal theory that a particular range value will fall outside the limits when all is well with the process. Clearly the upper line is much more important than the lower one, which may be dispensed with.

A question which suggests itself is the behavior of the control charts for non-normal data. Here the relative stability of the ratio $\mathscr{E}(W_n/\sigma)$ under many kinds of non-normality (see Section 7.3) is reassuring for the mean chart whose width is determined by d_n. Since the coefficient of variation of range is less stable, the range chart is more sensitive to non-normality. Of course, adjustments can be made to the control lines with the help of estimates of the β_2 values of the underlying population, as suggested by Cox. However, even without such refinements control charts have proved of great practical benefit by giving an easily understood visual account of the production output.

In the setting up of mean and range charts it has traditionally been assumed that accurate estimates of μ and σ are available. However, in the early stages of quality control for a new process this is not the case. Hillier (1964, 1967) has examined the errors incurred when the sampling fluctuations of $\bar{X}$ and $\hat{\sigma}_{\bar{w}}$ are ignored, and has shown how suitably modified control charts can be started even with quite limited data. Thus, in the case of the mean chart, the exact probability (on normal theory) that a particular sample mean $\bar{X}$ falls outside the control lines constructed as above is

$$P = 1 - \Pr\{\bar{\bar{X}} - A_2\bar{W} < \bar{X} < \bar{\bar{X}} + A_2\bar{W}\},$$

where now $\bar{\bar{X}}$, $\bar{W}$ are based on k previous samples of n. We have

$$P = 1 - \Pr\left\{-A_2 < \frac{\bar{X} - \bar{\bar{X}}}{\bar{W}} < A_2\right\}.$$

From Patnaik's approximation (Section 7.3) it follows that $(\bar{X} - \bar{\bar{X}})/\bar{W}$ is distributed approximately as

$$\frac{Z(1/n + 1/nk)\sigma}{c\chi_\nu\sigma/\nu^{1/2}} = \frac{1}{cnk}(k+1)t_\nu,$$

where Z is $N(0, 1)$, t_ν is a t variate with ν DF, ν and c being given by Table 7.3.2. Hillier (1964) shows that for $n = 5$ the value of P is increased from its supposed value 0.0027 (corresponding to $k = \infty$) to, for example, 0.0044, 0.0067, 0.012 for $k = 20, 10, 5$, respectively. Reciprocally, Hillier gives, for $n = 5$ and various k, values A_2^* which make

$$P^* = 1 - \Pr\{\bar{\bar{X}} - A_2^*\bar{W} < \bar{X} < \bar{\bar{X}} + A_2^*\bar{W}\}$$

approximately equal to $\alpha = 0.001, 0.0027, 0.01, 0.025, 0.05$. Similar results can be obtained for range charts (Hillier, 1967).

Control charts for largest and smallest values have been considered by Howell (1949) and for means, ranges, and sequential runs by Weiler (1954). The operating characteristic of the control chart for sample means has been studied by King (1952) for the case of random shifts in the true process mean. For systematic shifts the noncentral range results of Section 7.7 become relevant.

For a production process in which it takes some time to generate a single item for measurement, moving averages and the corresponding moving ranges (in samples of n) provide simple current measures of location and dispersion. The mean of such moving ranges may be regarded as a generalization of the mean successive difference (the case $n = 2$). The efficiency of the mean moving range as an estimator of σ is examined by David (1955), and its bias under trend or other systematic variations in process mean by Shimada (1957).

Tolerance Intervals for Normal Distributions Utilizing the Range

Related to statistical quality control is the following question: From k samples of n $N(\mu, \sigma^2)$ variates representing some production process can we find a (random) interval (L, V) such that a given high proportion γ, say 99%, of the output lies in the interval with specified probability β? A nonparametric solution, not taking advantage of normality, has been given in Section 2.6. Mitra (1957) has shown that approximate tolerance intervals are given by $(\bar{\bar{X}} - c\bar{W}, \bar{\bar{X}} + c\bar{W})$ and, using Patnaik's approximation to $\bar{W}$, has tabulated c for various k, n, γ, and β. Mathematically c satisfies approximately the equation

$$\Pr\left\{\frac{1}{(2\pi)^{1/2}\sigma}\int_{\bar{\bar{X}}-c\bar{W}}^{\bar{\bar{X}}+c\bar{W}} e^{-\frac{1}{2}(t-\mu)2/\sigma^2}dt > \gamma\right\} = \beta.$$

Of course, at the expense of convenience, the sample standard deviation could be used in place of the range with some gain in efficiency (expressed in shorter *average* length of tolerance interval).

EXERCISES

7.1.1. If W and S are the range and the standard deviation, respectively, of the same normal sample of n, prove that

(a) $\rho(W, S) = \dfrac{\sigma(S)/\mathscr{E}S}{\sigma(W)/\mathscr{E}W} = (\text{eff } \hat{\sigma}_w)^{1/2}$;

(b) the regression of W on S is linear;

(c) the variance of W, given $S = s$, is proportional to s^2.

For what statistics, in addition to the range, do similar results hold?
[Hint: For (a) write $\mathscr{E}(WS) = \mathscr{E}[(W/S)S^2]$ and use the independence of W/S and S^2 (Section 5.2).]

(Hartley, 1955; David and Perez, 1960)

7.2.1. By expanding the standard normal cdf $\Phi(x)$ about zero, show that

$$4\Phi(x)[1 - \Phi(x)] = e^{-2x^2/\pi}\left[1 + \frac{2(\pi - 3)}{3\pi^2}x^4 - \cdots\right].$$

Hence show, in normal $N(0, 1)$ samples of size $n = 2s + 1$ $(s = 0, 1, 2, \ldots)$ that the pdf of the median M is approximately proportional to

$$e^{-2sx^2/\pi}\left[1 + \frac{2(\pi - 3)s}{3\pi^2}x^4\right]e^{-\frac{1}{2}x^2},$$

and also that the variance and the β_2 value of M are given by the approximate formulae

$$\text{var } M \doteq \frac{\pi}{\pi + 4s}\left[1 + \frac{8(\pi - 3)s}{(\pi + 4s)^2}\right],$$

$$\beta_2(M) \doteq 3 + \frac{16(\pi - 3)s}{(\pi + 4s)^2}.$$

(Cadwell, 1952)

7.3.1. Show that for the extreme-value distribution

$$p(x) = \exp(-x - e^{-x}) \qquad -\infty \le x \le \infty,$$

the cdf and the expected value of the range W in samples of n are, respectively,

$$F(w) = n\sum_{i=1}^{n}(-1)^{i-1}\binom{n-1}{i-1}[i + (n-i)e^{-w}]^{-1},$$

$$\mathscr{E}W = \sum_{i=1}^{n}(-1)^i\binom{n}{i}\log\left(\frac{i}{n}\right).$$

(David, 1954)

7.3.2. Let (X_i, Y_i) $(i = 1, 2, \ldots, n)$ be a random sample from a bivariate normal distribution with unit variances and correlation coefficient ρ. Show that the correlation $\rho_w(n, \rho)$ between range X_i and range Y_i is, for $n = 2, 3$,

$$\rho_w(2, \rho) = \frac{\psi(\rho)}{\psi(1)}, \quad \rho_w(3, \rho) = \frac{\psi(\rho) + 2\psi(\frac{1}{2}\rho)}{\psi(1) + 2\psi(\frac{1}{2})}$$

where $\psi(\rho) = (2/\pi)[\rho \sin^{-1}\rho - 1 + (1 - \rho^2)^{\frac{1}{2}}]$.
[Hint: For $n = 3$ express range X_i as

$$\tfrac{1}{2}(|X_1 - X_2| + |X_2 - X_3| + |X_3 - X_1|), \text{ etc.}]$$

(Kurtz et al., 1966)

7.3.3. Let (X_i, Y_i) $(i = 1, 2, \ldots, n)$ be a random sample from a continuous bivariate distribution with joint cdf $H(x, y)$ and marginal cdf's $F(x)$, $G(y)$ $(a \le x \le b, c \le y \le d)$.

(a) By integration by parts show that

$$\text{cov}\,(X_i,\,Y_i) = \int_a^b \int_c^d (H - FG)\,dx\,dy$$

and hence that

$$\text{cov}\,(X_{(n)},\,Y_{(n)}) = \int_a^b \int_c^d (H^n - F^n G^n)\,dx\,dy.$$

(b) If $V = X_{(n)} - X_{(1)}$, $W = Y_{(n)} - Y_{(1)}$, use
$\text{cov}\,(V,\,W) = \text{cov}\,(X_{(n)},\,Y_{(n)}) + \text{cov}\,(X_{(1)},\,Y_{(1)}) - \text{cov}\,(X_{(n)},\,Y_{(1)}) - \text{cov}\,(X_{(1)},\,Y_{(n)})$
to show that

$$\text{cov}\,(V,\,W) = \int_a^b \int_c^d [H^n + (F - H)^n + (G - H)^n + (1 - F - G + H)^n$$

$$- F^n G^n - F^n(1 - G)^n - G^n(1 - F)^n - (1 - F)^n(1 - G)^n]\,dx\,dy.$$
(Mardia, 1967)

7.4.1. Establish the algebraic identity

$$\sum_{i=1}^n [i - \tfrac{1}{2}(n + 1)]x_{(i)} = \tfrac{1}{4} \sum_{i,\,j=1}^n |x_i - x_j|$$

to show that "σ" of (7.4.1) and G of (7.4.2) are linked by

$$\text{``}\sigma\text{''} = \tfrac{1}{2}\pi^{1/2}G.$$

Noting that $\mathscr{E}G = \mathscr{E}|X_1 - X_2|$, show that "$\sigma$" is an unbiased estimator of σ in normal samples and that for a distribution with cdf $P(x)$

$$\mathscr{E}\text{``}\sigma\text{''} = 2\pi^{1/2} \int_{-\infty}^{\infty} x[P(x) - \tfrac{1}{2}]\,dP(x).$$
(Nair, 1936; David, 1968)

7.5.1. If the regression line of Y on X is given by

$$\mathscr{E}(Y \,|\, x) = \alpha + \beta x,$$

show that an unbiased estimator of $\sigma_{xy} = \text{cov}\,(X,\,Y)$ is

$$\hat{\sigma}_{xy}' = \frac{(X_{(n)} - X_{(1)})(Y_{[n]} - Y_{[1]})}{c_n^2},$$

where $c_n^2 = \mathscr{E}W_x^2/\sigma_x^2$ $(W_x = X_{(n)} - X_{(1)})$.
If, in addition, var $(Y \,|\, x) = \sigma_y^2$, independent of x, show also that

$$\text{var}\,\hat{\sigma}_{xy}' = \left[\mathscr{E}\left(\frac{W_x^4}{\sigma_x^4 c_n^4} - 1\right)\rho^2 + \frac{2}{c_n^2}(1 - \rho^2)\right]\sigma_x^2\sigma_y^2.$$
(Tsukibayashi, 1962)

7.6.1. Derive the results stated in (7.6.8)–(7.6.13).

(Ogawa, 1962)

7.6.2. Verify that for a normal population the best linear estimators μ_0^*, σ_0^* of μ, σ, based on a common two-point spacing and minimizing var $\mu_0^* + c$ var σ_0^* for $c = 1, 2, 3$ are, together with their efficiencies, as in the following table:

c	μ_0^*	σ_0^*	Efficiency
1	$\frac{1}{2}(X_{(.1525n)} + X_{(.8475n)})$		.729
		$.4875(X_{(.8475n)} - X_{(.1525n)})$	.552
2	$\frac{1}{2}(X_{(.1274n)} + X_{(.8726n)})$		.683
		$.4391(X_{(.8726n)} - X_{(.1274n)})$	.594
3	$\frac{1}{2}(X_{(.1147n)} + X_{(.8853n)})$		.654
		$.4160(X_{(.8853n)} - X_{(.1147n)})$	.614

(Eisenberger and Posner, 1965)

7.7.1. Show that for model (7.7.3) the standard F test has power

$$\Pr\left\{F > \frac{F_\alpha}{1 + n\zeta^2}\right\},$$

where F_α is the upper α significance point of F with $k - 1$, $k(n - 1)$ DF, and $\zeta = \sigma'/\sigma$. By appropriate choice of ζ the power may be made equal to a specified value $1 - \beta$. Show that the power of the corresponding Q' test, for this value of ζ, is

$$\Pr\{(1 + n\zeta^2)^{\frac{1}{2}}Q_{k,v} > q_\alpha\} = \Pr\{Q_{k,v} > q_\alpha(F_{1-\beta}/F_\alpha)^{\frac{1}{2}}\}$$

with v given by Table 7.3.2. Evaluate the power for $\alpha = 0.05$, $\beta = 0.10$, $k = 8$, $n = 6$. (*Ans.* 0.87)

(David, 1953)

7.7.2. If $X_i = \mu + \alpha_i + Z_i$ ($i = 1, 2, \ldots, n$), then $W' = $ range X_i may be called the noncentral range. From Ex. 2.3.2 show that the cdf of W' in normal samples with $\sigma = 1$ is given by

$$\Pr\{W' \leq w\} = \sum_{i=1}^{n} \int_{-\infty}^{\infty} \phi(x_i - \alpha_i)\left\{\prod_{\substack{j=1 \\ j \neq i}}^{n} [\Phi(x_i - \alpha_j + w) - \Phi(x_i - \alpha_j)]\right\} dx_i,$$

and that the cdf of the studentized noncentral range $Q' = W'/S_v$ is

$$\Pr\{Q' \leq q\} = \int_0^\infty \Pr\{W' \leq s\}f(s)\,ds,$$

where $f(s)$ is the pdf of S_v.

7.8.1. Let $X_1, X_2, \ldots, X_n$ be a random sample from a $N(\mu, \sigma^2)$ population. Consider the statistic

$$W^* = \frac{\left(\sum_{i=1}^{n} a_i X_{(i)}\right)^2}{\sum_{i=1}^{n} (X_i - \bar{X})^2},$$

where the a_i are the normalized (i.e., $\Sigma\, a_i^2 = 1$) coefficients of the BLUE of σ. Write $b = \Sigma\, a_i X_{(i)}$ and $T^2 = \Sigma(X_i - \bar{X})^2$. Show that

(a) $\mathscr{E} W^{*r} = \mathscr{E} b^{2r} / \mathscr{E} T^{2r}$;

(b) the maximum value of W^* is 1;

(c) the minimum value of W^* is $n a_1^2 / (n - 1)$;

(d) for $n = 3$, the pdf of W^* is

$$\frac{3}{\pi} (1 - W^*)^{-\frac{1}{2}} W^{*-\frac{1}{2}} \qquad \tfrac{3}{4} \le W^* \le 1.$$

(Compare Section 5.2. W^* has been found, by empirical sampling methods, to provide a versatile test of normality.)

(Shapiro and Wilk, 1965, 1968; Shapiro *et al.*, 1968; Wilk and Shapiro, 1968)

CHAPTER 8

The Treatment of Outliers

8.1. THE PROBLEM OF OUTLIERS AND SLIPPAGE

The proper treatment of outlying observations has long been a subject for study. However, no historical account need be attempted here in view of several fairly recent publications with a historical introduction (Anscombe, 1960; Ferguson, 1961a; Doornbos, 1966). Traditionally, the approach has been to discover outliers by means of tests of significance, usually devised from intuitive considerations. Such tests generally have one of the following aims:

(*a*) to screen data in routine fashion preparatory to analysis;
(*b*) to sound an alarm that outliers are present, thus indicating the need for closer study of the data-generating process;
(*c*) to pin-point observations which may be of special interest just because they are extreme.

In Section 8.2 we list a variety of test statistics usually involving extreme order statistics, and wherever possible indicate sources of tables of percentage points; nearly always normality is assumed for the underlying population from which all the observations have ideally been drawn. In many cases the necessary distribution theory has already been given in Chapter 5.

Slippage tests are discussed in Section 8.3. A word of explanation on their connection with tests for outliers is called for. Slippage tests are designed to test the equality of several, say n, populations with special emphasis on the alternative that one of the populations has "slipped," that is, is different from the remaining ones, which are identical. When each population is represented by a single observation, the aim reduces to the detection of an outlier as in (*c*) above; for $m > 1$ observations per population the problem is to discover an outlying group of observations often representing a superior "treatment." An important example of the latter type occurs in the preliminary testing of n new drugs for an incurable disease, m patients on each

170

drug, where one does not have much expectation that any drug will be effective but hopes that one drug may be.

The performance of a number of commonly used tests for outliers from the point of view of their capacity to achieve aims (b) and (c) is studied in Section 8.4. Here it is frequently assumed that only a single outlier is present, differing in mean and occasionally in variance from the remainder of the sample. Such an assumption may be a reasonable approximation to the truth when the probability of "contamination" of the sample by outliers is low. However, we provide also some discussion of the more general situation (cf. Dixon, 1962). It is worth keeping in mind at all times that moderate outliers may merely reflect a long-tailed rather than a normal underlying distribution.

So far we have not dealt specifically with aim (a) above. In routine screening of data, outliers are merely a nuisance standing in the way of satisfactory analysis. Traditionally, the remedy adopted has been simple: reject observations found to be outlying by some test, and *then* estimate the parameters of interest or carry out the desired main test of significance. This is the so-called area of "rejection of outliers." We have tried to indicate that the treatment of outliers is a much broader subject; in fact, even aim (a) is broader. If the principal purpose of the analysis is the estimation of parameters, the properties of the estimators used should play a central role. It may still be reasonable to reject outliers before forming estimates, but the outlier tests should be carried out, not at conventional levels of significance, but rather at a level which will produce estimates optimal in some sense. Actually, outlying observations need not be wholly rejected; often better estimators will result if outliers are merely given reduced weight. These issues are discussed in Section 8.5. One may go a step further still and, skipping any sort of significance test, immediately use estimators which are robust in the presence of outliers, the simplest example being the use of the sample median as a measure of location. This takes us back to Chapter 6.

It will be clear that there are many aspects to the treatment of outliers and that much more work needs to be done. When a common cause may explain the presence of several outliers, it may become appropriate to treat such contaminated data as a sample from a mixture of two different populations (see, e.g., Blischke, 1968). Greater complications arise when the data are no longer (ideally) a single sample but rather a designed experiment. The methods to be described cover, or may readily be adapted to, one-way classifications; however, anything more complex introduces a host of new problems. Thus, for patterned data, casual inspection may no longer indicate where the outliers are most likely to be. A fairly obvious extension is to regard as prime candidates those observations giving rise to the largest residuals, possibly standardized in some way. We will not pursue this topic, but refer the reader to Anscombe (1960, 1961), Anscombe and Tukey (1963), Bross (1961), Daniel (1960), and Srikantan (1961).

Some simple points in the handling of outliers should not be overlooked in this welter of problems. Statistics can deal with only part of the outlier story. It goes without saying that the happiest ending is to find a physical cause and cure for their presence, be it before or after some test. If particular observations are suspect, the right thing is to try and find out why, by reference to the experimenter. Only if such a check is not practicable would one wish to turn to purely statistical procedures. In any case, an analysis is incomplete and certainly not quite honest unless it makes mention of the treatment adopted for outliers, including the values of any rejected observations. See also Kruskal (1960).

A "Bayesian approach to the rejection of outliers" is sketched by de Finetti (1961). The title seems a trifle playful since Bayesians do not reject observations; however, the outliers may in effect have very little weight.

General reviews of the subject of outliers are given by Ferguson (1961a), Dixon (1962), Chew (1964), and Grubbs (1969).

8.2. TESTS FOR OUTLIERS

Let X_1, X_2, $\ldots$, X_n be n independent variates following a common $N(\mu, \sigma^2)$ law. This is the null situation. We are concerned with test statistics designed to be sensitive to various non-null patterns, primarily a shift in mean (possibly accompanied by a change in variance) of one or a few of the variates. Even in this simple case a hierarchy of situations, (a)–(d) below, may usefully be distinguished, depending on the degree of information available on μ and σ. Moreover, one may be interested in shifts in a specified direction or in either direction. For each of the resulting eight cases we list at least one statistic, and wherever possible a reference giving tables of its percentage points. The one-sided statistic appears on the left of the page; for brevity this is expressed only in the form suitable for detecting an outlier on the right of the sample. PH stands for Pearson and Hartley (1966), w_n denotes the range, $s^2 = \Sigma (x_i - \bar{x})^2/(n - 1)$, and s_v^2 is an independent mean square estimate of σ^2 based on ν DF. Also

$$\bar{x}_n = \frac{\sum_{i=1}^{n-1} x_{(i)}}{n - 1}, \quad \bar{x}_{n,n-1} = \frac{\sum_{i=1}^{n-2} x_{(i)}}{n - 2}.$$

(a) Both μ and σ known

$$A_1 = (x_{(n)} - \mu)/\sigma \quad \text{(PH, Table 24)} \qquad A_2 = \max |x_i - \mu|/\sigma$$

$$A_3 = \chi_n^2 = \Sigma(x_i - \mu)^2/\sigma^2$$

(b) Only σ known

$B_1 = (x_{(n)} - \bar{x})/\sigma$ (Grubbs, 1950) $B_2 = \max |x_i - \bar{x}|/\sigma$

$B_3 = (x_{(n)} - x_{(n-1)})/\sigma$ (Irwin, 1925) $B_4 = w_n/\sigma$ (PH, Table 22)

$$B_5 = \chi^2_{n-1} = \Sigma(x_i - \bar{x})^2/\sigma^2$$

(c) Both μ and σ unknown but an independent estimate of σ^2 available

$C_1 = (x_{(n)} - \bar{x})/s_v$ $C_2 = \max |x_i - \bar{x}|/s_v$

    (PH, Table 26)   (Halperin *et al.*, 1955)

$$C_3 = w_n/s_v \quad \text{(PH, Table 29)}$$

$$C_4 = \frac{(x_{(n)} - \bar{x})}{[\Sigma(x_i - \bar{x})^2 + vs_v^2]^{\frac{1}{2}}} \qquad C_5 = \frac{\max |x_i - \bar{x}|}{[\Sigma(x_i - \bar{x})^2 + vs_v^2]^{\frac{1}{2}}}$$

    (PH, Table 26a)   (PH, Table 26b)

(d) Both μ and σ unknown

$D_1 = (x_{(n)} - \bar{x})/s$ $D_2 = \max |x_i - \bar{x}|/s$

    (Grubbs, 1950[1])   (PH, Table 26b)

$$D_3 = w_n/s \quad \text{(PH, Table 29c)}$$

$$D_4 = \frac{n^{\frac{1}{2}}\Sigma(x_i - \bar{x})^3}{[\Sigma(x_i - \bar{x})^2]^{\frac{3}{2}}} \qquad |D_4|$$

    (PH, Table 34b)

$$D_5 = \frac{n\Sigma(x_i - \bar{x})^4}{[\Sigma(x_i - \bar{x})^2]^2} \quad \text{(PH, Table 34c)}$$

$$D_6 = \frac{\sum_{i=1}^{n-2}(x_{(i)} - \bar{x}_{n,n-1})^2}{\sum_{i=1}^{n}(x_i - \bar{x})^2}$$

    (Grubbs, 1950)

Dixon's r statistics (Dixon, 1951)

For the most part the foregoing statistics have obvious intuitive appeal, extreme order statistics being used and unknown parameters being replaced by statistics sufficient for them. Indeed, in many instances various optimality properties have been discovered, usually long after the statistic made its first appearance. Note that C_1, D_1, and C_4 are all studentized versions of B_1, the studentization being, in the terminology of Chapter 5, external, internal, and

---

[1] Note that Grubbs takes $s = [\Sigma (x_i - \bar{x})^2/n]^{\frac{1}{2}}$.

pooled, respectively. The corresponding two-sided statistics C_2, D_2, and C_5 bear the same relation to B_2. Except for multiplicative constants, D_1 and D_2 are the special cases $\nu = 0$ of C_4 and C_5, respectively. Also included in the list are a few statistics (A_3, B_5, D_4, D_5) not primarily designed for the detection of outliers but nevertheless quite effective under appropriate conditions. Although these statistics do not single out the extremes, they are included here for comparative purposes. In regard to B_5, it is clear that a statistic focusing on a particular feature of the data should be preferable, under the circumstances for which it has been designed, to this overall statistic. We may mention in passing that this remark applies not only to a comparison of B_2 with the other B statistics but also in some other cases (Bodmer, 1959; David and Newell, 1965). But the edge which the specialized statistics have over B_2 may be slight and should be investigated in each case.

The greatest variety of statistics is available in situation (d), the most important case. The statistic D_6 was designed to be effective in the presence of two outliers, both to the right. The method of construction of D_6 suggests further generalizations, but no percentage points appear to have been tabulated. The simpler statistic

$$D_0 = \frac{\sum_{i=1}^{n-1}(x_{(i)} - \bar{x}_n)^2}{\sum_{i=1}^{n}(x_i - \bar{x})^2}$$

gives us nothing new since $D_0 = 1 - nD_1^2/(n - 1)^2$.

Dixon's r statistics are all ratios of differences of order statistics designed *ad hoc* to be especially effective under the following conditions:

(i) For a single outlier $x_{(n)}$ $\qquad\qquad r_{10} = \dfrac{x_{(n)} - x_{(n-1)}}{x_{(n)} - x_{(1)}}$

(ii) For outlier $x_{(n)}$, avoiding $x_{(1)}$ $\qquad r_{11} = \dfrac{x_{(n)} - x_{(n-1)}}{x_{(n)} - x_{(2)}}$

(iii) For outlier $x_{(n)}$, avoiding $x_{(1)}$, $x_{(2)}$ $\qquad r_{12} = \dfrac{x_{(n)} - x_{(n-1)}}{x_{(n)} - x_{(3)}}$

(iv) For outlier $x_{(n)}$, avoiding $x_{(n-1)}$ $\qquad r_{20} = \dfrac{x_{(n)} - x_{(n-2)}}{x_{(n)} - x_{(1)}}$

(v) For outlier $x_{(n)}$, avoiding $x_{(n-1)}$, $x_{(1)}$ $\qquad r_{21} = \dfrac{x_{(n)} - x_{(n-2)}}{x_{(n)} - x_{(2)}}$

(vi) For outlier $x_{(n)}$, avoiding $x_{(n-1)}$, $x_{(1)}$, $x_{(2)}$ $\qquad r_{22} = \dfrac{x_{(n)} - x_{(n-2)}}{x_{(n)} - x_{(3)}}$

As they stand, these are all one-sided statistics. Ferguson (1961b) has also considered a two-sided version of r_{10}, namely,

$$r_{10}^{(2)} = \max (r_{10}, r_{10}'), \quad \text{where} \quad r_{10}' = \frac{x_{(2)} - x_{(1)}}{x_{(n)} - x_{(1)}}. \tag{8.2.1}$$

Since for any constant $c > 0$

$$\Pr \{R_{10}^{(2)} > c\} = \Pr \{R_{10} > c\} + \Pr \{R_{10}' > c\} - \Pr \{R_{10} > c, R_{10}' > c\},$$

we have in the null case, due to symmetry,

$$\Pr \{R_{10}^{(2)} > c\} = 2 \Pr \{R_{10} > c\} - \Pr \{R_{10} > c, R_{10}' > c\}. \tag{8.2.2}$$

Thus

$$\Pr \{R_{10}^{(2)} > c\} \le 2 \Pr \{R_{10} > c\}, \tag{8.2.3}$$

and for sufficiently large c the RHS of (8.2.3) is a good approximation to its LHS since the last term of (8.2.2) is then small. The upper α significance point of R_{10} serves then also as an approximate (and strictly conservative) upper 2α significance point of $R_{10}^{(2)}$. This kind of result (King, 1953) is, of course, equally applicable to the other statistics in (a)–(d).

Indeed, the question may be raised whether we should not always use a two-sided test, since applying a one-sided test in the direction indicated as most promising by the sample at hand is clearly not playing fair. This criticism of what is often done in practice is valid enough, except that sticking to a precise level of significance may not be crucial in exploratory work, frequently the purpose of tests for outliers. Strictly speaking, one-sided tests should be confined to the detection of outliers in cases where only those in a specified direction are of interest, or to situations such as the repeated determination of the melting point of a substance, where outliers due to impurities must be on the low side since impurities depress the melting point. Similar arguments show that it is equally incorrect to pick one's outlier test after inspection of the data. The results of the next two sections, although necessarily incomplete, provide some guidance on the choice among competing statistics.

Example 8.2 (Quesenberry and David, 1961). Squibs are small devices for igniting the rocket motors of missiles. Watertightness and shock resistance are important characteristics of squibs. In order to study these characteristics, a random sample of size 48 was drawn from a large batch. The sample was randomly subdivided into 3 equal groups. The first group was used as a control unit and received no treatment, the second group was submerged in water, and the third group was dropped from a fixed height. Each squib in the entire sample was tested by having a current of 5 amperes passed through it and its time to failure recorded.

Table A

Control	(x_{1i})	Watertightness	(x_{2i})	Shock	(x_{3i})
.38	.51	.53	.39	.51	.35
.26	.55	.35	.74	.63	.41
.41	.53	.38	.32	.46	.49
.33	.41	.45	.74	.47	.40
.33	.47	1.09	.48	.42	.58
.37	.49	.46	.37	.45	.46
.54	.42	.57	.52	.41	.38
.76	.34	.47	.44	.39	.48

(Data furnished by Ordnance Missile Laboratories, ARGMA, AOMC, Redstone Arsenal, Alabama)

$$\Sigma x_{1i} = 7.10, \qquad \Sigma x_{2i} = 8.30, \qquad \Sigma x_{3i} = 7.29,$$

$$\bar{x}_1 = 0.4438, \qquad \bar{x}_2 = 0.5188, \qquad \bar{x}_3 = 0.4556,$$

$$\Sigma x_{1i}^2 = 3.3686, \qquad \Sigma x_{2i}^2 = 4.8768, \qquad \Sigma x_{3i}^2 = 3.4021,$$

$$\frac{(\Sigma x_{1i})^2}{16} = 3.1506, \qquad \frac{(\Sigma x_{2i})^2}{16} = 4.3056, \qquad \frac{(\Sigma x_{3i})^2}{16} = 3.3215,$$

$$SS(1) = 0.2180, \qquad SS(2) = 0.5712, \qquad SS(3) = 0.0806.$$

From investigation of a large series of similar data it has been found that these delay times are approximately normally distributed, but that for reasons not fully understood extremely large delay times occasionally occur. Consequently, an outlier test was used on each subgroup to pick out such divergent observations. The variance of the bulk of normal observations was assumed to be constant throughout the experiment. The data are given in Table A.

In groups 1 and 2 the largest observations, 0.76 and 1.09, respectively, are easily seen to be outliers. Thus for group 1

$$(n-1)^{\frac{1}{2}} D_1 = \frac{x_{(n)} - \bar{x}}{[\Sigma(x_i - \bar{x})^2]^{\frac{1}{2}}} = \frac{0.76 - 0.4438}{(0.2180)^{\frac{1}{2}}} = 0.677.$$

Interpolation for $n = 16$, $\nu = 0$ in Table 26a of PH gives the upper 5% point approximately as 0.630. Use of r_{10} leads to the same result:

$$r_{10} = \frac{x_{(n)} - x_{(n-1)}}{x_{(n)} - x_{(1)}} = \frac{0.76 - 0.55}{0.76 - 0.26} = 0.420,$$

which almost reaches the upper 1% point 0.426 (Sarhan and Greenberg, 1962, p. 332).

Although we have in this example a clear mandate for using a one-sided test, it may be noted that significance at the 5% level continues even for the two-sided test D_2:

$$(n - 1)^{1/2} D_2 = \frac{\max |x_{(i)} - \bar{x}|}{[\Sigma(x_i - \bar{x})^2]^{1/2}} = 0.677 \qquad \text{(upper 5\% pt.} \doteq 0.666),$$

but not quite for D_3:

$$D_3 = \frac{0.76 - 0.26}{0.4669}(15)^{1/2} = 4.15 \qquad \text{(upper 5\% pt.} = 4.24).$$

In the original paper the use of the statistic C_4 is illustrated. One has, again for group 1,

$$C_4 = \frac{x_{(n)} - \bar{x}}{[\text{SS}(1) + \text{SS}(2) + \text{SS}(3)]^{1/2}} = 0.339.$$

This is actually not significant ($n = 16$, $v = 30$; upper 5% pt. $\doteq 0.384$), but a similar test rejects the value 1.09 in group 2. Once this is removed (making $v = 29$) and C_4 recomputed for group 1, the resulting value 0.438 is clearly significantly large. No other outliers are found by a continuation of this process (see the original reference, where further analyses are also given).

Certain tests for outliers in non-normal underlying distributions have been considered by Darling (1952b), who finds the distribution of $\sum_{i=1}^{n} X_i/X_{(n)}$ when the X_i have a uniform (Ex. 5.4.6) or a χ^2 distribution with r (even) DF. The latter case—which is closely related to tests for slippage of a variance among normal populations (Comment 7 of Section 8.3)—is a generalization of Fisher's (1929) test for the largest harmonic (the case $r = 2$; see Section 5.4). Laurent (1963) and Basu (1965) give tests for outliers among two-parameter exponential variates, with one or both parameters unknown (Exs. 8.2.2 and 8.2.3).

Wilks (1963) tackles the outlier problem in the multivariate normal case, using as his basic statistics (with obvious notation) the ratio of determinants

$$r_\zeta = |a_{ij\zeta}|/|a_{ij}| \qquad i, j = 1, 2, \ldots, k; \, \zeta = 1, 2, \ldots, n,$$

where

$$a_{ij} = \sum_{\alpha=1}^{n}(x_{i\alpha} - \bar{x}_i)(x_{j\alpha} - \bar{x}_j),$$

and $a_{ij\zeta}$ is the corresponding sum with the ζth observation omitted. Then R_ζ is known to have a $\beta(\frac{1}{2}(n - k - 1), \frac{1}{2}k)$ distribution. The observation giving rise to the smallest value of r_ζ, namely, $r_{(1)}$, is the primary outlier candidate. Wilks uses the first Bonferroni inequality:

$$\Pr\{R_{(1)} < r\} \leq n \Pr\{R_\zeta < r\}$$

to obtain values r_α making the RHS $= \alpha$. For $k = 1$ the ratio $R_{(1)}$ is equivalent to the statistic $D_2 = \max |x_i - \bar{x}|/s$. If one is satisfied with the first Bonferroni inequality, the case of $r > 1$ outliers can be handled similarly.

8.3. SLIPPAGE TESTS

The first example of slippage tests is due to Mosteller (1948), who considered the following problem: Given samples of m from each of n continuous populations, to test the null hypothesis that all populations are identical against the alternative that one of them (we do not know which) has slipped to the right. To this end, Mosteller's nonparametric procedure is to pick the sample with the largest observation and to count the number of observations in this sample which exceed *all* observations of all other samples. If this number is large enough, the null hypothesis is rejected. The procedure is advanced by Mosteller as simple and reasonable, without any claims that it is in any sense best. Extensions to unequal sample sizes are easily made (Mosteller and Tukey, 1950). See also Bofinger (1965). If X_i denotes an observation in the ith sample ($i = 1, 2, \ldots, n$), slippage to the right may be generalized to

$$\Pr\{X_i > X_j\} > \tfrac{1}{2} \quad \text{for } some\ i, \quad j = 1, 2, \ldots, i - 1, i + 1, \ldots, n,$$

with the X_j iid variates.

Another simple nonparametric slippage test which may be expected to be better in general is as follows. Rank all the nm observations as for the Kruskal-Wallis test and form the rank sums T_i for each sample. If $\max T_i$ exceeds its critical value, declare the corresponding population to have slipped to the right. Such a procedure was first proposed by Doornbos and Prins (1958), who suggested Bonferroni-type approximations to the critical values. Exact tables are given by Odeh (1967). Paulson (1961) replaces ranks by normal scores. Similar results apply for the corresponding two-way Friedman rank analysis (or equivalently the "method of m rankings"). This case has also been considered by Doornbos and Prins (1958), and in more detail by Youden (1963) and by Thompson and Willke (1963), who give tables.

For a more formal approach we can do no better than follow closely the development given by Paulson (1952) in an early application to the case of normal populations. Let X_{ij} ($i = 1, 2, \ldots, n; j = 1, 2, \ldots, m$) be mutually independent $N(\mu_i, \sigma^2)$ variates as in a one-way analysis of variance. We say that population π_i has slipped to the right by an amount Δ (> 0) if

$$\mu_1 = \mu_2 = \cdots = \mu_{i-1} = \mu_{i+1} = \cdots = \mu_n \quad \text{and} \quad \mu_i = \mu_1 + \Delta. \quad (8.3.1)$$

Let $\mathcal{D}_0$ be the decision that the n means are equal, and $\mathcal{D}_i$ ($i = 1, 2, \ldots, n$)

the decision that $\mathscr{D}_0$ is incorrect and furthermore that π_i is the slipped population. The problem is to find a decision procedure for choosing one of the $n + 1$ decisions $(\mathscr{D}_0, \mathscr{D}_1, \ldots, \mathscr{D}_n)$ which will be in some sense optimal in detecting slippage to the right. With this aim we now impose the following restrictions:

(a) When all means are equal, $\mathscr{D}_0$ should be selected with probability $1 - \alpha$.

(b) The procedure should be invariant under the transformation $y = ax + b$, where a and b are constants with $a > 0$.

(c) The procedure should be symmetric in the sense that the probability of making the correct decision when (8.3.1) holds is to be the same for all i.

Since (a) fixes the probability of correctly choosing $\mathscr{D}_0$, an obvious optimality property to aim for is maximization of the probability of making the correct decision when one of the populations has slipped to the right. It will be shown that this is achieved by the following procedure, in which $\bar{x}_M$ stands for the largest of the n sample means $\bar{x}_i$:

$$\text{If } \frac{m(\bar{x}_M - \bar{x})}{\left[\sum_{i=1}^{n}\sum_{j=1}^{m}(x_{ij} - \bar{x})^2\right]^{1/2}} \begin{cases} < b_\alpha, & \text{choose } \mathscr{D}_0. \\ > b_\alpha, & \text{choose } \mathscr{D}_M, \end{cases} \tag{8.3.2}$$

Here b_α is just the upper α significance point of the statistic on the left, which is the special case $\nu = n(m - 1)$ of C_4 in Section 8.2. Note that the expression in brackets is the total (and not the error) sum of squares. As before, we shall denote the error mean square by

$$s_\nu{}^2 = \frac{1}{\nu}\sum_{i=1}^{n}\sum_{j=1}^{m}(x_{ij} - \bar{x}_i)^2.$$

Derivation of Optimum Procedure

Without loss of generality we may confine consideration to procedures depending only on $\bar{X}_1, \bar{X}_2, \ldots, \bar{X}_n$, and $S_\nu{}^2$ since these constitute a set of sufficient statistics for the unknown parameters $\mu_1, \mu_2, \ldots, \mu_n$, and σ^2. In fact, in view of (b) it is clear that any allowable decision procedure will depend only on the $n - 1$ ratios $(\bar{X}_1 - \bar{X}_n)/S_\nu, (\bar{X}_2 - \bar{X}_n)/S_\nu, \ldots, (\bar{X}_{n-1} - \bar{X}_n)/S_\nu$ (which are a maximal invariant). Let

$$W_t = \frac{\bar{X}_t - \bar{X}_n}{S_\nu} \quad \text{and} \quad \delta_t = \frac{\mu_t - \mu_n}{\sigma} \quad \text{for} \quad t = 1, 2, \ldots, n - 1.$$

Then the joint distribution of the W_t depends only on the δ_t. Also, $\mathscr{D}_0$

becomes the decision $\delta_1 = \delta_2 = \cdots = \delta_{n-1} = 0$; $\mathcal{D}_t$ that

$$\delta_1 = \delta_2 = \cdots = \delta_{t-1} = \delta_{t+1} = \cdots = \delta_{n-1} = 0, \quad \delta_t = \Delta/\sigma;$$

and $\mathcal{D}_n$ that

$$\delta_1 = \delta_2 = \cdots = \delta_{n-1} = -\Delta/\sigma.$$

To find the joint pdf of the W_t, write $V_t = (\bar{X}_t - \bar{X}_n)/\sigma$, so that $W_t = V_t\sigma/S_v$. The V_t are easily seen to have a $(n-1)$-dimensional multinormal distribution with

$$\mathscr{E}V_t = \delta_t, \quad \text{var } V_t = 2/m,$$

$$\text{corr } (V_t, V_{t'}) = \tfrac{1}{2} \quad t' = 1, 2, \ldots, n-1; t' \neq t.$$

Thus the joint pdf of the V_t is

$$C \exp\left\{-\tfrac{1}{2}\left[A\sum_{t=1}^{n-1}(v_t - \delta_t)^2 + B\sum_{t \neq t'}(v_t - \delta_t)(v_{t'} - \delta_{t'})\right]\right\},$$

where $A = (n-1)m/n$, $B = -m/n$, and C (as well as C' below) is a constant whose value is not needed. Upon studentization we have for the joint pdf of the W_t

$$f(w_1, w_2, \ldots, w_{n-1}) = C'\int_0^\infty y^{\nu+n-2}$$

$$\times \exp\left\{-\tfrac{1}{2}\left[\nu y^2 + A\sum_{t=1}^{n-1}(w_t y - \delta_t)^2 + B\sum_{t \neq t'}(w_t y - \delta_t)(w_{t'} y - \delta_{t'})\right]\right\} dy.$$

$$(8.3.3)$$

Let f_h $(h = 0, 1, \ldots, n)$ denote this pdf when $\mathcal{D}_h$ is the correct decision. Then by an extension of the Neyman-Pearson fundamental lemma[2] we select $\mathcal{D}_0$ for all points in the $w_1 w_2 \cdots w_{n-1}$ space for which $f_1 < \lambda f_0, f_2 < \lambda f_0, \ldots,$ $f_n < \lambda f_0$, where the constant λ is determined by restriction (a); for a point making $f_i > \lambda f_0$ for one or more i we choose $\mathcal{D}_i$ if $f_i = \max (f_1, f_2, \ldots, f_n)$. With the help of (8.3.3) it is now easy to calculate for each h the region where $\mathcal{D}_h$ is selected. For example, $\mathcal{D}_1$ will be chosen if $f_1 > \lambda f_0, f_1 > f_2, \ldots,$ $f_1 > f_n$. Now $f_1 > \lambda f_0$ if

$$\int_0^\infty y^{\nu+n-2} \exp\left[-\frac{y^2}{2}\left(\nu + A\sum_{t=1}^{n-1}w_t^2 + B\sum_{t \neq t'}w_t w_{t'}\right)\right]$$

$$\times \left(\exp\left(-\frac{A\Delta^2}{2\sigma^2}\right)\exp\left\{\left[(A-B)\frac{\Delta}{\sigma}w_1 + B\frac{\Delta}{\sigma}\sum_{t=1}^{n-1}w_t\right]y\right\} - \lambda\right) dy > 0.$$

---

[2] Here we depart briefly from the proof given by Paulson, who shows that (8.3.2) is the Bayes solution when equal prior probabilities are assigned to the decisions $\mathcal{D}_1, \mathcal{D}_2, \ldots, \mathcal{D}_n$,

With a change of variable this is equivalent to

$$\int_0^\infty t^{v+n-2} \exp\left(-\tfrac{1}{2}t^2\right)$$

$$\times \left\{ \exp\left(-\frac{A\Delta^2}{2\sigma^2}\right) \exp\left[\frac{\Delta}{\sigma}g(w_1, w_2, \ldots, w_{n-1})t\right] - \lambda \right\} dt > 0, \quad (8.3.4)$$

where

$$g(w_1, w_2, \ldots, w_{n-1}) = \frac{(A-B)w_1 + B\sum_{t=1}^{n-1} w_t}{\left(v + A\sum_{t=1}^{n-1} w_t^2 + B\sum_{t \neq t'} w_t w_{t'}\right)^{\frac{1}{2}}}.$$

The integrand on the left of (8.3.4) is for all t a monotonically increasing function of $g(w_1, w_2, \ldots, w_{n-1})$, so the region where $f_1 > \lambda f_0$ must be of the type $g(w_1, w_2, \ldots, w_{n-1}) > C''$, where C'' is a constant depending on Δ/σ and λ. Similarly it can be shown that $f_1 > f_{i'}$ ($i' = 2, 3, \ldots, n-1$) if and only if $w_1 > w_{i'}$, and that $f_1 > f_n$ iff $w_1 > 0$—results which are intuitively obvious. Thus $\mathscr{D}_1$ will be chosen if $w_1 > 0$, $w_1 > \max(w_2, w_3, \ldots, w_{n-1})$ and

$$(A-B)w_1 + B\sum_{t=1}^{n-1} w_t > C''\left(v + A\sum_{t=1}^{n-1} w_t^2 + B\sum_{t \neq t'} w_t w_{t'}\right)^{\frac{1}{2}}.$$

Recalling the definitions of A, B, C'', and w_t, we see that $\mathscr{D}_1$ is selected if $\bar{x}_1 > \max(\bar{x}_2, \bar{x}_3, \ldots, \bar{x}_n)$ and

$$m(\bar{x}_1 - \bar{x}) > C''\left[\sum_{i=1}^{n}\sum_{j=1}^{m}(x_{ij} - \bar{x})^2\right]^{\frac{1}{2}}.$$

Since corresponding results must hold for $\mathscr{D}_2, \mathscr{D}_3, \ldots, \mathscr{D}_n$, the decision procedure is just that of (8.3.2). Note that the constant C'' becomes b_α and does not in fact depend on Δ or σ. Thus, for given n, m, and α, the optimality property of (8.3.2) holds uniformly in Δ (> 0) and σ.

Comments and Extensions

1. As pointed out by Kudô (1956b), the optimality property of procedure (8.3.2) continues to hold when the shift in mean of one of the populations is accompanied by a decrease in variance (there being no change in either mean or variance for the other populations).

2. Even with the foregoing slight generalization the optimality property applies only against somewhat artificial alternatives. However, use of (8.3.2) is intuitively reasonable for other alternatives not too far removed from the slippage model. Power functions and related measures of performance are obtained in some limited instances in Section 8.4. Of course, $\mathscr{D}_0$ will always be chosen with probability $1 - \alpha$ if true. Moreover, Kapur (1957) establishes

the following unbiasedness property of Paulson's procedure for a general configuration of the μ_i. Define $\mathscr{D}_0$ as before and let $\mathscr{D}_i$ $(i = 1, 2, \ldots, n)$ be the decision that $\mu_i = \max(\mu_1, \mu_2, \ldots, \mu_n)$. Then the probability that any one of the decisions $\mathscr{D}_0, \mathscr{D}_1, \ldots, \mathscr{D}_n$ is chosen correctly is never less than the probability that it is chosen incorrectly.

3. If $m = 1$, the procedure may be termed the "slippage formulation of the outlier problem" (for a single outlier on the right). In this special case, (8.3.2) reduces at once (but with some slight repair work in the proof—see Kudô, 1956a) to the following:

$$\text{If } \frac{x_{(n)} - \bar{x}}{\left[\displaystyle\sum_{i=1}^{n}(x_i - \bar{x})^2\right]^{1/2}} \begin{cases} < b_\alpha, & \text{declare no outlier,} \\ > b_\alpha, & \text{declare } x_{(n)} \text{ an outlier.} \end{cases} \qquad (8.3.5)$$

This is essentially the test of Pearson and Chandra Sekar (1936). In line with Comment 2 we may wish to test whether $x_{(n-1)}$ is perhaps also an outlier, a question deferred to Section 8.4.

4. In the preceding outlier problem there may be available, in addition to the internal sum of squares $\Sigma(x_i - \bar{x})^2$, an external estimate s_ν of σ such that $\nu s_\nu^2/\sigma^2 \frown \chi_\nu^2$. Nair's (1948) studentized extreme deviate uses s_ν alone as the divisor of $x_{(n)} - \bar{x}$. The optimal procedure for correct detection of a single outlier on the right is seen to be (8.3.5) with the denominator replaced by $[\Sigma(x_i - \bar{x})^2 + \nu s_\nu^2]^{1/2}$. In fact, the original slippage problem may be regarded as the special case of this procedure for which $\nu = n(m - 1)$.

5. When slippage in either direction is of interest, that is, the sign of Δ in (8.3.1) is unspecified, Paulson's formulation can be retained with only the slight modifications that in restriction (b) the sign of a is unspecified and in (c) the probability of a correct decision is the same for $-\Delta$ as for Δ in (8.3.1). The resulting optimum procedure becomes (Kudô, 1956a):

$$\text{If } \frac{\max\limits_{i=1,2,\ldots,n} |\bar{x}_i - \bar{x}|}{\left[\displaystyle\sum_{i=1}^{n}\sum_{j=1}^{m}(x_{ij} - \bar{x})^2\right]^{1/2}} \begin{cases} < b_\alpha^*, & \text{choose } \mathscr{D}_0, \\ > b_\alpha^*, & \text{choose } \mathscr{D}_M, \end{cases}$$

where M now denotes the i maximizing $|\bar{x}_i - \bar{x}|$. Tables of b_α^* for $\alpha = 0.05$, 0.01 are given by Quesenberry and David (1961).

6. For unequal sample sizes m_i $(i = 1, 2, \ldots, n)$ Paulson's procedure may be adapted by replacing

$$\frac{m(\bar{x}_i - \bar{x})}{\left[\displaystyle\sum_{i=1}^{n}\sum_{j=1}^{m}(x_{ij} - \bar{x})^2\right]^{1/2}} \quad \text{by} \quad y_i = \frac{m_i(\bar{x}_i - \bar{x})}{\left[\displaystyle\sum_{i=1}^{n}\sum_{j=1}^{m_i}(x_{ij} - \bar{x})^2\right]^{1/2}}.$$

Pfanzagl (1959) shows that the following procedure is locally (small $\Delta > 0$)

optimum. With b_α' as the upper α significance point of $\max\limits_{i=1,2,\ldots,n} Y_i$, choose $\mathcal{D}_0$ if $\max y_i < b_\alpha'$ and choose D_M if $y_M = \max y_i$. In general, the b_α' have not been tabulated, but approximate values may be found with the help of the first Bonferroni bound (cf. Doornbos and Prins, 1958, but note that these authors use $m_i^{-\frac{1}{2}} y_i$ in place of y_i).

7. Corresponding procedures have been derived for the detection of slippage of a variance among n $N(\mu_i, \sigma_i^2)$ populations. Here $\mathcal{D}_0$ is the decision that

$$\sigma_1^2 = \sigma_2^2 = \cdots = \sigma_n^2$$

and $\mathcal{D}_i$ that

$$\sigma_1^2 = \sigma_2^2 = \cdots = \sigma_{i-1}^2 = \sigma_{i+1}^2 = \cdots = \sigma_n^2 = \sigma^2 \text{ (unspecified)}, \sigma_i^2 = \lambda'^2 \sigma^2.$$

Under the same restrictions as needed in Comment 2 and paralleling Paulson's proof, Truax (1953) shows that for $\lambda'^2 > 1$ (and samples of equal size m) the optimal procedure is based on Cochran's (1941) statistic

$$\frac{s_{\max}^2}{\sum\limits_{i=1}^{n} {}_i s^2}, \quad \text{where} \quad {}_i s^2 = \frac{\sum\limits_{j=1}^{m} (x_{ij} - \bar{x}_i)^2}{m-1}.$$

Pfanzagl (1959) generalizes this result also to the case of unequal m_i and obtains the locally (λ'^2 near 1) optimal decision criterion

$$\max (m_i - 1) \left(\frac{{}_i s^2}{s_\nu^2} - 1 \right),$$

where s_ν^2 is the usual error mean square. For $\lambda'^2 < 1$ (and equal m_i) the relevant statistic is $s_{\min}^2 / \Sigma_i s^2$. Doornbos (1956) obtains some approximate tables of the lower 5% points with the help of Bonferroni inequalities. A similar approach for handling slippage of a scale parameter in gamma distributions is used by Doornbos and Prins (1956).

Closely related is the problem of testing the significance of the n main effects in a 2^n factorial experiment in circumstances where it is known a priori that no more than a few of the effects are likely to be significant (see Birnbaum, 1959).

8. A general Bayesian formulation of slippage problems is given by Karlin and Truax (1960), who re-derive Paulson's results as a special case of their approach. They also consider the corresponding slippage problem when, in addition to the n treatment groups of size m, there is a control group of m independent $N(\mu_0, \sigma^2)$ variates. The alternative to the null hypothesis of equality of all $n+1$ means is now that one of the n treatment means has

slipped away from μ_0, and the test statistic becomes

$$\max_{i=1,2,\ldots,n} \frac{\bar{x}_i - \bar{\bar{x}}}{\sum_{i=0}^{n}\sum_{j=1}^{m}(x_{ij} - \bar{\bar{x}})^2} \quad \text{with} \quad \bar{\bar{x}} = \frac{\sum_{i=0}^{n}\sum_{j=1}^{m}x_{ij}}{(n + 1)m}.$$

9. Karlin and Truax also give a brief discussion of multivariate slippage problems. The multivariate outlier problem is considered by Kudô (1957) and Siotani (1959); see (5.3.11).

8.4. THE PERFORMANCE OF TESTS FOR OUTLIERS

Two main models have been advanced (Grubbs, 1950; Dixon, 1950) to provide "outlier alternatives" to the null hypothesis H_0 that the sample at hand has been drawn randomly from some normal parent. Both models assume that $X_1, X_2, \ldots, X_n$ are independent variates and that $n - k$ of the X's (we do not know which) have a common $N(\mu, \sigma^2)$ distribution. Under model A the remaining k variates have means $\mu + \lambda_i\sigma$ $(i = 1, 2, \ldots, k)$ and common variance σ^2; under model B these k variates have common mean μ and variances $\lambda_i'^2\sigma^2$ $(i = 1, 2, \ldots, k)$. Thus model A is concerned purely with shifts in location for some of the variates, model B purely with changes in dispersion. Evidently the true situation may not be so pure, but we have to start somewhere; in fact, further specialization of the models is usually necessary. For μ and σ categories (a)–(d) of Section 8.2 still apply. The λ_i and λ_i' are unknown parameters except that for one-sided tests (with, say, right-handed alternatives) we will take $\lambda_i > 0$ and $\lambda_i' > 1$, respectively.

Clearly slippage tests correspond to the special case $k = 1$, for which we write $\lambda_1 = \lambda$ and $\lambda_1' = \lambda'$. Then $\lambda\sigma$ is Δ of Section 8.3. In addition to the various optimality properties there stated for the statistics D_1 and D_2, it is of interest to note Ferguson's (1961b) result that D_2 continues to be optimal under model B for $\lambda' > 1$. Thus, out of all unbiased tests of size α invariant under changes of location and scale, the test based on D_2 maximizes the probability of rejecting the outlier (a) when the outlier differs in mean only, (b) when the outlier has increased variance, and (by Comment 1 of Section 8.3, which applies also to D_2), (c) when the outlier's shift in mean is accompanied by a decrease in variance. Together these make a strong case for the use of D_2 whenever this statistic is appropriate but still provide no safe guidance for $k > 1$. In fact, in the case of D_1 we have seen (Ex. 5.3.3) that only one of the ratios $(x_i - \bar{x})/s$ can exceed $D_{1,\alpha}$ for $n \leq 14$, $\alpha = 0.05$ or $n \leq 19$, $\alpha = 0.01$. This means that, if in this (n, α) region two observations happen to be equally outlying, *neither* can be detected by the use of D_1. If the two outliers come from a common $N(\mu + \lambda\sigma, \sigma^2)$ population, then, clearly, the probability P of detecting either outlier $\to 0$ as $\lambda \to \infty$, and P may be expected to be unsatisfactorily low for finite λ. This phenomenon was first pointed out by Pearson and Chandra Sekar (1936) and later termed the "masking effect" by Murphy (1951). Nor is the masking effect confined to D_1; when σ has to be estimated from the sample at hand, the effect is inevitable. To a smaller extent it is bound to creep also into C_4, especially for small value of ν (cf. Ex. 8.4.1).

Ferguson (1961b) also shows that the statistics D_4 and D_5 (more commonly denoted by g_1 and b_2) are locally most powerful among invariant tests, not only for $k = 1$ but also for $k < \frac{1}{2}n$ in the case of D_4 and for $k \le 0.31n$ in the case of D_5. More precisely, the alternative hypothesis for D_4 is that k of the X's have means $\mu + \lambda_i \delta \sigma$ with δ, $\lambda_i > 0$ $(i = 1, 2, \ldots, k)$. The power function is then expressible as $\beta_\omega(\delta\lambda_1, \delta\lambda_2, \ldots, \delta\lambda_k)$ for any invariant test ω. Of all these tests, D_4 maximizes the rate of increase of the power function at $\delta = 0$. Evidently such local optimality is of little practical value, nor are percentage points of the statistics available for small n. Note also that the criterion of optimality is the power of the test, which is not the same as the probability of correctly detecting the outlier(s).

We are therefore led rather naturally to inquire: What are appropriate measures of performance of a test for outliers, taking into account the three aims (a)–(c) of such tests, set out in Section 8.1? To this end we consider in some detail the case of model A for one outlier to the right. It is clear that any reasonable measure can then depend only on n and λ, and must be independent of which observation is the outlier. For convenience we take this to be x_1 and let H_1 denote the corresponding alternative hypothesis. Although the measures considered below could be generalized to apply to a wider set of statistics, we shall suppose also that the test statistic is of the form

$$v = \max_{i=1,2,\ldots,n} v_i.$$

This includes, in particular, the statistics A_1, B_1, C_1, C_4, and D_1, where for, e.g., C_4 we have

$$v_i = \frac{x_i - \bar{x}}{[\Sigma(x_i - \bar{x})^2 + \nu s_\nu^2]^{1/2}}.$$

With v_α standing for the upper α significance point of V, reasonable measures are:

1. Power function: $P_1 = \Pr\{V > v_\alpha \mid H_1\}$.
2. Probability that X_1 is significantly large: $P_2 = \Pr\{V_1 > v_\alpha \mid H_1\}$.
3. Probability that X_1 is significantly large *and* the largest in the sample:

$$P_3 = \Pr\{V_1 > v_\alpha; X_1 > X_2, X_3, \ldots, X_n \mid H_1\}.$$

4. Probability that only X_1 is significant:

$$P_4 = \Pr\{V_1 > v_\alpha; V_2, V_3, \ldots, V_n < v_\alpha \mid H_1\}.$$

5. Probability that X_1 is significantly large, given that it is the largest in the sample:

$$P_5 = \Pr\{V_1 > v_\alpha \mid X_1 > X_2, X_3, \ldots, X_n; H_1\}.$$

P_1 measures the probability of significance for any reason whatever and is thus especially suitable for sounding a general alarm—aim (b) of Section 8.1. P_2, P_3, and P_4 focus with increasing severity on the correct detection of the outlier—aim (c); only P_4 specifically excludes the possibility that good observations might be significant in addition to X_1. We see that

$$P_1 \geq P_2 \geq P_3 \geq P_4. \tag{8.4.1}$$

P_5 (Dixon, 1950) is related to P_3 by

$$P_5 = \frac{P_3}{\Pr\{X_1 > X_2, X_3, \ldots, X_n \mid H_1\}},$$

where the probability in the denominator has been tabulated by Teichroew (1955) for $n \leq 10$.

Of the five measures, P_2 is generally the easiest to evaluate numerically, P_3 and P_4 being quite difficult. However, under certain conditions only one of $V_1, V_2, \ldots, V_n$ can exceed v_α (as has been noted, this is so in the case of D_1 for $n \leq 14$, $\alpha = 0.05$, and $n \leq 19$, $\alpha = 0.01$). In this situation $V_1 > v_\alpha$ implies both $X_1 > X_2, X_3, \ldots, X_n$ and $V_2, V_3, \ldots, V_n < v_\alpha$, so that $P_2 = P_3 = P_4$.

The power function P_1 may be usefully bounded with the help of the first two Bonferroni inequalities (5.3.3). Identifying A_i with the event $V_i > v_\alpha$ ($i = 1, 2, \ldots, n$), we have $P_1 = \Pr\{\bigcup A_i\}$ and hence

$$P_1 \leq \Pr\{V_1 > v_\alpha \mid H_1\} + (n-1)\Pr\{V_j > v_\alpha \mid H_1\} \quad j = 2, 3, \ldots, n,$$

$$= P_2 + (n-1)\beta \tag{8.4.2}$$

where

$$\beta = \Pr\{V_j > v_\alpha \mid H_1\},$$

and

$$P_1 \geq P_2 + (n-1)\beta - (n-1)\Pr\{V_1 > v_\alpha, V_2 > v_\alpha \mid H_1\}$$

$$- \binom{n-1}{2}\Pr\{V_2 > v_\alpha, V_3 > v_\alpha \mid H_1\}. \tag{8.4.3}$$

Since the bivariate probabilities in (8.4.3) are usually difficult to obtain, it is worth noting that from (8.4.1) and (8.4.2)

$$P_2 \leq P_1 \leq P_2 + (n-1)\beta. \tag{8.4.4}$$

For given n the probability β is clearly a decreasing function of λ (> 0). Its upper bound β_0, corresponding to $\lambda = 0$, satisfies (8.4.3) with $H_1 = H_0$, so that

$$\alpha > n\beta_0 - \binom{n}{2} \text{Pr} \{V_1 > v_\alpha, V_2 > v_\alpha \mid H_0\}.$$

Now for each of the statistics A_1, B_1, C_1, C_4, and D_1, it is known that[3]

$$\text{Pr} \{V_1 > v_\alpha, V_2 > v_\alpha \mid H_0\} \leq \beta_0{}^2.$$

Hence it may easily be shown that $\beta_0 < \beta'$, where

$$\beta' = \frac{1}{n-1} \left[1 - \left(1 - 2\alpha \frac{n-1}{n} \right)^{\frac{1}{2}} \right], \qquad (8.4.5)$$

and that

$$\beta' < \frac{\alpha}{n-1} \quad \text{provided} \quad \alpha < \frac{2}{n}. \qquad (8.4.6)$$

This last condition is frequently satisfied, and (8.4.4) may then be replaced by the very simple but weaker inequalities

$$P_2 \leq P_1 \leq P_2 + \alpha. \qquad (8.4.7)$$

The power function P_1 of B_1 has been tabulated by David and A. S. Paulson (1965), who also give comparative figures for the power of the χ^2 test B_5. As expected, B_1 is always superior. For given λ, the balance in favor of B_1 increases with n. In the same paper, charts (Fig. 8.4) compare the performance, as measured by P_2, of B_1, C_1, C_4, and D_1. For C_1 and C_4, P_2 was computed for $\nu = 5, 10, 20$, but some curves were omitted so as not to crowd the graphs. The figure shows *inter alia* how, for given n and α, P_2 increases with increased knowledge of σ, and also just how much is added to the value of P_2 by the use of C_4 rather than C_1 in the present case of a single true outlier. Of course, the latter gain is highest when the internal information on σ^2 is large compared to the external information, that is, when $n - 1$ is large compared to ν. However, there are indications that internal degrees of freedom are less valuable than external ones. Thus for $n = 6$ the solid curve $\nu = 5$ lies well above the dotted curve $\nu = 0$, although in both cases there is a total of 5 DF.

Large-scale sampling experiments were carried out by Ferguson (1961b) to compare the power function P_1 for various competing statistics in case (d) of Section 8.2, namely, for the one-sided statistics D_1, D_4, r_{10}, and the two-sided D_2, D_5, $r_{10}^{(2)}$. Ferguson proceeded by chopping up the same 25,000 random normal deviates into samples of $n = 5, 10, 15, 20$, and 25 (i.e., 5000 samples of 5, 2500 samples of 10, etc.), adding successively a constant $\lambda = 0, 1, 2, \ldots$,

---

[3] For C_1 the result may break down for small values of ν (see Hume, 1965).

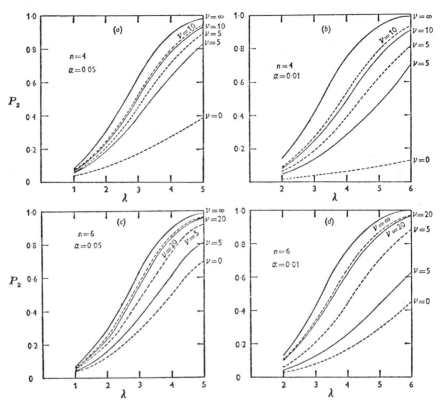

Fig. 8.4. Probability P_2 that the outlier X_1 is detected when B_1, C_1, C_4, and D_1 are used at level α and H_1 holds. B_1: $\nu = \infty$; D_1: $\nu = 0$; C_1: ———$\nu = 5, 10, 20$; C_4: - - -$\nu = 5, 10, 20$, (with permission of the Managing Editor of *Biometrika*).

15 to a fixed member of each sample, and noting the proportion of cases, for each n and λ, in which the statistic calculated from the sample exceeded its upper α significance point ($\alpha = 0.1, 0.05, 0.01$). A complicating factor was that, since no percentage points of D_4 and D_5 were available in the above range of n (except for D_4 when $n = 25$), Ferguson had to estimate the percentage points themselves by experimental sampling. This was also done for *all* two-sided statistics. Subject to these limitations, Ferguson's results may be outlined as follows:

(*a*) For small n there is no difference to approximately two decimal places between the power functions of the three one-sided tests or between the power functions of the three two-sided tests.

(*b*) The power functions pull increasingly apart as n increases. For $n = 25$, D_1 is best among the one-sided statistics, whereas D_2 and D_5 are

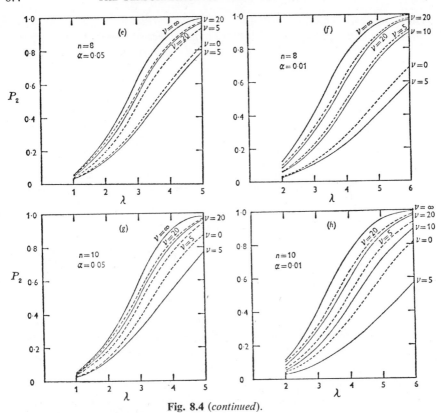

Fig. 8.4 (*continued*).

about equally good two-sided statistics. The numerical results for $\alpha = 0.05$ are set out in Table 8.4.

(*c*) The figures in parentheses in Table 8.4 provide a comparison of D_5 and D_2 when there are two outliers from $N(\mu + \lambda\sigma, \sigma^2)$. As foreshadowed above, D_2 performs unsatisfactorily and D_5 is distinctly better. Note that for D_5 the power function is higher than for a single outlier, whereas for D_2 it is lower. However, this does not mean that D_5 is exempt from the masking effect, which is more evident for smaller n. In fact, for $n = 5$ and 10, $P_1 \rightarrow 0$ as $\lambda \rightarrow \infty$ for both D_2 and D_5 even when α is as large as 0.10.

The most extensive investigation of the performance of outlier tests is still Dixon's (1950) sampling study. Dixon used essentially the same methods later employed by Ferguson but measured performance by the proportion of samples in which the outlying population provided the largest value $x_{(n)}$ in the sample *and* the statistic used led to significance. Any sample for which $x_{(n)}$ in fact came from a $N(\mu, \sigma^2)$ population was omitted from the calculations. Thus his measure of performance is an estimate of P_5 above. Dixon

Table 8.4.   Power of six tests ($\alpha = 0.05$) *for outliers based on* 1000 *normal samples of* $n = 25$, *when 1 or 2 (entries in parentheses) observations are from a* $N(\mu + \lambda\sigma, \sigma^2)$ *population and the others are* $N(\mu, \sigma^2)$

λ	1	2	3	4	5	6	7	8
D_4	.06	.17	.38	.69	.89	.98	1.00	
D_1	.06	.16	.42	.75	.94	.99	1.00	
r_{10}	.06	.13	.36	.67	.90	.98	.99	1.00
	(.06)	(.14)	(.37)	(.71)	(.95)	(1.00)		
D_5	.05	.12	.36	.68	.89	.98	1.00	
	(.06)	(.13)	(.34)	(.58)	(.81)	(.96)	(.99)	(1.00)
D_2	.05	.13	.35	.69	.91	.99	1.00	
$r_{10}^{(2)}$	.05	.11	.29	.60	.84	.96	.99	1.00

(From Ferguson, 1961b.)

deals with the following statistics (for all of which upper percentage points were available to him): B_1, B_2, B_4, B_5, C_1, C_3 (with $v = 9$), D_1, D_6, and his r statistics. Sample sizes are restricted to $n = 5$ or 15; α is usually 5%, but situations with two outliers as well as one are considered; and results are obtained for model B as well as model A. Each point on his graphs of performance as a function of λ is based on 66–200 determinations. We will not attempt to summarize Dixon's results here (see Dixon, 1962). They provide most useful indications but it is worth remembering that for a typical point the ordinate, being a binomial proportion based on (say) 100 trials, is subject to a standard error of 0.05 if $P_5 = 0.5$ and of 0.03 if $P_5 = 0.1$. However, intercomparisons of the performance of different tests will be more reliable due to the matching resulting from the multiple use of each sample of n.

Further Remarks on Dealing with More than One Outlier

There is an interesting theoretical result due to Murphy (1951) generalizing the optimality property of the statistic D_1 in the presence of a single outlier: if k observations are known to have come from a $N(\mu + \lambda\sigma, \sigma^2)$ population ($\lambda > 0$), then the optimal invariant test is to reject for large values of

$$D^{(k)} = \frac{1}{s}\left(x_{(n)} + x_{(n-1)} + \cdots + x_{(n-k+1)} - k\bar{x}\right).$$

No tables exist for $k > 1$; and, in any case, although the assumption of at most one outlier required for the optimality of D_1 may be a reasonable approximation to the truth, one will seldom be comfortable in assuming that there are exactly two outliers or none at all, as required for the optimality of $D^{(2)}$.

Quite generally, when it is suspected that k may well be greater than 1, we may ideally wish to proceed sequentially as follows:

Apply a certain test statistic to the sample of n. If significance is obtained, eliminate the most extreme observation and apply the same test statistic to the reduced sample of $n - 1$, adjusting the significance point to the new sample size. If significance holds again, repeat the procedure until the test statistic ceases to be significantly large.

Such a procedure has obvious appeal, since the same set of tables may be referred to repeatedly. However, it is unsuitable for statistics severely prone to the masking effect. For some theoretical studies see McMillan (1968).

8.5. THE EFFECT OF OUTLIER REJECTION ON THE ESTIMATION OF PARAMETERS

In discussing tests for outliers we have so far emphasized their role in sounding a general alarm and in pin-pointing extreme observations [aims (b) and (c) of Section 8.1]. Now we turn briefly to the difficult problem of assessing the effect of such tests on subsequent estimation of parameters. It is clear that there must be some effect whether observations have actually been rejected or not, although a two-sided test with two equal critical regions will introduce no bias in the estimation of the mean if the underlying population is symmetric. Systematic investigation of the subject appears to have begun with Dixon (1953), who based his work mainly on the sampling experiments just described. We shall follow the approach put forward by Anscombe (1960) and extended by Guttman and Smith (1966, 1969).

For n independent normal variates with common known variance σ^2 and hopefully common mean μ, let M be the serial number of the observation having the largest residual in magnitude. A simple rule is:

Reject x_M if $|x_M - \bar{x}| > c\sigma$, for some chosen constant c; otherwise reject no observation. Estimate μ by $\hat{\mu}$, the mean of the retained observations. Thus

$$\begin{aligned}\hat{\mu} &= \bar{x} & \text{if } |x_M - \bar{x}| \leq c\sigma, \\ &= \bar{x} - \frac{x_M - \bar{x}}{n - 1} & \text{if } |x_M - \bar{x}| > c\sigma.\end{aligned} \qquad (8.5.1)$$

The rule involves, of course, the statistic B_2 of Section 8.2, but c need not be one of the usual upper percentage points. In fact, one of Anscombe's main points is that the whole concept of the significance level of the associated test is not appropriate in the present estimation context. Rather, c should be chosen so as to make $\hat{\mu}$ a good estimator in some sense. Using the analogy of insurance, Anscombe suggests that we may be prepared to pay a *premium* in the form of a slightly raised mean square error when all is well in order to receive *protection* when something is very wrong. More precisely (and these definitions are clearly of wider applicability), we have

$$\text{premium} = \frac{\text{var } \hat{\mu} - \text{var } \bar{X}}{\text{var } \bar{X}}$$

$$\text{protection} = \frac{\mathscr{E}(\bar{X} - \mu)^2 - \mathscr{E}(\hat{\mu} - \mu)^2}{\mathscr{E}(\bar{X} - \mu)^2}.$$

Thus for a homogeneous sample

$$\text{protection} = -\text{premium} \leq 0,$$

but with increasing heterogeneity the protection will become positive.

Suppose now that a single $N(\mu + \lambda\sigma, \sigma^2)$ outlier is present. Write $z_i = x_i - \bar{x}$ ($i = 1, 2, \ldots, n$), and corresponding to (8.5.1) introduce the variate T defined by

$$\begin{aligned} T &= 0 & \text{if } |Z_M| \leq c\sigma, \\ &= \frac{-Z_M}{n-1} & \text{if } |Z_M| > c\sigma. \end{aligned} \right\} \tag{8.5.1'}$$

Then

$$\hat{\mu} = \bar{X} + T, \quad \text{where} \quad \bar{X} \frown N\left(\mu + \frac{\lambda\sigma}{n}, \frac{\sigma^2}{n}\right).$$

Also, because of the independence of $\bar{X}$ and T (a function of the Z_i),

$$\mathscr{E}(\hat{\mu} - \mu)^2 = \mathscr{E}\left(\bar{X} - \mu - \frac{\lambda\sigma}{n}\right)^2 + \mathscr{E}\left(T + \frac{\lambda\sigma}{n}\right)^2$$

$$= \frac{\sigma^2}{n} + \mathscr{E}\left(T + \frac{\lambda\sigma}{n}\right)^2.$$

Thus we have in this case

$$\text{premium} = n\mathscr{E}\left(\frac{T}{\sigma}\right)^2 \tag{8.5.2}$$

$$\text{protection} = \frac{-n^2\mathscr{E}[T(T + 2\lambda\sigma/n)]}{\sigma^2(n + \lambda^2)}. \tag{8.5.3}$$

Unfortunately, these quantities are difficult to calculate, and Monte Carlo methods have been used for $n > 3$ after analytic simplification of the integrals involved (Guttman and Smith, 1966). For $n = 3$ we have, taking $\sigma = 1$ for simplicity,

$$\text{premium} = 3\mathscr{E}T^2$$

$$= 3\int_{R_1} \frac{z_3^2}{4} f_{1,3}(z_1, z_3)\, dz_1\, dz_3 + 3\int_{R_2} \frac{z_1^2}{4} f_{1,3}(z_1, z_3)\, dz_3\, dz_1,$$

where

$$R_1 = \{z_1, z_3 \mid c < z_3 < \infty, \ -z_3 < z_1 < -\tfrac{1}{2}z_3\}, \\ R_2 = \{z_1, z_3 \mid -\infty < z_1 < -c, \ -\tfrac{1}{2}z_1 < z_3 < -z_1\},$$

(8.5.4)

and $f_{1,3}(z_1, z_3)$, the joint pdf of $Z_{(1)}$ and $Z_{(3)}$, is

$$f_{1,3}(z_1, z_3) = \frac{3\sqrt{3}}{\pi} \exp\left[-(z_1^2 + z_1 z_3 + z_3^2)\right]$$

over the region

$$\{z_1, z_3 \mid -\tfrac{1}{2}z_1 < z_3 < -2z_1\}.$$

(8.5.5)

By symmetry, the two integrals are equal. Setting

$$z_1 = -w^{1/2}v^{1/2}, \quad z_3 = w^{1/2}v^{-1/2},$$

we have

$$\text{premium} = \frac{9\sqrt{3}}{4\pi} \int_1^2 \int_{c^2/v}^{\infty} w \exp\left[-w\left(v - 1 + \frac{1}{v}\right)\right] dw \, dv$$

$$= \frac{9\sqrt{3}}{4\pi} \int_1^2 \frac{c^2(v^2 - v + 1) + v^2}{(v^2 - v + 1)^2} \exp\left[-\frac{c^2}{v}(v^2 - v + 1)\right] dv$$

$$= \frac{6}{\pi} \int_0^{1/\sqrt{3}} \frac{c^2[\tfrac{3}{4}(t^2 + 1)] + 1}{(1 + t^2)^2} \exp\left[-\tfrac{3}{4}c^2(1 + t^2)\right] dt,$$

where

$$t = \frac{2}{\sqrt{3}}\left(\frac{1}{2} - \frac{1}{v}\right).$$

By numerical integration c may now be determined so as to give the premium some acceptable value, such as 5% or 1%. The corresponding protection may then be found as a function of λ. Figure 8.5 (from Guttman and Smith, 1969) shows the protection corresponding to a 5% premium not only for the rule (8.5.1) but also for the following two rules, whose aim is to tame rather than eliminate the wild observation. We state these rules for general n.

(a) *Winsorization*: If $|z_M| > c\sigma$, set x_M equal to the observation closest to it (i.e., $x_{(2)}$ for $n = 3$) before averaging to obtain $\hat{\mu}_W$. Thus

$$\hat{\mu}_W = \bar{x} \qquad\qquad\qquad \text{if } |z_M| \leq c\sigma, \\ = \bar{x} - \frac{x_{(n)} - x_{(n-1)}}{n} \quad \text{if } |z_M| > c\sigma \ \text{ and } \ M = n, \\ = \bar{x} + \frac{x_{(2)} - x_{(1)}}{n} \quad\ \text{if } |z_M| > c\sigma \ \text{ and } \ M = 1.$$

(8.5.6)

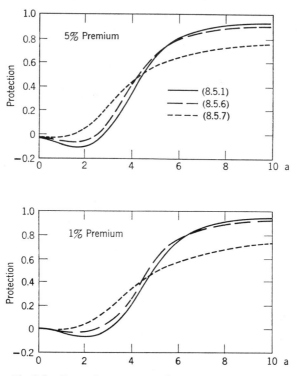

Fig. 8.5. Protections corresponding to premiums of 5% and 1% when a spurious observation from $N(\mu + a\sigma, \sigma^2)$ is present in a sample of size 6 and σ^2 is known (from Guttman and Smith, 1969, with permission of the authors and the Editor of *Technometrics*).

(*b*) *Modified Winsorization*: If $|z_M| > c\sigma$, set x_M equal to the closer of the two boundary values $\bar{x} \pm c\sigma$ before averaging to obtain $\mu_{W'}$. Thus

$$
\left.
\begin{aligned}
\hat{\mu}_{W'} &= \bar{x} && \text{if } |z_M| \leq c\sigma, \\[2mm]
&= \bar{x} - \frac{z_{(n)} - c\sigma}{n} && \text{if } |z_M| > c\sigma \text{ and } M = n, \\[2mm]
&= \bar{x} - \frac{z_{(1)} + c\sigma}{n} && \text{if } |z_M| > c\sigma \text{ and } M = 1.
\end{aligned}
\right\} \quad (8.5.7)
$$

Note that, for a fixed premium, c will change with the rule used.

Figure 8.5 shows that rule (8.5.7) is best for small λ, but is overtaken by rule (8.5.1) when $\lambda \doteq 4$. Of course, for very small λ simple averaging (with zero protection) is best of all. The Winsorization rule is intermediate but closer to Anscombe's than to the modified rule. Rather similar results hold for a premium of 1 %, and indeed for larger values of n, although differences in protection become less marked as n increases. Guttman and Smith (1966, 1969) also deal with the model B situation of a single outlying $N(\mu, \lambda'^2\sigma^2)$ variate. The results are qualitatively not very different, but rule (8.5.6) now is slightly superior to (8.5.1) for large λ'.

The case of several outliers differing in location from the remaining observations has also been examined by Anscombe and Barron (1966), who suggest a rule somewhat similar to (8.5.7). However, they set x_i equal to the closer of $\hat\mu \pm c\sigma$ for *each* x_i for which the residual $|x_i - \hat\mu|$ (not $|x_i - \bar x|$) exceeds $c\sigma$. Implementation of this rule may require several steps. Although the authors confine detailed analysis to the case $n = 3$ and a single outlier, they come up with the following appealing general recommendation for a two-stage procedure, which is in line with other findings we have described:

1. Apply the appropriate test for outliers at a very stringent level of significance, so stringent that good observations will very seldom be rejected. The purpose of this is to get rid only of wild observations very far removed from the main stream.

2. Apply the same outlier test again to the reduced data but now at quite a moderate level of significance. This time, unlike the preceding stage, do not reject the outlying observations but rather give them reduced weight in the subsequent estimation of parameters.

Except for this recommendation, which obviously leaves a number of questions up in the air, all the rules so far discussed assume σ known. For $n = 3$ Guttman and Smith (1966) have obtained some results for unknown σ and find modified Winsorization far better than the other two rules. (In the rules as stated above, σ is now to be replaced by s.)

Although we have of necessity been much concerned with the case $n = 3$, we have not alluded to the practice of estimating μ by the average of the two closest observations. There is nothing to be said in favor of this (Seth, 1950; Lieblein, 1952, 1962; Willke, 1966).

Bayesian estimators and the associated risks, particularly in the model B situation, are considered in some detail by Gebhardt (1964, 1966).

EXERCISES

8.2.1. Suggest tests for outliers in the normal case when σ is unknown but μ is known. Indicate what tables are available for both one-sided and two-sided alternatives.

(Cf. Chew, 1964)

8.2.2. Let $X_1, X_2, \ldots, X_n$ be independent variates with pdf

$$p(x) = \frac{1}{\theta_2} e^{-(x-\theta_1)/\theta_2} \qquad x \geq \theta_1, \theta_2 > 0,$$

and let

$$U_{(i)} = \frac{X_{(i)} - X_{(1)}}{\sum\limits_{i=1}^{n} (X_{(i)} - X_{(1)})} = \frac{X_{(i)} - X_{(1)}}{nY} \qquad i = 2, 3, \ldots, n.$$

Note that $X_{(1)}$, Y are complete sufficient statistics for θ_1, θ_2 and that the distribution of $U_{(i)}$ is free of parameters. From this it follows (D. Basu, 1955) that $U_{(i)}$ is statistically independent of $X_{(1)}$ and Y.

Hence show that, for $u \geq 0$ and $k \leq 1/u$,

$$\Pr\{U_{(i)} \leq u\} = 1 - \sum_{r=n-i+1}^{k} (-1)^{r+n-i+1} \binom{n}{r}\binom{r-1}{n-i}\left(1 - \frac{r}{n}\right)(1 - ru)^{n-2}.$$

(Laurent, 1963; Kabe, 1968)

8.2.3. Discuss tests for outliers (on the right) in the exponential case of Ex. 8.2.2. Distinguish cases in which (a) both θ_1 and θ_2 are known; (b) only θ_1 is known; (c) only θ_2 is known; (d) both θ_1 and θ_2 are unknown.

For (d) show that an appropriate statistic is $U_{(n)}$ of Ex. 8.2.2 and that

$$\Pr\{U_{(n)} \leq u\} = 1 - \sum_{j=1}^{k} (-1)^{j+1}\binom{n-1}{j}(1 - ju)^{n-2} \qquad k \leq 1/u.$$

(Laurent, 1963; A. P. Basu, 1965)

8.3.1. In Mosteller's problem with n samples of size m_i ($i = 1, 2, \ldots, n$) exactly y observations in one sample are greater than all observations in all other samples. Show that, if the samples are from a common continuous population, then

$$\Pr\{Y > y\} = \frac{\Sigma m_i^{(y)}}{(\Sigma m_i)^{(y)}}.$$

(Mosteller and Tukey, 1950)

8.3.2. Let T_i ($i = 1, 2, \ldots, m$) denote the Kruskal-Wallis rank sum for the ith of n samples of m taken from continuous populations, and let

$$V_i = \frac{T_i - \tfrac{1}{2}m(nm + 1)}{[nm^2(nm + 1)/12]^{1/2}}.$$

Show that asymptotically (as $m \to \infty$) $\max\limits_{i=1,2,\ldots,n} V_i$ is distributed as $\max\limits_{i=1,2,\ldots,n}$ $(X_i - \bar{X})$, where the X_i are independent unit normal variates.

(Cf. Odeh, 1967)

8.3.3. With n groups of m normal $N(\mu_i, \sigma^2)$ variates X_{ij} ($i = 1, 2, \ldots, n; j = 1, 2, \ldots, m$), as considered by Paulson, let $\mathscr{D}_{00}$ be the decision that all n means μ_i are equal and let $\mathscr{D}_{ii'}$ denote the decision that $\mathscr{D}_{00}$ is incorrect and $\mu_i = \mu_{\min}, \mu_{i'} = \mu_{\max}$ (minimum taken over $i = 1, 2, \ldots, n$, etc.). The pair $(\mu_i, \mu_{i'})$ may be said to have slipped by Δ (> 0) if

$$\mu_1 = \mu_2 = \cdots = \mu_{i-1} = \mu_{i+1} = \cdots = \mu_{i'-1} = \mu_{i'+1} = \cdots = \mu_k = \mu \qquad (\mu \text{ unspecified})$$

and

$$\mu_i = \mu - \Delta, \qquad \mu_{i'} = \mu + \Delta.$$

Under the restrictions that (a) when all μ_i are equal, $\mathscr{D}_{00}$ should be selected with probability $1 - \alpha$, (b) the decision procedure should be invariant under location and scale transformations of the variates, and (c) the procedure should be symmetric in the sense that the probability of making the correct decision when $(\mu_i, \mu_{i'})$ has slipped by Δ must be the same for all i, i', show that the optimal procedure when one of the pairs has slipped is as follows:

If
$$g = \frac{m(\bar{x}_{\max} - \bar{x}_{\min})}{\left[\sum_{i=1}^{n}\sum_{j=1}^{m}(x_{ij} - \bar{x})^2\right]^{1/2}} < g_\alpha, \qquad \text{choose } \mathscr{D}_{00};$$

if $g > g_\alpha$ and if $\bar{x}_{\min} = \bar{x}_i$, $\bar{x}_{\max} = \bar{x}_{i'}$,   choose $\mathscr{D}_{ii'}$.

(Ramachandran and Khatri, 1957)

(Note: Ramachandran and Khatri mistakenly regard g as the studentized range. The g_α have not been tabulated but could be obtained approximately with the help of the moments of G. As in Section 5.2, the kth raw moment of G is the ratio of the kth raw moment of the numerator to the kth raw moment of the denominator of G.)

8.4.1. In the situation of case (c) of Section 8.3 let
$$y_i = \frac{x_i - \bar{x}}{[\Sigma(x_i - \bar{x})^2 + \nu s_\nu^2]^{1/2}} \qquad i = 1,2,\ldots,n.$$

Prove that the joint pdf of Y_1, Y_2 is given by
$$f(y_1, y_2) = \left(\frac{n}{n-2}\right)^{1/2} \frac{n+\nu-3}{2\pi}\left(1 - \frac{n-1}{n-2}y_1^2 - \frac{2y_1 y_2}{n-2} - \frac{n-1}{n-2}y_2^2\right)^{1/2(n+\nu-5)}$$

over the ellipse
$$\frac{n-1}{n-2}y_1^2 - \frac{2y_1 y_2}{n-2} + \frac{n-1}{n-2}y_2^2 \le 1.$$

Hence show that, if $c \ge [(n-2)/2n]^{1/2}$, only one of the y_i can exceed c.

(Quesenberry and David, 1961)

8.4.2. Prove (8.4.5) and (8.4.6).

8.5.1. Show that corresponding to the premium (8.5.4)
$$\text{protection} = \frac{9}{3+\lambda^2}\int_{R_1} \frac{z_3}{2}\left(\frac{2\lambda}{3} - \frac{z_3}{2}\right) f_{1,3}(z_1, z_3; \lambda)\, dz_1\, dz_3$$
$$+ \frac{9}{3+\lambda^2}\int_{R_2} \frac{z_1}{2}\left(\frac{2\lambda}{3} - \frac{z_1}{2}\right) f_{1,3}(z_1, z_3; \lambda)\, dz_3\, dz_1,$$

where
$$f_{1,3}(z_1, z_3; \lambda) = \frac{\sqrt{3}}{\pi}\exp\left\{-\left[\left(z_1 + \frac{\lambda}{3}\right)^2 + \left(z_1 + \frac{\lambda}{3}\right)\left(z_3 + \frac{\lambda}{3}\right) + \left(z_3 + \frac{\lambda}{3}\right)^2\right]\right\}$$
$$+ \frac{\sqrt{3}}{\pi}\exp\left\{-\left[\left(z_1 - \frac{2\lambda}{3}\right)^2 + \left(z_1 - \frac{2\lambda}{3}\right)\left(z_3 + \frac{\lambda}{3}\right) + \left(z_3 + \frac{\lambda}{3}\right)^2\right]\right\}$$
$$+ \frac{\sqrt{3}}{\pi}\exp\left\{-\left[\left(z_1 + \frac{\lambda}{3}\right)^2 + \left(z_1 + \frac{\lambda}{3}\right)\left(z_3 - \frac{2\lambda}{3}\right) + \left(z_3 - \frac{2\lambda}{3}\right)^2\right]\right\}$$

over the region (8.5.5).

(Guttman and Smith, 1969)

8.5.2. Show that for rule (8.5.6)

$$\text{premium} = n\mathscr{E}(W/\sigma)^2,$$

where

$$W = \hat{\mu}_W - \overline{X}.$$

Hence show that, for $n = 3$, the expression for the premium analogous to (8.5.4) is

$$\text{premium} = \frac{2}{3}\int_{R_1} (2z_3 + z_1)^2 f_{1,3}(z_1, z_3)\, dz_1\, dz_3$$

(Guttman and Smith, 1966)

CHAPTER 9

Asymptotic Theory

9.1. INTRODUCTION

The asymptotic theory of order statistics is concerned with the distribution of $X_{r:n}$, suitably standardized, as $n \to \infty$. In the first instance it is usually assumed that $X_{r:n}$ is the rth order statistic in a random sample of n from some population with cdf $P(x)$. However, as we shall see, many kinds of dependence among $X_1, X_2, \ldots, X_n$ do not disturb the forms of the limiting distributions, a feature which adds greatly to the usefulness of the theory. If $r/n \to \lambda$ as $n \to \infty$, fundamentally different results are obtained according as $0 < \lambda < 1$ or $\lambda = 0, 1$. In the former case, $X_{r:n}$ is a sample quantile and (subject to certain regularity conditions) has an asymptotic normal distribution. The latter case includes the extremes $X_{1:n}$, $X_{n:n}$ and, more generally, the mth extremes $X_{m:n}$, $X_{n-m+1:n}$ with m fixed. These have nonnormal limiting distributions.

Asymptotic results have been alluded to frequently in previous chapters. In particular, asymptotic estimation by a limited number of quantiles has arisen in problems of "optimal spacing" (Section 7.6). In the next section we give the distribution theory underlying this application and, following Mosteller (1946), establish the asymptotic joint normality of quantiles.

The rest of this chapter deals with extreme-value theory (Sections 9.3–9.5) and the asymptotic distribution of linear functions of order statistics as well as their use in asymptotic estimation (Sections 9.6 and 9.7). Here, to a greater extent than elsewhere in this book, we confine ourselves to a summary of the very extensive literature available, giving proofs only of certain basic results.

The most striking result of extreme-value theory is by now classical: if $X_{n:n}$, suitably standardized, has a limiting distribution, then this must be one of the three types given in (9.3.1). Applications of these extreme-value distributions have been legion. For example, the humble premise that a chain is no stronger than its weakest link leads at once to $X_{1:n}$ as the strength of a chain (of n similar links), and thence to an impressive theory of breaking strength. Lieblein (1954b) traces this idea back to work by W. S. Chaplin in 1860. The most useful distribution describing breaking strength has been the so-called Weibull distribution having cdf

$$F(y) = 1 - \exp\left[-\left(\frac{y - \gamma}{\delta}\right)^{\alpha}\right] \quad \gamma < y < \infty, \delta > 0, \alpha > 0. \quad (9.1.1)$$

Here γ may be interpreted as a guaranteed minimum strength and δ as a scale factor. Clearly $X = -(Y - \gamma)/\delta$ has cdf $\Lambda_2(x)$; that is, the Weibull distribution is simply the second of the three types applied to the *smallest* rather than the largest variate.[1] In life tests or fatigue tests y may stand for *time* to failure. Again, the distribution of floods and of other extreme meteorological phenomena is often well represented in form by $\Lambda_3(x)$. The reader is referred to the book by Gumbel (1958), where other applications are indicated and numerous references are given.

Gumbel also discusses at length various methods for the estimation of parameters such as γ and δ in (9.1.1), the data being a set of n (not necessarily large) observed maxima or minima. Graphical methods are widely resorted to, especially probability plotting (cf. Section 7.8). Insofar as (9.1.1) is (for given α) a distribution depending on location and scale parameters, estimation by order statistics (Section 6.2) is, of course, also possible. In this connection we may mention the work of Maritz and Munro (1967), who deal with the estimation by order statistics of all three parameters of the "generalized extreme-value distribution"

$$F(y) = \exp\left\{-\left[\frac{1 - (y - \gamma)}{\delta\beta}\right]^{\beta}\right\},$$

where

$$-\infty < y < \gamma + \delta\beta \quad \beta > 0,$$
$$\gamma + \delta\beta < y < \infty \quad \beta < 0.$$

Taking $\gamma + \delta\beta = 0$, we obtain Λ_1 for $\delta\beta = 1$, $\beta = -\alpha$, and Λ_2 for $\delta\beta = 1$, $\beta = \alpha$. With $x = (y - \gamma)/\delta$ and $\beta \to \infty$, we have $\Lambda_3(x)$.

Other references supplementing Gumbel's book include his article on estimating the endurance limit in Sarhan and Greenberg (1962), as well as Gumbel (1961) on breaking strength and fatigue failure, Gumbel (1963) on forecasting droughts, Pike (1966) on cancer regarded as failure of the weakest cell, Barnett and Lewis (1967) on low-temperature probabilities, Epstein (1967) on bacterial extinction times, and Mann (1968) on estimation procedures.

---

[1] Note, however, that the three types are often labeled differently: $\Lambda_1(x)$, $\Lambda_2(x)$, $\Lambda_3(x)$ are referred to as the cdf's of the second, third, first asymptote (or even: Type), respectively.

The concluding section (9.7) of this chapter is concerned with the derivation of estimators which are asymptotically optimum for a particular parent distribution depending on location and scale parameters only. Closely related are methods yielding estimators which, although not necessarily optimal for the particular parent of most interest, have good properties over a chosen set of distributions. Such robust estimators have been discussed in Section 6.5, although primarily for small samples.

9.2. THE ASYMPTOTIC JOINT DISTRIBUTION OF QUANTILES

Let X_i ($i = 1, 2, \ldots, n$) be a random sample from a continuous distribution with pdf $p(x)$. We consider the asymptotic joint distribution of the k sample quantiles $X_{(n_j)}$ ($j = 1, 2, \ldots, k$), where $n_j = [n\lambda_j] + 1$ and $0 < \lambda_1 < \lambda_2 < \cdots < \lambda_k < 1$.

Theorem 9.2 (Mosteller, 1946). *If $p(x)$ is differentiable in the neighborhoods of the population quantiles ξ_{λ_j} and $p(\xi_{\lambda_j}) \neq 0$ ($j = 1, 2, \ldots, k$), then the joint distribution of $X_{(n_1)}, X_{(n_2)}, \ldots, X_{(n_k)}$ tends to a k-dimensional normal distribution with means $\xi_{\lambda_1}, \xi_{\lambda_2}, \ldots, \xi_{\lambda_k}$ and covariances*

$$\mathrm{cov}\,(X_{(n_j)}, X_{(n_{j'})}) = \frac{\lambda_j(1 - \lambda_{j'})}{np(\xi_{\lambda_j})p(\xi_{\lambda_{j'}})} \qquad j \leq j'.$$

Proof. To begin with, we assume that $p(x)$ is uniform $R(0, 1)$, since by the inverse of the probability integral transformation we may obtain any $p(x)$ satisfying the conditions of the theorem.
The joint pdf of the $X_{(n_j)}$ is (cf. 2.2.3)

$$h = \frac{n!}{\prod_{j=0}^{k} (n_{j+1} - n_j - 1)!} \prod_{j=0}^{k} \left(\int_{x_{(n_j)}}^{x_{(n_{j+1})}} dt_j \right)^{n_{j+1} - n_j - 1}$$

$$= C \prod_{j=0}^{k} (x_{(n_{j+1})} - x_{(n_j)})^{n_{j+1} - n_j - 1},$$

where $n_0 = 0$, $n_{k+1} = n + 1$, $x_{(n_0)} = 0$, $x_{(n_{k+1})} = 1$, and C denotes a generic constant.
Since $\mathscr{E} X_{(n_j)} = n_j/(n + 1)$, we put

$$y_j = \left(x_{(n_j)} - \frac{n_j}{n + 1} \right) n^{1/2} \qquad j = 0, 1, \ldots, k + 1.$$

Then $y_0 = 0$, $y_{k+1} = 0$, and

$$h = C \prod_{j=1}^{k+1} \left[1 + \frac{(y_j - y_{j-1})(n + 1)}{n^{1/2}(n_j - n_{j-1})} \right]^{n_j - n_{j-1} - 1}.$$

Thus, since Y_j is of order 1 in probability (briefly Y_j is $O_p(1)$), we have

$$\log h = C + \frac{n+1}{n^{\frac{1}{2}}} \sum_{j=1}^{k+1} \frac{(n_j - n_{j-1} - 1)(y_j - y_{j-1})}{n_j - n_{j-1}}$$

$$- \frac{1}{2} \frac{(n+1)^2}{n} \sum_{j=1}^{k+1} \frac{(n_j - n_{j-1} - 1)(y_j - y_{j-1})^2}{(n_j - n_{j-1})^2} + O_p\left(\frac{1}{n^{\frac{1}{2}}}\right).$$

Now

$$\sum_{j=1}^{k+1} \frac{(n_j - n_{j-1} - 1)(y_j - y_{j-1})}{n_j - n_{j-1}} = \sum_{j=1}^{k+1} (y_j - y_{j-1}) - \sum_{j=1}^{k+1} \left(\frac{y_j - y_{j-1}}{n_j - n_{j-1}}\right)$$

$$= 0 - O_p\left(\frac{1}{n}\right);$$

and, since $n_j = n\lambda_j + O(1)$,

$$n \sum_{j=1}^{k+1} \frac{(n_j - n_{j-1} - 1)(y_j - y_{j-1})^2}{(n_j - n_{j-1})^2} = \sum_{j=1}^{k+1} \frac{(y_j - y_{j-1})^2}{\lambda_j - \lambda_{j-1}} + O_p\left(\frac{1}{n}\right).$$

It follows that

$$\log h = C - \frac{1}{2} \sum_{j=1}^{k+1} \frac{(y_j - y_{j-1})^2}{\lambda_j - \lambda_{j-1}} + O_p\left(\frac{1}{n^{\frac{1}{2}}}\right). \tag{9.2.1}$$

Since

$$\sum_{j=1}^{k+1} \frac{(y_j - y_{j-1})^2}{\lambda_j - \lambda_{j-1}} = \sum_{j=1}^{k} \left(\frac{y_j^2}{\lambda_j - \lambda_{j-1}} + \frac{y_j^2}{\lambda_{j+1} - \lambda_j}\right) - 2\sum_{j=2}^{k} \frac{y_j y_{j-1}}{\lambda_j - \lambda_{j-1}}$$

$$= \sum_{j=1}^{k} y_j^2 \frac{\lambda_{j+1} - \lambda_{j-1}}{(\lambda_{j+1} - \lambda_j)(\lambda_j - \lambda_{j-1})} - 2\sum_{j=2}^{k} \frac{y_j y_{j-1}}{\lambda_j - \lambda_{j-1}}, \tag{9.2.2}$$

we see from (9.2.1) that the Y_j have asymptotically a k-variate normal distribution with zero means. The matrix of coefficients in the quadratic form (9.2.2) is $(A_{jj'})$, say, where

$$A_{jj} = \frac{\lambda_{j+1} - \lambda_{j-1}}{(\lambda_{j+1} - \lambda_j)(\lambda_j - \lambda_{j-1})}, \quad A_{j,j-1} = A_{j-1,j} = \frac{-1}{\lambda_j - \lambda_{j-1}},$$

and $A_{jj'} = 0$ for $|j - j'| > 1$. The inverse of this matrix, namely, the covariance matrix of the Y_j, may be verified to have elements

$$\text{cov}\,(Y_j, Y_{j'}) = \lambda_j(1 - \lambda_{j'}) \quad j \leq j'.$$

Since

$$X_{(n_j)} = \frac{Y_j}{n^{\frac{1}{2}}} + \frac{n_j}{n+1},$$

it follows that asymptotically the $X_{(n_j)}$ are also k-variate normal, with

$$\mathscr{E} X_{(n_j)} = \xi_{\lambda_j}(= \lambda_j) \quad \text{and} \quad \text{cov}\,(X_{(n_j)}, X_{(n_{j'})}) = \frac{\lambda_j(1 - \lambda_{j'})}{n},$$

which proves the theorem for the uniform case.   ▶

To obtain the general result we use the following lemma:

If the random variables $t_j(X_1, X_2, \ldots, X_n)$ $(j = 1, 2, \ldots, k)$ follow asymptotically a k-variate normal distribution with means θ_j, variances σ_j^2 which tend to 0 as $n \to \infty$, and covariances $\rho_{jj'}\sigma_j\sigma_{j'}$, and if $g_j(t_j)$ are single-valued functions with nonvanishing continuous derivatives $g_j{}'(t_j)$ in the neighborhood of $t_j = \theta_j$, then the $g_j(t_j)$ themselves have a k-variate normal distribution with means $g_j(\theta_j)$ and covariances $\rho_{jj'}\sigma_j\sigma_{j'}g_j{}'(\theta_j)g_{j'}{}'(\theta_{j'})$.

With $t_j = X_{(n_j)}$, $\theta_j = \lambda_j$, the transformation $g_j(t_j) = P^{-1}(X_{(n_j)})$ clearly satisfies the conditions of the lemma. The theorem follows when we note that

$$g_j(\theta_j) = P^{-1}(\lambda_j) = \xi_{\lambda_j}$$

and

$$g_j{}'(\theta_j) = \frac{dg_j(\theta_j)}{d\lambda_j} = \frac{1}{d\lambda_j/d\xi_{\lambda_j}} = \frac{1}{p(\xi_{\lambda_j})},$$

where ξ_{λ_j} now refers to the general population with pdf $p(x)$.

Remark 1.   The means, variances, and covariances given by the theorem correspond to the first terms of (4.5.3)–(4.5.5), respectively.

Remark 2.   The results of the theorem have already been used in the problem of optimal spacing of order statistics (Section 7.6).

For an alternative approach to Theorem 9.2 see Ex. 9.2.2 and also Kiefer (1967). Yet another rather simple but rigorous proof via characteristic functions has recently been given by Walker (1968). Using the Bahadur representation of Ex. 9.2.2, Sen (1968) establishes the asymptotic normality (under suitable conditions) of sample quantiles for m-dependent, not necessarily stationary, processes, that is, for random vectors $(X_1, \ldots, X_i)$ and $(X_j, X_{j+1}, \ldots)$ which are stochastically independent if $j - i > m$, $m = 0, 1, 2, \ldots$. Smirnov (1966, 1967) discusses the behavior of $X_{(k)}$ when k is a function of n and states conditions under which $X_{(k(n))}$ tends to normality as $n \to \infty$ when $k(n) \to \infty$ but $k(n)/n \to 0$. See also Cheng (1964) and the direct arguments of van der Vaart (1961a). The large-sample estimation of nonunique quantiles is considered by Feldman and Tucker (1966); see Ex. 9.2.1 for one of their results. The asymptotic joint normality of sample quantiles for a multivariate distribution is established by Weiss (1964) under mild conditions.

Mosteller's theorem assumes that $0 < \lambda_1 < \cdots < \lambda_k < 1$, that is, the λ's are strictly increasing. In contrast, starting from the joint pdf of $X_{(i-k)}$, $X_{(i)}$, and $X_{(i+l)}$ with $i/n \to \lambda$ and k, l of $o(n)$, and transforming to

$$U = \frac{n^{\frac{1}{2}}(X_{(i)} - \xi_\lambda)p_\lambda}{[\lambda(1 - \lambda)]^{\frac{1}{2}}},$$

$$U_1 = \frac{n}{k}(X_{(i)} - X_{(i-k)}), \quad U_2 = \frac{n}{l}(X_{(i+l)} - X_{(i)}),$$

Siddiqui (1960) shows that asymptotically U, U_1, and U_2 are independently distributed, U being of course $N(0, 1)$. Moreover, $2kp_\lambda U_1$ and $2lp_\lambda U_2$ are distributed as χ^2 variates with $2k$, $2l$ DF, respectively. Thus asymptotically,

$$2np_\lambda(X_{(i+l)} - X_{(i-k)}) \frown \chi^2_{2(k+l)} \qquad (9.2.3)$$

with distribution independent of that of $X_{(i)}$. These results hold whether or not k and l are constants. If k, l are of order n^α ($0 < \alpha < 1$), then (9.2.3) is equivalent (cf. Bloch and Gastwirth, 1968) to

$$\frac{np_\lambda(X_{(i+l)} - X_{(i-k)}) - (k + l)}{(k + l)^{\frac{1}{2}}} \frown N(0, 1).$$

Bloch and Gastwirth are concerned with the estimation of the reciprocal of the density function. Somewhat similar problems arise in the estimation of the mode from some function of the first and the last of those s (say) chosen consecutive order statistics which are closest together (e.g., Venter, 1967).

9.3. THE ASYMPTOTIC DISTRIBUTION OF THE EXTREME

The asymptotic behavior of $X_{(n)}$, the largest in a random sample of n from a population with cdf $P(x)$, has provided a challenge to many able mathematical statisticians. Noteworthy contributions were made by Dodd (1923), von Mises (1923, 1936), Fréchet (1927), Fisher and Tippett (1928), de Finetti (1932), Gumbel (from 1935 on, summarized 1958) and finally Gnedenko (1943), who gave the most complete and rigorous discussion of the problem. See also Barndorff-Nielsen (1963) for a summary of these and related results, as well as Dwass (1964) and Lamperti (1964) for a stochastic process approach.

Some of the main findings are as follows. For an arbitrary parent distribution, $X_{(n)}$, even after suitable standardization, will not in general possess a limiting distribution (ld). However, if $P(x)$ is such that a ld exists, then this ld must be one of just three types,[2] namely,

$$\begin{aligned}
\Lambda_1(x) &= 0 & x \le 0, \alpha > 0, \\
&= \exp\left(-x^{-\alpha}\right) & x > 0; \\
\Lambda_2(x) &= \exp\left[-(-x)^\alpha\right] & x \le 0, \alpha > 0, \\
&= 1 & x > 0; \\
\Lambda_3(x) &= \exp\left(-e^{-x}\right) & -\infty < x < \infty.
\end{aligned} \qquad (9.3.1)$$

More formally, we have the following theorem (Gnedenko):

The class of ld's for $P^n(a_n x + b_n)$, where $a_n > 0$ and b_n are suitably chosen constants, contains only laws of the types $\Lambda_k(x)$ ($k = 1, 2, 3$).

We shall not prove this theorem but instead give the ingenious key idea already used by Fisher and Tippett: Since the largest in a sample of mn may be regarded as the largest member of a sample of n maxima in samples of m, and since, if a limiting form $\Lambda(x)$ exists, both of the distributions will tend to $\Lambda(x)$ as $m \to \infty$, it follows that $\Lambda(x)$ must be such that

$$\Lambda^n(a_n x + b_n) = \Lambda(x), \qquad (9.3.2)$$

that is, the largest in a sample of n drawn from a distribution with cdf $\Lambda(x)$ must, upon the same standardization as above, itself have limiting cdf $\Lambda(x)$. The solutions for $\Lambda(x)$ in this functional equation give all the possible limiting forms.

Now, if in (9.3.2) $a_n \ne 1$, then $x = a_n x + b_n$ when $x = b_n/(1 - a_n) = x_0$ (say), and $\Lambda^n(x_0) = \Lambda(x_0)$, that is, $\Lambda(x_0) = 0$ or 1. Under the assumption that a ld $\Lambda(x)$ exists, x_0 must be a constant which may be taken as zero without loss of generality. Then, since $x_0 = 0$ implies $b_n = 0$, the solutions fall into three classes:

(1) $\Lambda(x) = 0$ $x \le 0$, $\Lambda^n(a_n x) = \Lambda(x)$ $x > 0$;
(2) $\Lambda^n(a_n x) = \Lambda(x)$ $x \le 0$, $\Lambda(x) = 1$ $x > 0$;
(3) $\Lambda^n(x + b_n) = \Lambda(x)$.

These classes correspond evidently to $a_n > 1$, $a_n < 1$, and $a_n = 1$. From standard mathematical results it follows that the only solutions of the functional equations (1)–(3) are respectively of the form $\Lambda_1(x)$ to $\Lambda_3(x)$.

Let us now take a closer look at $\Lambda_3(x)$. This, in view of its pre-eminent position among the three types, is often called *the* extreme-value distribution although, of course, all three fit the term. Clearly, the maximum in a sample of n drawn from $\Lambda_3(x)$ as parent distribution has a cdf differing from $\Lambda_3(x)$

---

[2] See the footnote in Section 9.1.

only in a displacement b_n to the right, where b_n is given by

$$\exp\left(-ne^{-x}\right) = \exp\left(-e^{-(x-b_n)}\right),$$

that is,

$$b_n = \log n.$$

The pdf $\Lambda_3'(x) = \exp\left(-x - e^{-x}\right)$ is represented in Fig. 9.3. By considering the cumulant generating function it is easy to show that

$$\mu = \gamma \text{ (Euler's constant)} = 0.5772\ldots,$$

$$\mu_2 = \sum_{n=1}^{\infty} \frac{1}{n^2} = \frac{\pi^2}{6} = 1.6449\ldots,$$

$$\beta_1 = 1.2986\ldots, \quad \beta_2 = 5.4.$$

Gnedenko (1943) gives necessary and sufficient conditions for a distribution $P(x)$ to belong to the "domain of attraction" of one of the three limiting forms:

(1) $P(x)$ belongs to the domain of attraction of $\Lambda_1(x)$ if and only if

$$\lim_{x \to \infty} \frac{1 - P(x)}{1 - P(kx)} = k^{\alpha}$$

for every $k > 0$.

(2) $P(x)$ belongs to the domain of attraction of $\Lambda_2(x)$ if and only if
(a) there exists an x_0 such that

$$P(x_0) = 1 \quad \text{and} \quad P(x_0 - \varepsilon) < 1$$

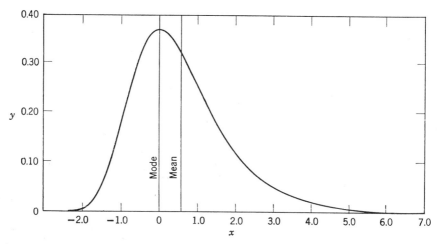

Fig. 9.3. $y = \Lambda_3'(x) = \exp\left(-x - e^{-x}\right).$

for every $\varepsilon > 0$;

$$(b) \qquad \lim_{x \to -0} \frac{1 - P(kx + x_0)}{1 - P(x + x_0)} = k^{\alpha}$$

for every $k > 0$.

$P(x)$ is clearly unlimited on the right in (1) and limited in (2). Gnedenko points out that $\Lambda_3(x)$ may arise in either case (Ex. 9.3.1). Instead of citing his rather complex necessary and sufficient conditions we shall, following von Mises (1936), prove a necessary condition convenient when $P(x)$ is un-limited on the right:

Suppose $P(x)$ is less than 1 for every finite x, is twice differentiable at least for all x greater than some value x_0, and is such that

$$\lim_{x \to \infty} \frac{d}{dx} \left[\frac{1 - P(x)}{p(x)} \right] = 0. \qquad (9.3.3)$$

Then

$$\lim_{n \to \infty} \Pr \left\{ (X_{(n)} - l_n) n p(l_n) \leq u \right\} = \Lambda_3(u)$$

holds uniformly for every u in $(-\infty, \infty)$, where l_n is given by

$$P(l_n) = \frac{n - 1}{n} .$$

Proof.   First note that

$$\Pr \left\{ X_{(n)} \leq l_n \right\} \equiv F_n(l_n) = P^n(l_n) = \left(1 - \frac{1}{n} \right)^n$$

so that

$$\lim_{n \to \infty} P_n(l_n) = e^{-1}.$$

Now put

$$x = l_n + \frac{u}{c_n}, \qquad (9.3.4)$$

where the c_n are constants at our disposal. Then $P^n(l_n + u/c_n)$ has a definite limit for $u = 0$, and we shall consider its limit when u is any constant.

The definition of l_n gives

$$P(x) = 1 - \frac{1}{n} \frac{1 - P(x)}{1 - P(l_n)} .$$

If now, as a result of some hypothesis on x, the ratio $[1 - P(x)]/[1 - P(l_n)]$ tends to a finite limit as $n \to \infty$, we have

$$-\log F_n(x) = -n \log P(x) = \frac{1 - P(x)}{1 - P(l_n)} + O\left(\frac{1}{n} \right),$$

so that

$$\log\,[-\log F_n(x)] = \log\,\{n[1 - P(x)]\} + O\Big(\frac{1}{n}\Big),$$

or, with $G_n(x) = \log\,\{n[1 - P(x)]\}$,

$$\lim_{n\to\infty} \log\,[-\log F_n(x)] = \lim_{n\to\infty} G_n(x)$$

$$= -\lim_{n\to\infty} \int_{l_n}^{x} g(t)\,dt, \qquad (9.3.5)$$

since $G_n(l_n) = 0$, where

$$g(x) = -G_n{}'(x) = \frac{p(x)}{1 - P(x)}\,.$$

Now specify the c_n in (9.3.4) by

$$c_n = g(l_n) = np(l_n)$$

and express $G_n(x)$ as

$$G_n(x) = \int_x^{l_n} g(t)\,dt = (l_n - x)g(\xi) \qquad x < \xi < l_n.$$

Then

$$G_n\Big(l_n + \frac{u}{c_n}\Big) = -u\,\frac{g(\xi)}{g(l_n)}\,. \qquad (9.3.6)$$

We show that $g(\xi)/g(l_n) \to 1$ as $n \to \infty$, when u lies in the finite interval $(-u_0, u_0)$. First note that $\xi \to \infty$ as $n \to \infty$; for by (9.3.4) $\xi > l_n - u_0/g(l_n)$ and the RHS is the value for $x = l_n$ of the function $x - u_0/g(x)$, whose derivative tends to 1 as $x \to \infty$ by (9.3.3). Therefore the function itself increases without limit as x runs through the values $l_1, l_2, l_3, \ldots$.

Expanding $1/g(\xi)$ in a Taylor series about $\xi = l_n$ and multiplying through by $g(l_n)$, we obtain

$$\frac{g(l_n)}{g(\xi)} = 1 + g(l_n)(\xi - l_n)\frac{d}{dx}\Big[\frac{1}{g(x)}\Big]_{x=\xi_1} \qquad \xi < \xi_1 < l_n. \qquad (9.3.7)$$

Now $|g(l_n)(\xi - l_n)| < u_0$ by (9.3.4). Also, as n increases, $\xi_1 \to \infty$ since $\xi \to \infty$, so that by hypothesis the last term of (9.3.7) tends to zero. Hence $g(l_n)/g(\xi) \to 1$ as $n \to \infty$. In conjunction with (9.3.6) and (9.3.5) this shows that

$$\lim_{n\to\infty} \log\Big[-\log F_n\Big(l_n + \frac{u}{np(l_n)}\Big)\Big] = -u$$

uniformly for every u in $(-u_0, u_0)$, or

$$\lim_{n\to\infty} F_n\Big(l_n + \frac{u}{np(l_n)}\Big) = e^{-e^{-u}},$$

which proves the theorem; for the uniform convergence may be extended to $(-\infty, \infty)$ since every cdf $F_n(\cdot)$ and also $e^{-e^{-u}}$ tend uniformly to 0 and 1, respectively, as $u \to -\infty, +\infty$.

Example 9.3.1. For the exponential distribution $p(x) = e^{-x}$ ($x \geq 0$) we have *exactly* $l_n = \log n$ and $np(l_n) = 1$. Condition (9.3.3) is obviously satisfied, so that $X_{(n)} - \log n$ has limiting cdf $\Lambda_3(x)$. This is also easily established from first principles (Ex. 2.1.2). In the present context all distributions in the domain of attraction of $\Lambda_3(x)$ are said to be of *exponential type*.

Example 9.3.2. For the normal pdf $p(x) = (2\pi)^{-\frac{1}{2}}e^{-\frac{1}{2}x^2}$ it is well known that asymptotically for large x

$$\frac{1 - P(x)}{p(x)} \sim \frac{1}{x}, \tag{9.3.8}$$

so that (9.3.3) is satisfied and the theorem applies.

For given n, l_n may be evaluated from tables of the normal cdf. However, note that by (9.3.8)

$$\frac{1}{n} = 1 - P(l_n) \sim \frac{e^{-\frac{1}{2}l_n{}^2}}{l_n(2\pi)^{\frac{1}{2}}},$$

giving as a first approximation (cf. Ex. 9.3.2)

$$l_n \sim (2 \log n)^{\frac{1}{2}}. \tag{9.3.9}$$

Also

$$np(l_n) \sim (2 \log n)^{\frac{1}{2}},$$

so that asymptotically $(2 \log n)^{\frac{1}{2}}[X_{(n)} - (2 \log n)^{\frac{1}{2}}]$ has cdf $\Lambda_3(x)$.

As first pointed out by Fisher and Tippett (1928), the tendency toward the asymptotic form is exceedingly slow in the normal case. This is in contrast to the situation in, for example, exponential and logistic parents. However, even in the normal case the agreement is already good for $n = 100$, except in the tails (see Gumbel, 1958, p. 222). The penultimate behavior of the distribution of the extreme for exponential-type parents has been studied by Uzgören (1954). More general results are given by Dronkers (1958). See also Haldane and Jayakar (1963), who point out that for a normal parent the cdf of $X_{(n)}^2$ (suitably standardized) approaches $\Lambda_3(x)$ much more quickly than does the cdf of $X_{(n)}$ itself.

Similar results hold, of course, for the asymptotic distribution of the standardized minimum. The three possible limiting types corresponding to

(9.3.1) are as follows:

$$\Lambda_1'(x) = 1 - \exp\left[-(-x)^{-\alpha}\right] \qquad x \leq 0, \alpha > 0,$$
$$= 1 \qquad x > 0;$$
$$\Lambda_2'(x) = 0 \qquad x \leq 0, \alpha > 0, \qquad (9.3.10)$$
$$= 1 - \exp\left(-x^\alpha\right) \qquad x > 0;$$
$$\Lambda_3'(x) = 1 - \exp\left(-e^x\right) \qquad -\infty < x < \infty.$$

We will not enter into such questions as whether the sequence of extremes is stable in some (technical) sense or whether the moments of the extremes tend to the moments of the appropriate asymptotic distribution except to refer the reader to, for example, Geffroy (1958), Barndorff-Nielsen (1963), Sen (1959, 1961, 1964), McCord (1964) and Pickands (1967b, 1968).

9.4. EXTREME-VALUE THEORY—GENERALIZATIONS FOR INDEPENDENT, IDENTICALLY DISTRIBUTED VARIATES

We turn now to some generalizations for iid variates of the results in Section 9.3, confining ourselves for the most part to a statement of the principal results.

Distribution of the mth Extreme

In direct generalization (Gumbel, 1935; Smirnov, 1952) of the results in (9.3.1) there are again just three possible limiting distributions for a suitably standardized form of the mth extreme $X_{n-m+1:n}$ as $n \to \infty$ with m a fixed non-negative integer. The three types are:

$$\Lambda_1^{(m)}(x) = 0 \qquad\qquad x \leq 0, \alpha > 0,$$
$$= \frac{1}{(m-1)!} \int_{x^{-\alpha}}^{\infty} e^{-t} t^{m-1}\, dt \qquad x > 0;$$
$$\Lambda_2^{(m)}(x) = \frac{1}{(m-1)!} \int_{(-x)^\alpha}^{\infty} e^{-t} t^{m-1}\, dt \qquad x \leq 0, \alpha > 0,$$
$$= 1 \qquad\qquad x > 0;$$
$$\Lambda_3^{(m)}(x) = \frac{1}{(m-1)!} \int_{e^{-x}}^{\infty} e^{-t} t^{m-1}\, dt \qquad -\infty < x < \infty.$$

Smirnov shows also that Gnedenko's (Section 9.3) necessary and sufficient conditions for a distribution to belong to the domain of attraction of one of the above ld's continue to hold for $m > 1$.[3]

---

[3] Generalizations for independent but nonidentically distributed variates are given by Mejzler and Weissman (1969).

Joint Distribution of Extremes

We consider first the joint pdf of $U = nP(X_{r:n})$ and $V = n[1 - P(X_{s:n})]$, which from (2.2.1) is

$$f(u, v) = \frac{n!}{(r - 1)!(s - r - 1)!(n - s)!} \left(\frac{u}{n}\right)^{r-1} \left(1 - \frac{u}{n} - \frac{v}{n}\right)^{s-r-1} \left(\frac{v}{n}\right)^{n-s} \frac{1}{n^2}$$

$$u \geq 0, v \geq 0, u + v \leq n.$$

Letting $n \to \infty$ but keeping r and $n - s + 1 = t$ fixed, we see that the RHS becomes

$$\frac{1}{\Gamma(r)} u^{r-1} e^{-u} \cdot \frac{1}{\Gamma(t)} v^{t-1} e^{-v} \qquad u \geq 0, v \geq 0,$$

showing that U and V are independent $\gamma(r)$ and $\gamma(t)$ variates. Thus any "lower" extreme $X_{r:n}$ is asymptotically independent of any "upper" extreme $X_{n+1-t:n}$. This result is, of course, very useful in the derivation of the limiting distributions of statistics such as the range and the midrange.

Asymptotic Distribution of the Range

We consider symmetric parent distributions satisfying (9.3.3). Then $Y = (X_{(n)} - l_n)np(l_n)$ has limiting cdf $\Lambda_3(y)$, and by symmetry so does $-Z$, where Z is the reduced minimum

$$Z = (X_{(1)} - l_1)np(l_1) \qquad l_1 = -l_n.$$

By the asymptotic independence of Y and Z their joint asymptotic pdf is

$$\exp\left(-y - e^{-y} + z - e^z\right),$$

so that the reduced range $W' = (X_{(n)} - X_{(1)} - 2l_n)np(l_n)$ has limiting pdf

$$\int_{-\infty}^{\infty} \exp\left(-w' - e^{-w'-z} - e^z\right) dz = 2e^{-w'} K_0(2e^{-\frac{1}{2}w'}), \qquad (9.4.1)$$

where $K_0(x)$ is a modified Bessel function of the second kind. This result is due to Gumbel (1947) and Cox (1948), and is used by Gumbel (1949) in the construction of tables of both pdf and cdf of W'.

Various authors (Elfving, 1947; Cox, 1948; Cadwell, 1953a) have improved on this result as an approximation to the distribution of range W in finite samples. If we write the pdf of W as

$$f(w) = n(n - 1) \int_{-\infty}^{\infty} p(x - \tfrac{1}{2}w)p(x + \tfrac{1}{2}w)[P(x + \tfrac{1}{2}w) - P(x - \tfrac{1}{2}w)]^{n-2} dx$$

and take $p(x)$ symmetric about $x = 0$ and unimodal, then the integrand will have a maximum at $x = 0$ and will fall rapidly to zero on either side of

$x = 0$. This suggests, as in the method of steepest descent, expanding the integrand in powers of x.

It is easy to verify (Cadwell, 1953a) that

$$p(x - \tfrac{1}{2}w)p(x + \tfrac{1}{2}w) = p^2 \exp\left\{-\left[\left(\frac{p'}{p}\right)^2 - \frac{p''}{p}\right]x^2 + \cdots\right\},$$

$$P(x + \tfrac{1}{2}w) - P(x - \tfrac{1}{2}w) = (2P - 1) \exp\left(\frac{p'}{2P - 1}x^2 + \cdots\right),$$

where P, p, and its derivatives are evaluated at $x = \tfrac{1}{2}w$, and the dots denote a series in higher even powers of x. Thus the integrand may be written in the form

$$p^2(2P - 1)^{n-2}(1 + Ax^4 + Bx^6 + \cdots)\exp -\left[\left(\frac{p'}{p}\right)^2 - \frac{p''}{p} - \frac{(n - 2)p'}{2P - 1}\right]x^2,$$

and with the help of Watson's lemma (see, e.g., Jeffreys and Jeffreys, 1946, §17.03) may be integrated term by term to give an asymptotic series for $f(w)$. The first and dominant term of this series is clearly

$$f(w) \sim \frac{n(n - 1)\pi^{1/2}p^2(2P - 1)^{n-2}}{\left[\left(\frac{p'}{p}\right)^2 - \frac{p''}{p} - \frac{(n - 2)p'}{2P - 1}\right]^{1/2}}. \qquad (9.4.2)$$

When $p(x) = \phi(x)$, the unit normal pdf, (9.4.2) simplifies to

$$f(w) \sim \frac{n(n - 1)\pi^{1/2}\phi^2(2\Phi - 1)^{n-3/2}}{[2\Phi - 1 - (n - 2)\phi']^{1/2}}. \qquad (9.4.3)$$

Cadwell shows that this leading term approximation already gives good agreement for the first four moments of W when $n = 20$:

	Mean	SD	β_1	β_2
Exact value	3.7350	.7287	.1627	3.259
Error by use of (9.4.3)	.0086	.0025	−.0043	−.019

Further improvements can be effected by the use of additional terms. Cadwell deals also with quasi-ranges, for which the first approximation is even better (cf. Ex. 9.4.1). He handles the cdf of W separately (Cadwell, 1954).

From the joint asymptotic pdf of the extremes, Gumbel and Keeney (1950a, b) have derived, respectively, the asymptotic distributions of the "geometric range" $[(X_{(n)})(-X_{(1)})]^{1/2}$ and of the "extremal quotient" $X_{(n)}/-X_{(1)}$. Tables of the cdf of the latter are given by Gumbel and Pickands (1967).

Asymptotic Distribution of the Sample Spacings

Steepest descent methods similar to the above are used by Darwin (1957) in a study of the distribution of the sample spacings $X_{i+1:n} - X_{i:n}$ $(i = 1, 2, \ldots, n - 1)$. This topic has been reviewed for both small (cf. Section 5.4) and large samples by Pyke (1965), who gives many references. See also Weiss (1965), one of the more recent papers by a chief contributor to the subject, and Blumenthal (1966).

Other Independence Results

Not only are lower extremes asymptotically independent of upper extremes but also both are asymptotically independent of (a) central order statistics (Rossberg, 1963; Krem, 1963; Rosengard, 1964b) and (b) the sample mean (Rossberg, 1965b; Rosengard, 1964a). These results are subsumed in the following theorem by Rossberg (1965b):

Let $g_h(x_{1:n}, \ldots, x_{h:n})$ and $g_{h'}(x_{n+1-h':n}, \ldots, x_{n:n})$ be arbitrary Borel-measurable functions of the arguments indicated. If the iid variates $X_1, \ldots, X_n$ have finite variance σ^2 and

$$\lim_{n \to \infty} \frac{h}{n} = \lim_{n \to \infty} \frac{h'}{n} = 0,$$

then the random variables

$$g_h(X_{1:n}, \ldots, X_{h:n}), \quad n^{1/2}(\overline{X} - \mathscr{E}X)/\sigma, \quad g_{h'}(X_{n+1-h':n}, \ldots, X_{n:n})$$

are asymptotically independent.

Limiting Distribution of the Studentized Extreme Deviate

Let $X_1, X_2, \ldots$ be iid variates with $\mathscr{E}X_1 = 0$, $\mathscr{E}X_1^2 = 1$. Berman (1963) proves that, if for some sequences $\{a_n\}$ and $\{b_n\}$ $(a_n > 0)$ the normed extreme $(X_{(n)} - b_n)/a_n$ has ld $\Lambda(x)$, then the normed (internally) studentized extreme deviate

$$\frac{1}{a_n}\left(\frac{X_{(n)} - \overline{X}}{S} - b_n\right)$$

also has ld $\Lambda(x)$ under the condition that

$$\lim_{n \to \infty} \frac{b_n}{a_n n^{1/2}} = 0.$$

This condition obviously holds in the normal case when (Example 9.3.2) $a_n = (2 \log n)^{-1/2}$, $b_n = (2 \log n)^{1/2}$, and is in fact shown by Berman to be implied by the von Mises criterion (9.3.3). If $p(x)$ is symmetric, Berman proves also that under the above conditions

$$\frac{1}{a_n}\left(\max\left|\frac{X_i - \bar{X}}{S}\right| - b_n\right)$$

has ld $\Lambda^2(x)$.

Miscellaneous

The limiting distribution of the maximum of a *random* number of random variables has been studied by Berman (1962a), Barndorff-Nielsen (1964), and Richter (1964). Chernoff and Teicher (1965) consider the limit distributions of the minimax (or maximin) of doubly indexed iid variates, that is, of $\min_i \max_j X_{ij}$.

Bivariate Extensions

Several authors have dealt with bivariate (and multivariate) extremal distributions. Early results are reviewed by Tiago de Oliveira (1963). Gumbel and Mustafi (1967) consider explicitly two possible extremal pdf's:

$$\Lambda_{k,l}^{(1)}(x, y, a) = \Lambda_k(x)\Lambda_l(y)\exp\left\{a\left[\frac{1}{-\log\Lambda_k(x)} + \frac{1}{-\log\Lambda_l(y)}\right]^{-1}\right\}$$

and

$$\Lambda_{k,l}^{(2)}(x, y, m) = \exp\left(-\{[-\log\Lambda_k(x)]^m + [-\log\Lambda_l(y)]^m\}^{1/m}\right),$$

where $\Lambda_k(x)$, $\Lambda_l(y)$ ($k, l = 1, 2, 3$) are the three possible univariate ld's of (9.3.1), and a and m are parameters of association restricted by

$$0 \le a, \quad 1/m \le 1.$$

The cases $a = 0$ and $m = 1$ correspond to independence of X and Y.

See also Berman (1962b), Mardia (1964a), Srivastava *et al.* (1964), Srivastava (1967), and for two applications Gumbel and Goldstein (1964).

9.5. EXTREME-VALUE THEORY FOR DEPENDENT VARIATES

The first major result in extreme-value theory dealing with dependent variates appears to be due to Watson (1954), who showed that the limiting distributions of the maximum for stationary m-dependent variates are, under certain conditions, the same as for independence. Stationarity means that

$$\Pr\{X_i \le x_i, X_j \le x_j, \ldots\} = \Pr\{X_{i+1} \le x_i, X_{j+1} \le x_j, \ldots\},$$

and m-dependence is defined toward the end of Section 9.2. Watson's conditions are that the X's be unbounded above and that

$$\lim_{c \to \infty} \frac{1}{\Pr\{X_i > c\}} \max_{|i-j| \le m} \Pr\{X_i > c, X_j > c\} = 0,$$

conditions which hold, for example, when X_i, X_j are bivariate normal $N(0, 0, 1, 1, \rho)$ with $|\rho| < 1$. For a generalization see Newell (1964).

For stationary Gaussian processes with

$$
\left.
\begin{aligned}
\mathscr{E}X_i = 0, \quad \mathscr{E}X_i^2 = 1 \qquad i = 1, 2, \ldots, \\
\mathscr{E}X_iX_{i+N} = r_N \qquad\qquad N = 1, 2, \ldots,
\end{aligned}
\right\}
\tag{9.5.1}
$$

Berman (1964) has shown that the condition of m-dependence (namely, $r_N = 0$ for $N > m$) can be strengthened: provided $r_N \to 0$ sufficiently fast, the ld $\Lambda_3(x)$, true under independence, continues to apply. Specifically it is sufficient that

$$
\lim_{N \to \infty} r_N \log N = 0
$$

or that

$$
\sum_{N=1}^{\infty} r_N^2 < \infty.
$$

It is not sufficient that $\lim_{N \to \infty} r_N = 0$, although this will ensure (Pickands, 1967a) that

$$
(2 \log n)^{-\frac{1}{2}} X_{(n)} \to 1 \text{ a.s.}
$$

A simple but striking example (Berman, 1962c) of a case in which $\Lambda_3(x)$ fails to be the limiting cdf occurs for equicorrelated normal variates [i.e., (9.5.1) with $r_N = \rho$, all N]. To see this, note first [cf. (5.5.7)] that, for $\rho > 0$, X_i may be expressed as the sum of two independent normal variates $U_i + Y$ with

$$
\mathscr{E}U_i = 0 = \mathscr{E}Y,
$$

$$
\mathscr{E}U_i^2 = 1 - \rho, \quad \mathscr{E}Y^2 = \rho, \quad \mathscr{E}U_iU_j = 0 \qquad i \neq j.
$$

Thus

$$
X_{(n)} = \max_{i=1,2,\ldots,n} (U_i + Y) = U_{(n)} + Y,
\tag{9.5.2}
$$

so that the cdf of $X_{(n)}$ is simply the convolution of the distribution of $U_{(n)}$, the maximum of n *independent* normal variates, with the distribution of Y. Now (Gnedenko, 1943)

$$
\mathop{\mathrm{p\,lim}}_{n \to \infty} \{U_{(n)} - [2(1 - \rho) \log n]^{\frac{1}{2}}\} = 0,
$$

from which it follows in conjunction with (9.5.2) that the distribution of

$$
X_{(n)} - [2(1 - \rho) \log n]^{\frac{1}{2}}
$$

converges to the distribution of Y, which is *normal* $N(0, \rho)$.

For further generalizations see Pickands (1967a), who gives many references, and also Loynes (1965).

The asymptotic distribution of the range of *sums* of independent variates has been studied by Feller (1951) and Kemperman (1959).

9.6. THE ASYMPTOTIC DISTRIBUTION OF LINEAR FUNCTIONS OF ORDER STATISTICS

From the asymptotic joint normality of the quantiles established in Theorem 9.2 it is clear that a linear function of (a finite number of) such quantiles must be asymptotically normal under the conditions stated. On the other hand, the extremes in random samples ($X_{i:n}$ with $i/n \to 0$, 1) have non-normal limiting distributions (if they have ld's at all). Now the sample mean $\bar{X}$, which can be written as $\sum_{i=1}^{n} X_{i:n}/n$, is asymptotically normal (provided only X_i has finite variance σ^2) although it involves the extremes. The question arises therefore: Under what conditions does

$$T_n = \sum_{i=1}^{n} c_{in} X_{i:n}$$

have a limiting normal distribution? Apart from its theoretical interest, this question has been motivated by the work of Bennett (1952) and Jung (1955), who were concerned with finding optimal weights for T_n regarded as an estimator of location or scale.

As the above simple considerations illustrate, asymptotic normality of T_n will require suitable conditions on both the c_{in} and the form of the under-lying cdf $P(x)$. Various rather complicated sets of conditions have been developed, some severe on the c_{in} and weak on $P(x)$, others the reverse. However, a few simple remarks can be made. Since the sample mean is asymptotically normal if $\sigma^2 < \infty$, the trimmed mean

$$\sum_{i=[n\lambda]}^{[n(1-\lambda)]} \frac{X_{i:n}}{n(1 - 2\lambda)} \qquad 0 \leq \lambda < \tfrac{1}{2}$$

must be so a fortiori ([ ] denotes integral part). Likewise the Winsorized mean

$$\frac{1}{n} \left(n\lambda X_{[n\lambda]:n} + \sum_{i=[n\lambda]}^{[n(1-\lambda)]} X_{i:n} + n\lambda X_{[n(1-\lambda)]:n} \right),$$

in so far as it places less weight on the extremes than does $\bar{X}$, may be expected to be asymptotically normal; but this presupposes satisfactory behavior of $P(x)$ at ξ_λ and $\xi_{1-\lambda}$ in the spirit of Theorem 9.2 (cf. Bickel, 1967). Actually, asymptotic normality may hold even when the extremes carry much more weight. A simple example is provided by the statistic

$$``\sigma" = \frac{\pi^{1/2}}{n(n-1)} \sum_{i=1}^{n} (2i - n - 1) X_{i:n} \qquad (9.6.1)$$

of (7.4.1). This, being essentially Gini's mean difference, is a U statistic (Hoeffding, 1948) and is consequently asymptotically normal (provided only $\sigma^2 < \infty$).

For more general results the reader is referred to Bickel (1967), Chernoff *et al.* (1967), Govindarajulu (1968b), and Stigler (1969). A relatively simple form is given by Moore (1968). To this end, note that T_n may be written as

$$T_n = \frac{1}{n} \sum_{i=1}^{n} J\left(\frac{i}{n}\right) X_{i:n},$$

where $J(u)$ is a function of u ($0 \le u \le 1$) such that $J(i/n) = nc_{in}$. Alternatively, T_n may be expressed as a Stieltjes integral:

$$T_n = \int_{-\infty}^{\infty} x J(P_n(x)) \, dP_n(x),$$

where $P_n(x)$ is the empirical distribution function of X. Also let Q be any inverse of P. Then

Theorem 9.6. *If $\mathscr{E}\,|X| = \int_0^1 |G(u)|\,du < \infty$ and J is continuous on $[0, 1]$ except for jump discontinuities at $a_1, \ldots, a_M$, and J' is continuous and of bounded variation on $[0, 1] - \{a_1, \ldots, a_M\}$, then*

$$n^{1/2}\left[T_n - \int_{-\infty}^{\infty} x J(P(x)) \, dP(x) \right]$$

is asymptotically normal $N(0, \sigma^2)$ provided $\sigma^2 < \infty$, where

$$\sigma^2 = 2\iint_{s<t} J(P(s))J(P(t))P(s)[1 - P(t)]\,ds\,dt.$$

Example.   For "σ" in (9.6.1) we have

$$J\left(\frac{i}{n}\right) = \frac{\pi^{1/2}(2i - n - 1)}{n - 1},$$

so that we take $J(u) = \pi^{1/2}(2nu - n - 1)/(n - 1)$, giving

$$\int_{-\infty}^{\infty} x J(P(x)) \, dP(x) = \pi^{1/2} \int_{-\infty}^{\infty} x \frac{[2nP(x) - n - 1]}{n - 1} \, dP(x),$$

which is in asymptotic agreement with the known exact result (cf. David, 1968)

$$\text{``}\sigma\text{''} = \pi^{1/2} \int_{-\infty}^{\infty} x[2P(x) - 1] \, dP(x).$$

We may also mention that Stigler (1969) uses a procedure, applied by Hájek (1968) to linear rank statistics, to represent T_n as a linear combination $\hat{T}_n$ of *independent* variates plus a remainder term. The usual central-limit theory can be applied to $\hat{T}_n$, and the remainder can be shown, under quite general conditions, to converge to zero in mean square.

9.7. OPTIMAL ASYMPTOTIC ESTIMATION BY ORDER STATISTICS

In the preceding section we examined the asymptotic distribution of

$$T_n = \sum_{i=1}^{n} c_{in} X_{i:n},$$

where the c_{in} are given constants. We now turn to the question of how the c_{in} are to be chosen to make T_n a good estimator of some underlying parameter. In his remarkable unpublished Ph.D. thesis, C. A. Bennett considered this question already in 1952 for distributions depending on location and scale parameters only, that is, with pdf $p(x) = (1/\sigma)g((x - \mu)/\sigma)$ and cdf $G((x - \mu)/\sigma)$. Having first obtained (independently) essentially the results of Lloyd (1952), he proceeded naturally from the determination of optimal weights in small samples to the derivation of optimal asymptotic weights. Bennett's treatment allowed for multiply-censored samples, but we shall outline his approach only for double censoring (which includes single and no censoring as special cases). The interested reader is also referred to Chernoff et al. (1967) for extensions of some of Bennett's results.

We consider first the matrix $\mathbf{A'\Omega A}$ of (6.2.4), writing it more fully as

$$\begin{pmatrix} \Sigma\Sigma\beta^{ij} & \Sigma\Sigma\alpha_i\beta^{ij} \\ \Sigma\Sigma\alpha_i\beta^{ij} & \Sigma\Sigma\alpha_i\alpha_j\beta^{ij} \end{pmatrix}, \qquad (9.7.1)$$

where all sums extend over $i, j = 1, 2, \ldots, n$, and the β^{ij} are the elements of $\mathbf{\Omega}$. Now Section 4.5 gives to order $1/n$, for $r \le s$,

$$\operatorname{cov}(X_{r:n}, X_{s:n}) \equiv \sigma^2\beta_{rs} + \sigma^2 \cdot \frac{p_r q_s}{n+2} \cdot \frac{1}{g(H_r)g(H_s)},$$

where $H_r = G^{-1}(r/(n+1)) = (Q_r - \mu)/\sigma$, so that $g(H_r) = \sigma p(Q_r)$. The β^i are therefore given (cf. 9.2.3) by

$$\left.\begin{aligned} \beta^{ii} &= ng^2(H_i)\left(\frac{1}{p_{i+1} - p_i} + \frac{1}{p_i - p_{i-1}}\right), \\ \beta^{i,i-1} &= \beta^{i-1,i} = \frac{-ng(H_i)g(H_{i-1})}{p_i - p_{i-1}}, \\ \beta^{ij} &= 0 \qquad \text{otherwise.} \end{aligned}\right\} \qquad (9.7.2)$$

Note that these results hold not only when i, j runs through all the integers $1, 2, \ldots, n$ but also for any subset. In particular, for r_1 observations censored on the left and r_2 on the right, we obtain from (9.7.2), on taking $p_i = i/(n+1)$ $(i = r_1 + 1, r_1 + 2, \ldots, n - r_2)$ and $p_{r_1} = 0 = p_{n-r_2+1}$:

$$\frac{1}{n}\sum_{i=r_1+1}^{n-r_2}\sum_{j=r_1+1}^{n-r_2} \beta^{ij} = \sum_{i=r_1+1}^{n-r_2-1} \frac{[\Delta g(H_i)]^2}{\Delta p_i} + \frac{g^2(H_{r_1+1})}{p_{r_1+1}} + \frac{g^2(H_{n-r_2})}{1 - p_{n-r_2}},$$

where $\Delta g(H_i) = g(H_{i+1}) - g(H_i)$ and $\Delta p_i = 1/(n+1)$. Similarly, using $\alpha_i = H_i + O(1/n)$, we have, apart from terms of lower order,

$$\frac{1}{n}\Sigma\Sigma\alpha_i\beta^{ij} = \sum \frac{\Delta g(H_i)\,\Delta[H_i g(H_i)]}{\Delta p_i}$$

$$+ \frac{H_{r_1}g^2(H_{r_1+1})}{p_{r_1+1}} + \frac{H_{n-r_2}g^2(H_{n-r_2})}{1 - p_{n-r_2}},$$

$$\frac{1}{n}\Sigma\Sigma\alpha_i\alpha_j\beta^{ij} = \sum \frac{\{\Delta[H_i g(H_i)]\}^2}{\Delta p_i}$$

$$+ \frac{H_{r_1+1}^2 g^2(H_{r_1+1})}{p_{r_1+1}} + \frac{H_{n-r_2}^2 g^2(H_{n-r_2})}{1 - p_{n-r_2}}.$$

As $n \to \infty$, with $r_1/n \to \lambda_1$, $(n - r_2)/n \to \lambda_2$, we obtain

$$\sum_{i=r_1+1}^{n-r_2} \frac{[\Delta g(H_i)]^2}{\Delta p_i} \to \int_{\lambda_1}^{\lambda_2} \left[\frac{dg(H(u))}{du}\right]^2 du,$$

$$= \int_{\lambda_1}^{\lambda_2} \left[\frac{g'(H(u))}{g(H(u))}\right]^2 du,$$

since

$$\frac{dg(H(u))}{du} = \frac{dg(H(u))/dH(u)}{du/dH(u)} \quad \text{and} \quad u = G(H(u)).$$

For ease of writing put $y = H(u)$ and $\Psi(y) = g'(y)/g(y)$. Then in the notation of (7.6.1)

$$\lim_{n \to \infty} \frac{1}{n}\Sigma\Sigma\beta^{ij} = \int_{\lambda_1}^{\lambda_2} \Psi^2(y)\,du + \frac{f_1^2}{\lambda_1} + \frac{f_2^2}{1 - \lambda_2}, \tag{9.7.3a}$$

and similarly

$$\lim_{n \to \infty} \frac{1}{n}\Sigma\Sigma\alpha_i\beta^{ij} = \int_{\lambda_1}^{\lambda_2} \Psi(y)[1 + y\Psi(y)]\,du$$

$$+ \frac{\xi_{\lambda_1}f_1^2}{\lambda_1} + \frac{\xi_{\lambda_2}f_2^2}{1 - \lambda_2}, \tag{9.7.3b}$$

$$\lim_{n \to \infty} \frac{1}{n}\Sigma\Sigma\alpha_i\alpha_j\beta^{ij} = \int_{\lambda_1}^{\lambda_2} [1 + y\Psi(y)]^2\,du$$

$$+ \frac{\xi_{\lambda_1}^2 f_1^2}{\lambda_1} + \frac{\xi_{\lambda_2}^2 f_2^2}{1 - \lambda_2}. \tag{9.7.3c}$$

In the uncensored case ($\lambda_1 = 0$, $\lambda_2 = 1$) the last two terms on the right of (9.7.3a–c) all vanish provided that

$$\lim_{\lambda_1 \to 0} \frac{\xi_{\lambda_1}{}^2 f_1{}^2}{\lambda_1} = 0, \quad \lim_{\lambda_2 \to 1} \frac{\xi_{\lambda_2}{}^2 f_2{}^2}{1 - \lambda_2} = 0. \tag{9.7.4}$$

The inverse of the covariance matrix of the LS estimators μ^*, σ^* of μ, σ is just $\mathbf{A'\Omega A}/\sigma^2$. It is easy to show that this tends, under (9.7.4), to the information matrix. For example, since $p(x) = (1/\sigma)g(y)$, where $y = (x - \mu)/\sigma$ ($= H(u)$), we have

$$\mathscr{E}\left[\frac{\partial \log p(x)}{\partial \mu}\right]^2 = \mathscr{E}\left[\frac{\partial \log g(y)}{\partial \mu}\right]^2$$

$$= \frac{1}{\sigma^2}\mathscr{E}\left[\frac{\partial \log g(y)}{\partial y}\right]^2 = \frac{1}{\sigma^2}\int_0^1 \Psi^2(y)\, du,$$

in agreement with (9.7.3a). Thus our linear estimators of μ and σ are asymptotically efficient. This result can be shown to hold in the censored case also (see Chernoff *et al.*, 1967).

So far, the explicit form of the estimators themselves has not been needed. We now obtain this. According to (6.2.5) and (6.2.5′), adapted for censored samples (so that $\mathbf{A}$ has $n - r_1 - r_2$ rows, etc.), the LS estimators of μ and σ are given by

$$\mu^* = \sum_{i=r_1+1}^{n-r_2} \beta_i X_{i:n}, \quad \sigma^* = \sum_i \gamma_i X_{i:n},$$

where

$$\left.\begin{array}{l} \beta_i = \dfrac{1}{|\mathbf{A'\Omega A}|}\left(\displaystyle\sum_{j=r_1+1}^{n-r_2} \beta^{ij} \cdot \sum_i \sum_j \alpha_i\alpha_j\beta^{ij} - \sum_j \alpha_j\beta^{ij} \cdot \sum_i \sum_j \alpha_j\beta^{ij}\right), \\[16pt] \gamma_i = \dfrac{1}{|\mathbf{A'\Omega A}|}\left(\displaystyle\sum_j \alpha_j\beta^{ij} \cdot \sum_i \sum_j \beta^{ij} - \sum_j \beta^{ij} \cdot \sum_i \sum_j \alpha_j\beta^{ij}\right). \end{array}\right\} \tag{9.7.5}$$

From (9.7.2) we find

$$\frac{1}{n}\sum_{j=r_1+1}^{n-r_2} \beta^{ij} = -g(H_i)\frac{\delta^2 g(H_i)}{\Delta p} \qquad i = r_1 + 2, \ldots, n - r_2 - 1.$$

Also it is easily seen that

$$\frac{1}{n}\sum_{j=r_1+1}^{n-r_2} \beta^{r_1+1, j} = -g(H_{r_1+1})\left[\frac{\delta^2 g(H_{r_1+1})}{\Delta p} + \frac{g(H_{r_1+1}) - g(H_r)}{\Delta p} - \frac{g(H_{r_1+1})}{p_{r_1+1}}\right].$$

Then we have asymptotically, for $i = r_1 + 2, \ldots, n - r_2 - 1$,

$$\frac{1}{n}\sum_j \beta^{ij} \sim -g(\xi_{\lambda_i})\frac{d^2 g(\xi_{\lambda_i})}{d\lambda_i^2}\, d\lambda_i = a_1(\lambda_i)\, d\lambda_i \text{ (say)} \sim \frac{a_1(\lambda_i)}{n}$$

and

$$\frac{1}{n} \sum_j \beta^{r_1+1,j} \sim a_1(\lambda_{r_1+1})\, d\lambda_{r_1+1} + a_{1,r_1+1},$$

where

$$a_{1,r_1+1} = \frac{g^2(H_{r_1+1})}{p_{r_1+1}} - g'(H_{r_1+1});$$

and likewise for $a_{1,n-r_2}$. Also (for $i = r_1 + 2, \ldots, n - r_2 - 1$)

$$\frac{1}{n} \Sigma \alpha_j \beta^{ij} \sim g(\xi_{\lambda_i}) \frac{d^2(\xi_{\lambda_i} g(\xi_{\lambda_i}))}{d\lambda_i^2}\, d\lambda_i = a_2(\lambda_i)\, d\lambda_i.$$

Thus, except for the two most extreme order statistics in the sample, the a_1 and a_2 functions are of the form, with $y = H(u)$:

$$\left.\begin{aligned}
a_1(u) &= -g(y) \frac{d^2 g(y)}{du^2} = -\Psi'(y), \\
a_2(u) &= -g(y) \frac{d^2}{du^2}[yg(y)] = -[\Psi'(y) + y\Psi''(y)].
\end{aligned}\right\} \quad (9.7.6)$$

Correspondingly, from (9.7.5), the coefficients β_i, γ_i, respectively, are given asymptotically by setting $u = i/(n + 1)$ in the continuous weight functions

$$\left.\begin{aligned}
\beta(u) &= \frac{a_1(u)I_{22} - a_2(u)I_{12}}{n(I_{11}I_{22} - I_{12}^2)}, \\
\gamma(u) &= \frac{a_2(u)I_{11} - a_1(u)I_{12}}{n(I_{11}I_{22} - I_{12}^2)};
\end{aligned}\right\} \quad (9.7.7)$$

also

$$\beta_{r_1+1} = \beta(\lambda_{r_1+1}) + \frac{a_{1,r_1+1}I_{22} - a_{2,r_1+1}I_{12}}{n(I_{11}I_{22} - I_{12}^2)}, \quad \text{etc.} \quad (9.7.8)$$

Here I_{11}, I_{12}, I_{22} are given by the RHS's of (9.7.3a–c).

For a symmetrically censored sample from a symmetric distribution, (9.7.7) simplifies to

$$\beta(u) = \frac{a_1(u)}{nI_{11}}, \qquad \gamma(u) = \frac{a_2(u)}{nI_{22}}. \quad (9.7.9)$$

Example (Chernoff *et al.*, 1967). For an uncensored sample from a $N(\mu, \sigma^2)$ parent we have $I_{11} = 1$, $I_{12} = 0$, $I_{22} = 2$, giving

$$\sigma^* = \frac{1}{2n} \sum_{i=1}^n a_2\left(\frac{i}{n+1}\right) X_{i:n}.$$

Since $\Psi'(y) = -y$, we see from (9.7.6) that $a_2(u) = 2y$. Thus

$$\sigma^* = \frac{1}{n}\sum_{i=1}^{n}\Phi^{-1}\left(\frac{i}{n+1}\right)X_{i:n}.$$

The efficient estimator of μ is, of course, just $\bar{X}$.

Next, suppose we wish to estimate μ from a symmetrically censored sample, perhaps because of suspected outliers. Then, for $\lambda_1 = 1 - \lambda_2 = \lambda$,

$$I_{11} = \int_{\lambda}^{1-\lambda}\Psi^2(y)\,du + \frac{2f_\lambda^2}{\lambda}$$

$$= \int_{\Phi^{-1}(\lambda)}^{\Phi^{-1}(1-\lambda)} y^2\phi(y)\,dy + \frac{2\phi^2(\Phi^{-1}(\lambda))}{\lambda}$$

$$= 1 - 2\lambda + 2\Phi^{-1}(\lambda)\phi(\Phi^{-1}(\lambda)) + \frac{2\phi^2(\Phi^{-1}(\lambda))}{\lambda}.$$

Since $a_1(u) = 1$ the required estimator is of the form

$$\mu^* = c(X_{r+1:n} + X_{n-r:n}) + \frac{1}{nI_{11}}\sum_{i=r+1}^{n-r}X_{i:n},$$

where $r = [n\lambda]$, and c, the additional weight of the extreme variates retained, is by (9.7.8)

$$\frac{\phi^2(\Phi^{-1}(\lambda))/\lambda + \Phi^{-1}(\lambda)\phi(\Phi^{-1}(\lambda))}{I_{11}}.$$

μ or σ known

For brevity we consider the uncensored case. If μ is known, $X_{i:n}$ can be replaced by $X_{i:n} - \mu$, leaving Ω unchanged and giving $\gamma(u)$ as in (9.7.8). The resulting estimator

$$\sum_{i=1}^{n}\gamma\left(\frac{i}{n+1}\right)(X_{i:n} - \mu)$$

is asymptotically unbiased and has variance σ^2/nI_{22}. Note, however, that $\Sigma\gamma(i/(n+1))X_{i:n}$ is in general asymptotically biased for σ. Likewise, if σ is known, the appropriate estimator is

$$\sum_{i=1}^{n}\beta\left(\frac{i}{n+1}\right)(X_{i:n} - \alpha_i\sigma),$$

where α_i may be replaced by H_i.

A rather different approach to optimal asymptotic estimation is due to Jung (1955). This is summarized in Sarhan and Greenberg (1962).

EXERCISES

9.2.1. For $0 < \lambda < 1$ assume that there exist numbers $a < b$ such that

$$a = \inf \{x \mid P(x) = \lambda\} \quad \text{and} \quad b = \sup \{x \mid P(x) = \lambda\}.$$

Show that the limiting distribution of $X_{[n\lambda]:n}$ (where $[x]$ denotes the integral part of x) is given by

$$\lim_{n \to \infty} \Pr \{X_{[n\lambda]:n} \leq x\} = 0 \quad \text{if } x < a,$$
$$= \tfrac{1}{2} \quad \text{if } a \leq x < b,$$
$$= 1 \quad \text{if } x \geq b.$$

(Feldman and Tucker, 1966)

9.2.2. Bahadur (1966) shows that, if $P(\xi_\lambda) = \lambda$, P has at least two derivatives in some neighborhood of ξ_λ, $P''(x)$ is bounded in the neighborhood, and $P'(\xi_\lambda) = p(\xi_\lambda) > 0$, then $X_{[n\lambda]:n}$ may be expressed as

$$X_{[n\lambda]:n} = \xi_\lambda - \frac{P_n(\xi_\lambda) - \lambda}{p(\xi_\lambda)} + R_n,$$

where $P_n(\xi_\lambda)$ is the proportion of $X_i \leq \xi_\lambda$ $(i = 1, 2, \ldots, n)$ and R_n becomes negligible as $n \to \infty$.

Noting that, with

$$\delta(x) = \begin{matrix} 1 & x \geq 0, \\ 0 & x < 0 \end{matrix}$$

$P_n(x)$ may be written as

$$P_n(x) = \frac{1}{n} \sum_{i=1}^{n} \delta(x - X_i),$$

show that for $x \leq y$

$$\text{cov}\,(P_n(x), P_n(y)) = \frac{1}{n} P(x)[1 - P(y)]$$

and establish Theorem 9.2.

9.3.1. Let
(a) $P(x) = 1 - e^{-x^\alpha}$ $x \geq 0, \alpha > 0$;
(b) $P(x) = 1 - \exp[-x/(1 - x)]$ $0 < x \leq 1$.

Show from first principles or otherwise that

$$P^n(a_n x + b_n) \to e^{-e^{-x}},$$

where for (a)

$$a_n = \frac{1}{\alpha} (\log n)^{(1-\alpha)/\alpha}, \quad b_n = (\log n)^{1/\alpha},$$

and for (b)

$$a_n = (\log n)^{-2}, \quad b_n = \frac{\log n}{1 + \log n}.$$

(Gnedenko, 1943)

9.3.2. Show that, for a unit normal parent, a better approximation to l_n than (9.3.9) is

$$l_n \sim (2 \log n)^{1/2} - \tfrac{1}{2}(2 \log n)^{-1/2}(\log \log n + \log 4\pi).$$

[Hint: Improve (9.3.9) by Newton's method.]

(Cramér, 1946)

9.3.3. Let $X_1, X_2, X_3, \ldots$ be iid *positive* variates with cdf $P(x)$ and let

$$R_n = \frac{\displaystyle\sum_{i=1}^{n} X_{1:i}}{\log n}.$$

When the X_i are uniform $(0, 1)$ show that

(a) $\lim_{n \to \infty} \mathscr{E} R_n = 1,$

(b) $\mathrm{cov}\,(X_{1:n}, X_{1:n+k}) = \dfrac{n}{(n + 1)(n + k + 1)(n + k + 2)}$ $\quad k = 1, 2, \ldots,$

(c) $\lim_{n \to \infty} \mathrm{var}\, R_n = 0.$

Hence show that for general $P(x)$ as above, R_n converges in probability to $\lim_{t \downarrow 0} t/P(t)$ (assumed to exist as a finite or infinite number).

(Grenander, 1965)

9.4.1. Show that for the ith (i fixed) quasi-range

$$W_{(i)} = X_{(n-i+1)} - X_{(i)},$$

(9.4.2) generalizes to

$$f(w_{(i)}) \sim \frac{Cp^2(1 - P)^{2(i-1)}(2P - 1)^{n-2i}}{\left[\left(\dfrac{p'}{p}\right)^2 - \dfrac{p''}{p} + \dfrac{i - 1}{1 - P}\left(\dfrac{p^2}{1 - P} + p'\right) - \dfrac{(n - 2i)p'}{2P - 1} \right]^{1/2}},$$

where

$$C = \frac{n!\pi^{1/2}}{(i - 1)!(n - 2i)!(i - 1)!}.$$

(Cadwell, 1953a)

9.7.1. Show that the asymptotically efficient estimators of μ and σ for the logistic distribution

$$P(x) = \left\{ 1 + \exp\left[\frac{-(x - \mu)}{\sigma} \right] \right\}^{-1} \quad -\infty < x < \infty$$

are determined by the respective weight functions

$$\beta(u) = \frac{6u(1 - u)}{n},$$

$$\gamma(u) = \frac{9\{2u - 1 + 2u(1 - u) \log [u/(1 - u)]\}}{n(\pi^2 + 3)}.$$

(Gupta and Gnanadesikan, 1966; Chernoff et al., 1967)

APPENDIX

Guide to Tables

Sections in the Appendix are numbered to correspond to those in the main text.

PH: Pearson and Hartley (1966)—*Biometrika Tables I*
SG: Sarhan and Greenberg (1962)—*Contributions to Order Statistics*
D: decimal (e.g., to 3D = to 3 decimal places)
S: significant (e.g., to 4S = to 4 significant figures)
A3.1: Appendix section 3.1

2.1. The cdf of the extreme in samples from a unit normal parent was tabulated early by Tippett (1925) for $n = 3, 5, 10, 20, 30, 50, 100$ (100) 1000. Percentage points for $n \leq 30$ appear in Table 24 of PH. Percentage points of all normal order statistics are given by Gupta (1961) for $n \leq 10$ and by Govindarajulu and Hubacker (1964) for $n \leq 30$.

The last two authors deal also with the percentage points of order statistics from a uniform parent ($n = 1(1)30(5)60$), from a chi distribution with 1 DF (see A3.1 for definition of this distribution and others) and from three Weibull distributions. Gupta (1960) treats the chi-square distribution with even DF and, with Shah (1965), the logistic distribution.

Eisenhart *et al.* (1963) give percentage points of the median in samples from normal, double-exponential, uniform, Cauchy, sech, and sech^2 distributions.

2.3. The extensive tables by Harter and Clemm (1959) of cdf and percentage points of the range W in samples from a unit normal parent give $F(w)$ to 8D for w in steps of 0.01 and $n = 2(1)20(2)40(10)100$, and 23 different percentage points to 6D for each n. The percentage points appear also in Harter (1960). See also Pearson and Hartley (1942; 1966, Tables 22 and 23), who provide the first detailed exact tabulation.

Percentage points of W in samples from a rectangular parent of unit s.d. are given by Harter (1961c) to 6D for $n = 2(1)20(2)40(10)100$.

225

2.4. Percentage points of the range of n independent binomial $b(p, N)$ variates are given by Siotani and Ozawa (1958). See also Ishii and Yamasaki (1961).

2.5. MacKinnon (1964) tables the integer $I = r - 1$ making $(X_{(r)}, X_{(n-r+1)})$ a confidence interval for the median with confidence coefficient $\geq 1 - \alpha$ for $\alpha = 0.001, 0.01, 0.02, 0.05, 0.10, 0.50$ and $n = 1(1)1000$. See also Dixon and Massey (1957) and Owen (1962, p. 362).

2.6. See Murphy (1948), Somerville (1958), and Owen (1962, p. 317).

3.1. Variances and covariances of order statistics from the uniform distribution (as well as in some other cases) are tabled for $n \leq 10$ by Hastings *et al.* (1947). Expected values (selected $n \leq 100$) are given by Lieblein and Salzer (1957) for the extreme-value distribution with cdf

$$P(x) = e^{-e^{-x}} \qquad -\infty < x < \infty.$$

For $n \leq 6$ Lieblein and Zelen (1956) also tabulate the covariances (reproduced in SG, p. 405). All means and variances for $n \leq 20$ (and privately for $n \leq 100$) are given by White (1969); strictly, White deals with $-X$, which he calls a "reduced log-Weibull" variate. Lieblein (1955) presents procedures for the Weibull distributions with cdf's $(m > 0)$

$$\begin{aligned} P(x) &= 1 - \exp\left[-(-x)^{-m}\right] & x &\leq 0, \\ &= 1 & x &> 0, \end{aligned}$$

and

$$\begin{aligned} P(x) &= 0 & x &\leq 0, \\ &= 1 - e^{-x^m} & x &> 0. \end{aligned}$$

Closely related is the generalized extreme-value distribution (Maritz and Munro, 1967) with cdf

$$\begin{aligned} P(x) &= \exp\left[-(1 - \gamma x)^{1/\gamma}\right] & \gamma &> 0, \quad -\infty < x < 1/\gamma \\ & & \gamma &< 0, \quad 1/\gamma < x < \infty. \end{aligned}$$

For $5 \leq n \leq 10$, $\gamma = -0.10(0.05)0.40$, the authors give 3D tables of all expected values (see also Section 9.1).

For the gamma distribution $(r > 0)$

$$\begin{aligned} p(x) &= e^{-x} x^{r-1}/\Gamma(r) & x &\geq 0, \\ &= 0 & x &< 0, \end{aligned}$$

Gupta (1960) gives the first four moments for $r = 1(1)5$ and $n \leq 10$ and the moments of $X_{1:n}$ for $n \leq 15$. Breiter and Krishnaiah (1968) add the first four moments for $r = 0.5(1)10.5$ and $n \leq 9$. Malik (1966) tabulates to 4D means and covariances for $n \leq 8$ for the Pareto distribution

$$p(x) = va^v x^{-v-1} \qquad a > 0, v > 0, x \geq a,$$
$$= 0 \qquad \text{elsewhere,}$$

with $v = 2.5(0.5)5.0$. Sarhan (1954) tables means and covariances ($n \leq 5$) for the triangular distribution

$$p(x) = 4x \qquad 0 \leq x \leq \tfrac{1}{2},$$
$$= 4(1-x) \qquad \tfrac{1}{2} \leq x \leq 1,$$

and the double exponential

$$p(x) = \tfrac{1}{2} e^{-|x|} \qquad -\infty < x < \infty.$$

For the latter distribution all means and covariances are tabulated for $n \leq 20$ by Govindarajulu (1966). Govindarajulu and Eisenstat (1965) give all means and covariances for $n = 1(1)20(10)100$ for the chi distribution with 1DF, namely, for

$$p(x) = (2/\pi)^{1/2} e^{-\frac{1}{2}x^2} \qquad x \geq 0,$$
$$= 0 \qquad x < 0.$$

Exact values for $n \leq 5$ were derived earlier by Govindarajulu (1962). The first four moments for the standardized logistic distribution

$$p(x) = \frac{\pi}{\sqrt{3}} \frac{e^{-\pi x/\sqrt{3}}}{(1 + e^{-\pi x/\sqrt{3}})^2} \qquad -\infty < x < \infty$$

are given ($n \leq 10$) by Gupta and Shah (1965). Shah (1966b) also tables the covariances for $n \leq 10$, and Gupta et al. (1967) for $11 \leq n \leq 25$. Compare Tarter and Clark (1965).

For the Cauchy distribution

$$p(x) = \frac{1}{\pi(1 + x^2)} \qquad -\infty < x < \infty$$

$\mu_{r:n}$ does not exist for $r = 1, n$ and $\sigma_{r:n}^2$ does not exist for $r = 1, 2, n-1, n$. Barnett (1966) tabulates all existing means for $n \leq 20$ and $\sigma_{rs:n}$ for $r, s = 3$ to $n - 2$ with $n = 5(1)16(2)20$.

For use in robustness studies (Section 6.5) Gastwirth and Cohen (1968) give to 5D all means and covariances for $n \leq 20$ for the (scale-) contaminated normal distribution

$$p_{\gamma,\kappa}(x) = (2\pi)^{-1/2}\left[(1-\gamma)e^{-\frac{1}{2}x^2} + \left(\frac{\gamma}{\kappa}\right)e^{-\frac{1}{2}x^2/\kappa^2}\right]$$

with $\gamma = 0.01, 0.05, 0.10$, and $\kappa = 3$.

3.2. Harter (1961a) tabulates $\mu_{r:n}$ for a unit normal parent to 5D for all r and for $n = 2(1)100(25)250(50)400$. For $n \leq 20$ all $\mu_{r:n}$ and $\mu_{rs:n}$ are given to 10D by Teichroew (1956), abbreviated from unpublished computations to 20D; the corresponding $\sigma_{rs:n}$ to 10D appear in Sarhan and Greenberg (1956). Ruben (1954) tables the first ten moments of $X_{n:n}$ for $n \leq 50$, and Borenius (1966) the first two to 7D for $n \leq 120$.

Expected values and variances of the rth quasi-range ($r = 0, 1, \ldots, 8$) are tabled by Harter (1959) for $n \leq 100$. He also gives (1960) the mean, variance (10D), Pearson's $\beta_1^{1/2}$ (8D), and β_2 (6D) for the range W_n for $n \leq 100$. The means had already been computed to 5D for $n \leq 1000$ by Tippett in 1925.

5.2. Harter *et al.* (1959) give the cdf of $Q_{n,\nu}$ to 6D or 6S, whichever is less accurate, for $n = 2(1)20(2)40(10)100$ and $\nu = 1(1)20, 24, 30, 40, 60, 120$. Their percentage points to 4D or 4S, whichever is less accurate, corresponding to cdf's 0.001, 0.005, 0.01, 0.025, 0.05, 0.1(0.1)0.9, 0.95, 0.975, 0.99, 0.995, 0.999 are reproduced by Harter (1960). Less extensive upper percentage points were obtained earlier by Pearson and Hartley (see PH, Table 29).

5.3. Grubbs (1950) tabulates the cdf of $X_{(n)} - \bar{X}$ to 5D for $n = 2(1)25$ at intervals of 0.05. He also provides upper 10, 5, 1, and 0.5% points to 3D. Related statistics for which upper significance points have been tabulated wholly or in part by the method of this section include (normal parent distributions assumed throughout):

$(X_{(n)} - \bar{X})/S_\nu$ to 2D for $n = 2(1)10, 12; \nu \geq 5$ or 10;
$\alpha = 10, 5, 2.5, 1, 0.5, 0.1\%$

$\qquad\qquad\qquad\qquad\qquad\qquad\qquad\qquad$ (PH, Table 26)

$\max |X_{(i)} - \bar{X}|/S_\nu$ to 2D for $n = 3(1)10(5)20(10)60; \nu \geq 3$;
$\alpha = 5, 1\%$

$\qquad\qquad\qquad\qquad\qquad\qquad\qquad\qquad$ (Halperin *et al.*, 1955)

$(X_{(n)} - \bar{X})/[(n-1)S^2 + \nu S_\nu^2]^{1/2}$ $\left.\begin{array}{c} \\ \\ \end{array}\right\}$ to 3D for $n = 2(1)10, 12, 15, 20$;

$\max_i |X_{(i)} - \bar{X}|/[(n-1)S^2 + \nu S_\nu^2]^{1/2}$ $\quad 0 \leq \nu \leq 50, \alpha = 5, 1\%$

$\qquad\qquad\qquad\qquad\qquad\qquad\qquad\qquad$ (PH, Tables 26a, b)

W_n/S to 2D or 3D for $n = 3(1)20(5)100(50)200, 500, 1000$; upper and lower 10, 5, 2.5, 1, 0.5, 0% points

$\qquad\qquad\qquad\qquad\qquad\qquad\qquad\qquad$ (PH, Table 29c)

Siotani (1959) tabulates to 2D upper 5, 2.5, 1% points of $\chi^2_{\max \cdot D}$ for $n = 3(1)10(2)20(5)30$ and $p = 2, 3, 4$ and of $T^2_{\max \cdot D}$ for $n = 3(1)12, 14$; $p = 2; \nu \geq 20$.

Let $X_1, X_2, \ldots, X_n$ be n independent χ^2 variates with ν_1 DF and let X_0 be another independent χ^2 with ν_2 DF. Armitage and Krishnaiah (1964) give to 2D upper 10, 5, 2.5, 1% points of

$$F_{n,\alpha}^* = \nu_2 \cdot \max_{i=1,2,\ldots,n} X_i/\nu_1 X_0$$

for $n = 1(1)12$, $\nu_1 = 1(1)19$, $\nu_2 = 5$ or $6(1)45$. Published tables are confined to $\nu_1 = 1$ (Table 19 of PH with extensions by Chambers, 1967, but there are discrepancies between the two tables). Krishnaiah and Armitage (1964) give similar 4D tables of lower 10, 5, 2.5, 1% points of

$$\nu_2 \min X_i/\nu_1 X_0.$$

For $\nu_1 = \nu_2 = 2(2)50$, $k = 1(1)10$, Gupta and Sobel (1962) provide 4D tables of lower 25, 10, 5, 1% points.

5.4. Fisher gives upper 5 and 1% points of $Y_{(n)}$ (1950) and upper 5% points of $Y_{(n-1)}$ (1940), all for $n \leq 50$. PH, Table 31a, lists to 4D upper 5 and 1% points of $\max {}_j S^2/\sum_{j=1}^n {}_j S^2$ for $n = 2(1)10$, 12, 15, 20 and $\nu \geq 1$; Table 31b gives to 3D upper 5% points of $\max {}_j W/\Sigma_j W$ for the same n and $m = 2(1)10$.

5.5. Let $Y_1, Y_2, \ldots, Y_n$ be multinormal with zero means, unit variances, and equal correlations ρ. Gupta (1963a) tabulates $\Pr\{Y_{(n)} < y\}$ to 5D for $n = 1(1)12$, y in steps of 0.1, and $\rho = 0.1(0.1)0.9$; 0.125(0.125)0.875; $\frac{1}{3}$, $\frac{2}{3}$. Upper 10, 5, 2.5, 1, and 0.5% points to 3D of $Y_{(n)}$ for $n = 2(1)10$ and $\rho = 0(0.2)1.0$ were obtained earlier by Thigpen (1961), who also found the same percentage points of $\max |Y_i|$. The latter table is largely superseded by the work of Krishnaiah and Armitage (1965a), who give tables of $\Pr\{Y_{(n)}^2 \leq y\}$ to 6D for $n = 1(1)10$, y in steps of 0.1, and $\rho = 0.0(0.0125)0.85$, as well as to 3D the corresponding upper 10, 5, 2.5, and 1% points.

Let $Y_1, Y_2, \ldots, Y_n$ be multinormal with zero means, common unknown variance σ^2, and equal correlation ρ. Let S_ν be the usual rms estimator of σ independent of the Y_i. Krishnaiah and Armitage (1965b) tabulate to 2D upper 10, 5, 2.5, 1% points of

$$\max (Y_i/S_\nu) = Y_{(n)}/S_\nu$$

for $n = 1(1)10$, $\nu = 5(1)35$, and $\rho = 0.0(0.1)0.9$. Pillai and Ramachandran (1954) give upper and lower 5% points of $\max |Y_i/S_\nu|$ for $\rho = 0$, the studentized maximum modulus. See also the bibliography by Gupta (1963b).

6.1. Rider's (1951) tables of W_1/W_2 for a $R(0, \theta)$ parent provide two-sided 10, 5, 1% points for $n_1, n_2 \leq 10$. For the same sample sizes Hyrenius (1953) gives upper 10, 5, 1% points of his T, and upper and lower 10, 5, 1% points of his U and V.

For k independent samples of n from a $R(0, \theta)$ parent Khatri (1960) tabulates for $k = 2(1)5$ and $n = 4(1)10(5)20$ lower 5% points of $W_{(1)}/W_{(k)}$, the ratio of the smallest to the largest range, and for $k = 2(1)11$ and $n = 1(1)10(5)30, 40, 60, 100, 500, 1000$, lower 5% points of $Y_{(1)}/Y_{(k)}$, the ratio of the smallest to the largest sample maximum.

Harter (1961c) tabulates the reciprocals of many lower and upper percentage points w_α of the range in samples of $n = 2(1)20(2)40(10)100$ from a rectangular parent of unit s.d. Since $\Pr\{W/\sigma < w_\alpha\} = 1 - \alpha$ implies $\Pr\{\sigma > W/w_\alpha\} = 1 - \alpha$, it is only necessary to multiply an observed range by $1/w_\alpha$ to obtain a lower confidence bound on σ with confidence coefficient $1 - \alpha$.

6.3. Many of the references listed in this section deal with ML estimation, and some of them contain tables giving variances, covariances, and efficiencies of the estimators. We confine ourselves here to tables giving the coefficients of the best linear unbiased estimates (BLUE), the uncensored case being always included.

Sarhan and Greenberg (1962) table to 4D the coefficients of the BLUE of μ and σ for a $N(\mu, \sigma^2)$ population, covering all cases of (Type II) censoring for $n \leq 20$ (pp. 218–51) with variances, covariances, and efficiencies of the estimators (pp. 252–68). For $n \leq 5$ they deal similarly in a combined table with certain symmetrical populations, namely, U-shaped, rectangular, parabolic, triangular, normal, and double-exponential (pp. 391–5, where the populations are defined). Also considered are the extreme-value distribution for $n \leq 6$ and censoring on the right (pp. 404–5; note table headings are reversed) and the right-triangular distribution for $n \leq 10$ (uncensored, pp. 452–3). All cases of censoring on the right for $n \leq 20$ in the extreme-value distribution are covered by White (1964), who gives 7D tables of the coefficients.

Govindarajulu and Eisenstat (1965) deal with the folded normal distribution, namely,

$$p(x) = (2/\pi\sigma^2)^{1/2} \exp\{-[(x - \mu)/\sigma]^2\} \quad x > \mu,$$
$$= 0 \qquad\qquad\qquad\qquad\qquad \text{elsewhere.}$$

They table to 4D coefficients for many cases of censoring, mainly on the right, when $n \leq 20$. Govindarajulu (1966) treats symmetric censoring in the double exponential for $n \leq 20$. For the reader equipped with magnifying glasses Gupta et al. (1967) handle all cases of censoring for the logistic distribution for $n = 2, 5(5)25$.

7.3. Upper 10, 5, and 1% points of the range of rank totals in a two-way classification are given by Dunn-Rankin and Wilcoxon (1966).

7.4. Harter (1959) tabulates the means and variances of $W_{(i)}/\sigma$ in normal samples for $n \leq 100$ and $i \leq 9$ (including $i = 1$—his $r = 0$—the range). He also gives the efficiencies of $W_{(i)}$ for $n = 2(2)50(5)100$ and $i \leq 9$ and of estimators which are linear combinations of two $W_{(i)}$.

7.6. Ogawa (1962, pp. 278–82) tabulates the optimal linear k-point estimators of μ and σ (normal parent) for $k = 1(1)10$ (for $\mu_0{}^*$) and $k = 1(1)6$ (for $\sigma_0{}^*$). (Note that in the latter case his coefficients for $k = 2$ are twice the correct values.) Corresponding (correct) results are given by Eisenberger and Posner (1965) for $k = 2(2)20$. For the same k, these authors also give the estimators of μ and σ based on a common spacing and minimizing var $\mu^* + c$ var σ^* ($c = 1, 2, 3$). Efficiencies accompany all estimators.

7.7. Harter (1964b) gives to 6D, for $n = 2(1)20(2)40(10)100$, $\alpha = 0.005$, 0.01, 0.05, 0.1, the two values which on multiplication by the indicated $w_{(i)}$ give upper and lower $1 - \alpha$ confidence limits for σ.

Upper 50, 25, 10, 5, 2.5, 1, 0.5, and 0.1% points of $_1W/_2W$ are tabulated by Harter (1963) (also Table 29b of PH) to 4S for $n_1, n_2 \leq 15$.

Table 31 of PH gives upper 5 and 1% points of S^2_{max}/S^2_{min} to 3S and at least 2S, respectively, for $k = 2(1)12$ and $v = 2(1)10$, 12, 15, 20, 30, 60. Leslie and Brown (1966), for the same k and v, tabulate to 4S upper 5, 2.5, 1, and 0.5% points of W_{max}/W_{min}.

Moore (1957) (also Table 29a of PH) gives to 3D upper 10, 5, 2, and 1% points of $\frac{1}{2}R_2$ for $n_1, n_2 = 2(1)20$. Jackson and Ross (1955) give to 2D upper 10, 5, and 1% points of G_1 and G_2 for $n' = 2(1)15$ and $k, k_1, k_2 = 1(1)15$.

Upper 5 and 1% points of Q' are given to at least 2S for $k, n = 2(1)10$ (Beyer, 1968). For a two-way classification, calculation of the scale factor c and the equivalent degrees of freedom v presents a difficult problem in numerical integration. Hartley's (1950a) values have been revised by Mardia (1967) and printed in a form evidently meant to replace Table 30B of PH at the next reprinting. The practical effect of the revision is not great. The table gives c to 2D and v to 1D for $n = 2(1)10$ and all k.

7.8. For $n \leq 50$ Shapiro and Wilk (1965) give 4D tables of the coefficients a_i and 3D tables of 1, 2, 5, 10, 50, 90, 95, 98, 99% points of their statistic W^*, defined in Ex. 7.8.1.

7.9. Mitra (1957) tabulates c to 3D (but using the original form of Patnaik's approximation to $\overline{W}$) for $\gamma = 0.75, 0.90, 0.95, 0.99, 0.999$; $\beta = 0.75, 0.90, 0.95, 0.99$; $n = 2(1)20$, $k = 1$, and $n = 4, 5, k = 4(1)20(5)40$, 50, 75, 100, ∞.

8.2. Many of the available tables of percentage points for the statistics of (a)–(d) are reproduced (but not always in full) in PH and SG. In some cases details have already been given in A5.2 and A5.3.

A_1: 3D for $n = 1(1)30$, lower and upper 10,
5, 2.5, 1, 0.5, 0.1% points   (PH, SG p. 322)
B_1: See A5.3   (PH, SG p. 323)
B_2: Use C_2 for $v = \infty$ or Krishnaiah and
Armitage (1965a) as in A5.5
B_3: $\Pr\{X_{(n)} - X_{(n-1)} > \lambda\sigma\}$ for
$n = 2,\ 3,\ 10(10)100(100)1000;\ \lambda = 0.1$
$(0.1)5$ (SG p. 325)
C_1: See A5.3   (PH, SG p. 326)
C_2: See A5.3
C_3: See A5.2   (PH, SG pp. 114–5)
C_4, C_5 See A5.3   (PH)
D_1: Grubbs (1950) tabulates to 3D upper 10,
5, 2.5, 1% points of

$$\frac{x_{(n)} - \bar{x}}{\left[(1/n)\sum_{i=1}^{n}(x_i - \bar{x})^2\right]^{1/2}}$$

and to 4D the corresponding lower percentage points of the equivalent statistic

$$\frac{\sum_{i=1}^{n-1}(x_{(i)} - \bar{x}_n)^2}{\sum_{i=1}^{n}(x_i - \bar{x})^2}$$

(SG p. 330)
D_2: Use $D_2 = C_5(n-1)^{1/2}$ for $v = 0$ (PH)
D_3: See A5.3   (PH, SG pp. 328–9)
D_4: 3D for $n = 25(5)50(10)100(25)200(50)$
$1000(200)2000(500)5000$; upper 5, 1%
points   (PH)
D_5: 2D for $n = 50(25)150(50)700(100)1000$
$(200)2000(500)5000$; upper and lower 5,
1% points.   (PH)
D_6: 4D for $n = 4(1)20$, lower 10, 5, 2.5, 1%
points   (SG, p. 331)
Dixon's r statistics: 3D for $n = 3(1)30$ and the following percentiles: 99.5,
99, 98, 95, 90(10)10, 5.   (SG, pp. 332–7)

Laurent (1963) tabulates $\Pr\{U_{(n)} \leq u\}$, with $U_{(n)}$ as defined in Exs. 8.2.2 and
8.2.3, to 5D for $n = 3(1)10$ and $u = 0.1(0.1)1$. He also gives upper 10
and 1% points of $U_{(n)}$.

Wilks (1963) tabulates r_α to 5D for $n = 5(1)30(5)100(100)500$, $k = 1(1)5$;
$\alpha = 0.1, 0.05, 0.025, 0.01$.

8.3. Mosteller (1948) tabulates the probability $P_r = nm^{(r)}/(nm)^{(r)}$ that in n samples of m there is one sample having r or more observations larger than those of the other $n - 1$ samples for $m = 3, 5, 7, 10, 15, 20, 25, \infty$; $n = 2(1)5$ or 6; $r = 2(1)5$ or 6. Tables for max T_i, the largest Kruskal-Wallis rank sum in n groups of m, are given by Odeh (1967), who provides, in addition to critical values for various α and $n = 2(1)6$, $m = 2(1)8$, the cdf for $n \leq 5$, $m \leq 5$. Thompson and Willke (1963) tabulate the significance points with their associated exact significance levels for a two-sided rank sum test when n objects are ranked by each of m judges, for nominal significance levels $\alpha = 0.01, 0.03, 0.05$; $n = 3(1)15$; $m = 3(1)15$.

For b_α, $b_\alpha{}^*$ see A5.3.

For tables of $s_{max}^2/\sum_{i=1}^{n} {}_i s^2$ see A5.4.

8.4. David and Paulson (1965) tabulate P_1 of B_1 and B_5 to 3D for $\lambda = 1(1)5$; $n = 2(1)10, 12, 15, 20, 25$; $\alpha = 0.05, 0.01$. (Actually $n = 3, 4, 5, 6, 8, 12$ for B_5.)

References

The numbers in square brackets give the pages on which the corresponding reference is cited.

Abdel-Aty, S. H. (1954). Ordered variables in discontinuous distributions. *Statist. Neerlandica* **8**, 61–82. [23]

Aiyar, K. R. (1963). On uncorrelated linear functions of order statistics. *J. Amer. Statist. Ass.* **58**, 245–6. [132]

Ali, M. M. and Chan, L. K. (1964). On Gupta's estimates of the parameters of the normal distribution. *Biometrika* **51**, 498–501. [105–6]

Ali, M. M. and Chan, L. K. (1965). Some bounds for expected values of order statistics. *Ann. Math. Statist.* **36**, 1055–7. [63, 69]

Anis, A. A. (1955). The variance of the maximum of partial sums of a finite number of independent normal variates. *Biometrika* **42**, 96–101. [85]

Anis, A. A. (1956). On the moments of the maximum of partial sums of a finite number of independent normal variates. *Biometrika* **43**, 79–84. [85]

Anis, A. A. and Lloyd, E. H. (1953). On the range of partial sums of a finite number of independent normal variates. *Biometrika* **40**, 35–42. [85]

Anscombe, F. J. (1960). Rejection of outliers. *Technometrics* **2**, 123–47. [170–1, 191]

Anscombe, F. J. (1961). Examination of residuals. *Proc. 4th Berkeley Symp.* I, 1–36. [171]

Anscombe, F. J. and Barron, B. A. (1966). Treatment of outliers in samples of size three. *J. Res., Nat. Bur. Stand.* **70B**, 141–51. [195]

Anscombe, F. J. and Tukey, J. W. (1963). The examination and analysis of residuals. *Technometrics* **5**, 141–60. [171]

Armitage, J. V. and Krishnaiah, P. R. (1964). Tables for the studentized largest chi-square distribution and their applications. Aerospace Research Laboratories 64–188. [229]

Babik, S. (1968). Application of reliability theory to a reactor safety circuit. *Appl. Statist.* **17**, 137–56. [124]

Bahadur, R. R. (1966). A note on quantiles in large samples. *Ann. Math. Statist.* **37**, 577–80. [120]

Bain, L. J. and Antle, C. E. (1967). Estimation of parameters in the Weibull distribution. *Technometrics* **9**, 621–7. [101]

Bain, L. J. and Thoman, D. R. (1968). Some tests of hypotheses concerning the three-parameter Weibull distribution. *J. Amer. Statist. Ass.* **63**, 853–60. [124]

Bain, L. J. and Weeks, D. L. (1965). Tolerance limits for the generalized gamma distribution. *J. Amer. Statist. Ass.* **50**, 1142–52. [223]

Barlow, R. E. (1965). Bounds on integrals with applications to reliability problems. *Ann. Math. Statist.* **36**, 565–74. [62]

235

Barlow, R. E. and Gupta, S. S. (1966). Distribution-free life test sampling plans. *Technometrics* **8**, 591–613. [124]

Barlow, R. E., Madansky, A., Proschan, F., and Scheuer, E. M. (1968). Statistical estimation procedures for the "burn-in" process. *Technometrics* **10**, 51–62. [124]

Barlow, R. E., Marshall, A. W., and Proschan, F. (1963). Properties of probability distributions with monotone hazard rate. *Ann. Math. Statist.* **34**, 375–89. [62]

Barlow, R. E. and Proschan, F. (1965). *Mathematical Theory of Reliability.* Wiley, New York. [121]

Barlow, R. E. and Proschan, F. (1967). Exponential life test procedures when the distribution has monotone failure rate. *J. Amer. Statist. Ass.* **62**, 548–60. [124]

Barnard, G. A. (1953). Time intervals between accidents—a note on Maguire, Pearson and Wynn's paper. *Biometrika* **40**, 212–13. [124]

Barndorff-Nielsen, O. (1963). On the limit behaviour of extreme order statistics. *Ann. Math. Statist.* **34**, 992–1002. [204, 210]

Barndorff-Nielsen, O. (1964). On the limit distribution of the maximum of a random number of independent random variables. *Acta Mathematica* **15**, 399–403. [214]

Barnett, F. C., Mullen, K., and Saw, J. G. (1967). Linear estimates of a population scale parameter. *Biometrika* **54**, 551–4. [146–7]

Barnett, V. D. (1966). Order statistics estimators of the location of the Cauchy distribution. *J. Amer. Statist. Ass.* **61**, 1205–18. Correction **63**, 383–5. [40, 227]

Barnett, V. D. and Lewis, T. (1967). A study of low-temperature probabilities in the context of an industrial problem. *J. R. Statist. Soc. A* **130**, 177–206. [200]

Barr, D. R. (1966). On testing the equality of uniform and related distributions. *J. Amer. Statist. Ass.* **61**, 856–64. [100]

Bartholomew, D. J. (1957). A problem in life testing. *J. Amer. Statist. Ass.* **52**, 350–5. [124]

Bartholomew, D. J. (1963). The sampling distribution of an estimate arising in life testing. *Technometrics* **5**, 361–74. [123]

Barton, D. E. and Casley, D. J. (1958). A quick estimate of the regression coefficient. *Biometrika* **45**, 431–5. [149]

Barton, D. E. and David, F. N. (1956). Some notes on ordered random intervals. *J. R. Statist. Soc. B* **18**, 79–94. [89]

Barton, D. E. and David, F. N. (1959). Combinatorial extreme value distributions. *Mathematika* **6**, 63–76. (Slightly elaborated in Chapter 13 of David and Barton, 1962.) [76, 82, 90]

Basu, A. P. (1965). On some tests of hypotheses relating to the exponential distribution when some outliers are present. *J. Amer. Statist. Ass.* **60**, 548–59. [124, 177, 196]

Basu, A. P. (1967). On the large sample properties of a generalized Wilcoxon-Mann-Whitney statistic. *Ann. Math. Statist.* **38**, 905–15. [124]

Basu, A. P. (1968). On a generalized Savage statistic with applications to life testing. *Ann. Math. Statist.* **39**, 1591–604. [124]

Basu, D. (1955). On statistics independent of a complete sufficient statistic. *Sankhyā* **15**, 377–80. [71, 96, 196]

Bechhofer, R. E., Kiefer, J., and Sobel, M. (1968). *Sequential Identification and Ranking Procedures.* University of Chicago Press. [3]

Bell, C. B., Blackwell, D., and Breiman, L. (1960). On the completeness of order statistics. *Ann. Math. Statist.* **31**, 794–7. [94]

Belz, M. H. and Hooke, R. (1954). Approximate distribution of the range in the neighborhood of low percentage points. *J. Amer. Statist. Ass.* **49**, 620–36. [145]

Bennett, B. M. (1966). Note on confidence limits for a ratio of bivariate medians. *Metrika* **10**, 52–4. [23]

Bennett, B. M. and Nakamura, E. (1968). Percentage points of the range from a symmetric multinomial distribution. *Biometrika* **55**, 377–9. [160]

Bennett, C. A. (1952). Asymptotic properties of ideal linear estimators. Ph.D. Thesis, University of Michigan. [108, 216, 218]

Benson, F. (1949). A note on the estimation of mean and standard deviation from quantiles. *Suppl. J. R. Statist. Soc.* **11**, 91–100. [137]

Berman, S. M. (1962a). Limiting distribution of the maximum term in sequences of dependent random variables. *Ann. Math. Statist.* **33**, 894–908. [214]

Berman, S. M. (1962b). Convergence to bivariate limiting extreme value distributions. *Ann. Inst. Statist. Math.*, *Tokyo* **13**, 217–23. [214]

Berman, S. M. (1962c). Equally correlated random variables. *Sankhyā A* **24**, 155–6. [215]

Berman, S. M. (1963). Limiting distribution of the studentized largest observation. *Skand. Aktuarietidskr.* 1962, 154–61. [213]

Berman, S. M. (1964). Limit theorems for the maximum term in stationary sequences. *Ann. Math. Statist.* **35**, 502–16. [215]

Beyer, W. H. (Ed.) (1968). *Handbook of Probability and Statistics*, 2nd Ed. The Chemical Rubber Company, Cleveland. [3, 118, 159, 231]

Bhattacharjee, G. P. (1965). Distribution of range in non-normal samples. *Aust. J. Statist.* **7**, 127–41. [145]

Bickel, P. J. (1965). On some robust estimates of location. *Ann. Math. Statist.* **36**, 847–58. [129]

Bickel, P. J. (1967). Some contributions to the theory of order statistics. *Proc. 5th Berkeley Symp.* **I**, 575–91. [216–7]

Bickel, P. J. and Hodges, J. L., Jr. (1967). The asymptotic theory of Galton's test and a related simple estimate of location. *Ann. Math. Statist.* **38**, 73–89. [41, 129]

Birnbaum, A. (1959). On the analysis of factorial experiments without replication. *Technometrics* **1**, 343–57. [163, 183]

Birnbaum, A. (1961). A multi-decision procedure related to the analysis of single degrees of freedom. *Ann. Inst. Statist. Math.*, *Tokyo* **12**, 227–36. [163]

Birnbaum, A. and Laska, E. (1967). Optimal robustness: a general method, with applications to linear estimators of location. *J. Amer. Statist. Ass.* **62**, 1230–40. [127, 136]

Birnbaum, Z. W., Esary, J. D., and Saunders, S. C. (1961). Multi-component systems and structures and their reliability. *Technometrics* **3**, 55–77. [124]

Birnbaum, Z. W. and Saunders, S. C. (1958). A statistical model for life-length of materials. *J. Amer. Statist. Ass.* **53**, 151–60. [124]

Bland, R. P., Gilbert, R. D., Kapadia, C. H., and Owen, D. B. (1966). On the distributions of the range and mean range for samples from a normal distribution. *Biometrika* **53**, 245–8. [141]

Bland, R. P. and Owen, D. B. (1966). A note on singular normal distributions. *Ann. Inst. Statist. Math.*, Tokyo **18**, 113–6. [84]

Blischke, W. R. (1968). Mixtures of distributions. In: Sills, D. L. (Ed.). *International Encyclopedia of the Social Sciences*, Vol. 4, 235–41. Macmillan and Free Press, New York. [171]

Bliss, C. I., Cochran, W. G., and Tukey, J. W. (1956). A rejection criterion based upon the range. *Biometrika* **43**, 418–22. [81]

Bliss, C. I. and Stevens, W. L. (1937). The calculation of the time-mortality curve. *Ann. Appl. Biol.* **24**, 815–52. [162]

Bloch, D. (1966). A note on the estimation of the location parameter of the Cauchy distribution. *J. Amer. Statist. Ass.* **61**, 852–5. [154]

Bloch, D. A. and Gastwirth, J. L. (1968). On a simple estimate of the reciprocal of the density function. *Ann. Math. Statist.* **39**, 1083–5. [204]

Blom, G. (1958). *Statistical Estimates and Transformed Beta-Variables.* Almqvist and Wiksell, Uppsala, Sweden; Wiley, New York. [64, 66, 106, 132, 161]

Blom, G. (1962). Nearly best linear estimates of location and scale parameters. *SG*, 34–46. [106]

Blumenthal, S. (1966). Contributions to sample spacings theory, I: Limit distributions of sums of ratios of spacings. *Ann. Math. Statist.* **37**, 904–24. [213]

Bodmer, W. F. (1959). A significantly extreme deviate in data with a non-significant heterogeneity chi square. *Biometrics* **15**, 538–42. [174]

Bofinger, V. J. (1965). The k-sample slippage problem. *Aust. J. Statist.* **7**, 20–31. [178]

Borenius, G. (1959). On the distribution of the extreme values in a sample from a normal distribution. *Skand. Aktuarietidskr.* 1958, 131–66. [86]

Borenius, G. (1966). On the limit distribution of an extreme value in a sample from a normal distribution. *Skand. Aktuarietidskr.* 1965, 1–15. [86, 228]

Bose, R. C. and Gupta, S. S. (1959). Moments of order statistics from a normal population. *Biometrika* **46**, 433–40. [31]

Box, G. E. P. (1953). Non-normality and tests on variances, *Biometrika* **40**, 318–35. [156]

Breiter, M. C. and Krishnaiah, P. R. (1968). Tables for the moments of gamma order statistics. *Sankhyā B* **30**, 59–72. [226]

Brillinger, D. R. (1966). An extremal property of the conditional expectation. *Biometrika* **53**, 594–5. [25]

Bross, I. D. J. (1961). Outliers in patterned experiments: a strategic appraisal. *Technometrics* **3**, 91–102. [171]

Buckland, W. R. (1964). *Statistical Assessment of the Life Characteristic.* Griffin, London; Hafner, New York. [124]

Burr, I. W. (1955). Calculation of exact sampling distribution of ranges from a discrete population. *Ann. Math. Statist.* **26**, 530–2. Correction **38**, 280. [23]

Cacoullos, T. and DeCicco, H. (1967). On the distribution of the bivariate range. *Technometrics* **9**, 476–80. [148]

Cadwell, J. H. (1952). The distribution of quantiles of small samples. *Biometrika* **39**, 207–11. [139, 166]

Cadwell, J. H. (1953a). The distribution of quasi-ranges in samples from a normal population. *Ann. Math. Statist.* **24**, 603–13. [44, 146, 211–2, 224]

Cadwell, J. H. (1953b). Approximating to the distributions of measures of dispersion by a power of χ^2. *Biometrika* **40**, 336–46. [147, 156]

Cadwell, J. H. (1954). The probability integral of range for samples from a symmetrical unimodal population. *Ann. Math. Statist.* **25**, 803–6. [212]

Carlton, A. G. (1946). Estimating the parameters of a rectangular distribution. *Ann. Math. Statist.* **17**, 355–8. [21]

Chambers, C. (1967). Extension of the tables of percentage points of the largest variance ratio, s^2_{max}/s_0^2. *Biometrika* **54**, 225–7. [71, 229]

Chan, L. K. (1967a). Linear estimation of the location and scale parameters from type II censored samples from symmetric unimodal distributions. *Naval Res. Logist. Quart.* **14**, 135–45. [108]

Chan, L. K. (1967b). Remark on the linearized maximum likelihood estimate. *Ann. Math. Statist.* **38**, 1876–81. [66]

Chan, L. K. (1967c). On a characterization of distributions by expected values of extreme order statistics. *Amer. Math. Monthly* **74**, 950–1. [19]

Cheng, B. (1964). The limiting distributions of order statistics. *Acta Math. Sinica* **14**, 694–714. Translation in *Chinese Mathematics* (1965) **6**, 84–104. [203]

Chernoff, H., Gastwirth, J. L., and Johns, M. V., Jr., (1967). Asymptotic distribution of linear combinations of functions of order statistics with applications to estimation. *Ann. Math. Statist.* **38**, 52–72. [108, 217–8, 220–1, 224]

Chernoff, H. and Lieberman, G. J. (1954). Use of normal probability paper. *J. Amer. Statist. Ass.* **49**, 778–85. [105, 161–2]

Chernoff, H. and Lieberman, G. J. (1956). The use of generalized probability paper for continuous distributions. *Ann. Mat. Statist.* **27**, 806–18. [161]

Chernoff, H. and Teicher, H. (1965). Limit distributions of the minimax of independent identically distributed random variables. *Trans. Amer. Math. Soc.* **116**, 474–91. [214]

Chew, V. (1964). Tests for the rejection of outlying observations. *RCA Systems Analysis Tech. Rept. Memo.* 64–7. [172, 195]

Chu, J. T. (1955). On the distribution of the sample median. *Ann. Math. Statist.* **26**, 112–6. [139]

Chu, J. T. (1957). Some uses of quasi-ranges. *Ann. Math. Statist.* **28**, 173–80. [15]

Chu, J. T. (1968). Some statistical methods for large scale and preliminary data analyses. *Ann. Inst. Statist. Math., Tokyo* **20**, 489–99. [23]

Chu, J. T. and Hotelling, H. (1955). The moments of the sample median. *Ann. Math. Statist.* **26**, 593–606. [139]

Chu, J. T. and Ya'coub, K. (1968). Linear order estimates using subsamples. *SIAM J. Appl. Math.* **16**, 162–6. [109]

Churchman, C. W. and Epstein, B. (1946). Tests of increased severity. *J. Amer. Statist. Ass.* **41**, 567–89. [124]

Clark, C. E. and Williams, G. T. (1958). Distributions of the members of an ordered sample. *Ann. Math. Statist.* **29**, 862–70. [66]

Cochran, W. G. (1941). The distribution of the largest of a set of estimated variances as a fraction of their total. *Ann. Eugen.* **11**, 47–52. [81, 89, 183]

Cohen, A. C., Jr. (1954). Estimation of the Poisson parameter from truncated samples and from censored samples. *J. Amer. Statist. Ass.* **49**, 158–68. [120]

Cohen, A. C., Jr. (1955a). Restriction and selection in samples from bivariate normal distributions. *J. Amer. Statist. Ass.* **50**, 884–93. [119]

Cohen, A. C., Jr. (1955b). Maximum likelihood estimation of the dispersion parameter of a chi-distributed radial error from truncated and censored samples with applications to target analysis. *J. Amer. Statist. Ass.* **50**, 1122–35. [119]

Cohen, A. C., Jr. (1955c). Censored samples from truncated normal distributions. *Biometrika* **42**, 516–9. [119]

Cohen, A. C., Jr. (1957). Restriction and selection in multinormal distributions. *Ann. Math. Statist.* **28**, 731–41. [119]

Cohen, A. C., Jr. (1959). Simplified estimators for the normal distribution when samples are singly censored or truncated. *Technometrics* **1**, 217–37. [110]

Cohen, A. C., Jr. (1961). Tables for maximum likelihood estimates: singly truncated and censored samples. *Technometrics* **3**, 535–41. [110, 112–3, 115]

Cohen, A. C., Jr. (1963). Progressively censored samples in life testing. *Technometrics* **5**, 327–39. [124]

Cohen, A. C., Jr. (1965). Maximum likelihood estimation in the Weibull distribution based on complete and on censored samples. *Technometrics* **7**, 579–88. [120]

Cohen, A. C., Jr. (1966). Life testing and early failure. *Technometrics* **8**, 539–45. [124]

Cohn, R., Mosteller, F., Pratt, J. W., and Tatsuoka, M. (1960). Maximizing the probability that adjacent order statistics of samples from several populations form overlapping intervals. *Ann. Math. Statist.* **31**, 1095–104. [21]

Cole, R. H. (1951). Relations between moments of order statistics. *Ann. Math. Statist.* **22**, 308–10. [37]

Conover, W. J. (1965). A k-sample model in order statistics. *Ann. Math. Statist.* **36**, 1223–35. [24]

Cox, D. R. (1948). A note on the asymptotic distribution of range. *Biometrika* **35**, 310–15. [211]

Cox, D. R. (1949). The use of range in sequential analysis. *Suppl. J. R. Statist. Soc.* **11**, 101–14. [156]

Cox, D. R. (1954). The mean and coefficient of variation of range in small samples from non-normal populations. *Biometrika* **41**, 469–81. Correction **42**, 277. [29, 144]

Cox, D. R. (1956). A note on the theory of quick tests. *Biometrika* **43**, 478–80. [138]

Cox, D. R. (1959). The analysis of experimentally distributed life-times with two types of failure. *J. R. Statist. Soc.* B **21**, 411–21. [124]

Cox, D. R. (1964). Some applications of exponential ordered scores. *J. R. Statist. Soc.* B **26**, 103–10. [124]

Cox, D. R. (1968). Notes on some aspects of regression analysis. *J. R. Statist. Soc.* A **131**, 265–79. [163]

Cox, D. R. and Lauh, Elizabeth. (1967). A note on the graphical analysis of multi-dimensional contingency tables. *Technometrics* **9**, 481–8. [163]

Cox, D. R. and Lewis, P. A. W. (1966). *The Statistical Analysis of Series of Events.* Methuen, London; Wiley, New York. [121]

Craig, C. C. (1962). On the mean and variance of the smaller of two drawings from a binomial population. *Biometrika* **49**, 566–9. [36]

Cramér, H. (1946). *Mathematical Methods of Statistics*. Princeton University Press. [224]

Crow, E. L. and Siddiqui, M. M. (1967). Robust estimation of location. *J. Amer. Statist. Ass.* **62**, 353–89. [125–7]

Curnow, R. N. and Dunnett, C. W. (1962). The numerical evaluation of certain multivariate normal integrals. *Ann. Math. Statist.* **33**, 571–9. [91]

Daly, J. F. (1946). On the use of the sample range in an analogue of Student's *t*-test. *Ann. Math. Statist.* **17**, 71–4. [19, 157]

Daniel, C. (1959). Use of half-normal plots in interpreting factorial two-level experiments. *Technometrics* **1**, 311–41. [163]

Daniel, C. (1960). Locating outliers in factorial experiments. *Technometrics* **2**, 149–56. [171]

Daniels, H. E. (1952). The covering circle of a sample from a circular normal distribution. *Biometrika* **39**, 137–43. [148]

Darling, D. A. (1952a). The influence of the maximum term in the addition of independent random variables. *Trans. Amer. Math. Soc.* **73**, 95–107. [90]

Darling, D. A. (1952b). On a test for homogeneity and extreme values. *Ann. Math. Statist.* **23**, 450–6. Correction **24**, 135. [90, 177]

Darling, D. A. (1953). On a class of problems related to the random division of an interval. *Ann. Math. Statist.* **24**, 239–53. [80]

Darwin, J. H. (1957). The difference between consecutive members of a series of random variables arranged in order of size. *Biometrika* **44**, 211–8. [213]

David, F. N. and Barton, D. E. (1962). *Combinatorial Chance*. Griffin, London; Hafner, New York. [88]

David, F. N. and Johnson, N. L. (1954). Statistical treatment of censored data. I. Fundamental formulae. *Biometrika* **41**, 228–40. [28, 66, 147]

David, F. N. and Johnson, N. L. (1956). Some tests of significance with ordered variables (with discussion). *J. R. Statist. Soc. B* **78**, 1–31. [147, 155]

David, H. A. (1951). Further applications of range to the analysis of variance. *Biometrika* **38**, 393–409. [143, 159]

David, H. A. (1953). The power function of some tests based on range. *Biometrika* **40**, 347–53. [159–60, 168]

David, H. A. (1954). The distribution of range in certain non-normal populations. *Biometrika* **41**, 463–8. [166]

David, H. A. (1955). A note on moving ranges. *Biometrika* **42**, 512–5. [165]

David, H. A. (1956). Revised upper percentage points of the extreme studentized deviate from the sample mean. *Biometrika* **43**, 449–51. [86]

David, H. A. (1957). Estimation of means of normal populations from observed minima. *Biometrika* **44**, 282–6. [42, 124]

David, H. A. (1962). Order statistics in short-cut tests. *SG*, 94–128. [142, 144]

David, H. A. (1966). A note on "A *k*-sample model in order statistics" by W. J. Conover. *Ann. Math. Statist.* **37**, 287–8. [24]

David, H. A. (1968). Gini's mean difference rediscovered. *Biometrika* **55**, 573–5. [167, 217]

David, H. A., Hartley, H. O. and Pearson, E. S. (1954). The distribution of the ratio, in a single normal sample, of range to standard deviation. *Biometrika* **41**, 482–93. [72]

David, H. A. and Joshi, P. C. (1968). Recurrence relations between moments of order statistics for exchangeable variates. *Ann. Math. Statist.* **39**, 272–4. [84]

David, H. A. and Mishriky, R. S. (1968). Order statistics for discrete populations and for grouped samples. *J. Amer. Statist. Ass.* **63**, 1390–8. [117]

David, H. A. and Newell, D. J. (1965). The identification of annual peak periods for a disease. *Biometrics* **21**, 645–50. [174]

David, H. A. and Paulson, A. S. (1965). The performance of several tests for outliers. *Biometrika* **52**, 429–36. [187, 233]

David, H. A. and Perez, C. A. (1960). On comparing different tests of the same hypothesis. *Biometrika* **47**, 297–306. [138, 166]

David, H. T. (1962). The sample mean among the moderate order statistics. *Ann. Math. Statist.* **33**, 1160–6. [84]

David, H. T. (1963). The sample mean among the extreme normal order statistics. *Ann. Math. Statist.* **34**, 33–55. [34, 84]

Davis, R. C. (1951). On minimum variance in nonregular estimation. *Ann. Math. Statist.* **22**, 43–57. [95]

de Finetti, B. (1932). Sulla legge di probabilità degli estremi. *Metron* **9**, 127–38. [204]

de Finetti, B. (1961). The Bayesian approach to the rejection of outliers. *Proc. 4th Berkeley Symp.* **1**, 199–210. [172]

Dempster, A. P. and Kleyle, R. M. (1968). Distributions determined by cutting a simplex with hyperplanes. *Ann. Math. Statist.* **39**, 1473–8. [82]

Dixon, W. J. (1950). Analysis of extreme values. *Ann. Math. Statist.* **21**, 488–506. [184, 186, 189]

Dixon, W. J. (1951). Ratios involving extreme values. *Ann. Math. Statist.* **22**, 68–78. [173]

Dixon, W. J. (1953). Processing data for outliers. *Biometrics* **9**, 74–89. [191]

Dixon, W. J. (1957). Estimates of the mean and standard deviation of a normal population. *Ann. Math. Statist.* **28**, 806–9. [139, 146–7]

Dixon, W. J. (1960). Simplified estimation from censored normal samples. *Ann. Math. Statist.* **31**, 385–91. [129, 140, 146]

Dixon, W. J. (1962). Rejection of observations. *SG*, 299–342. [171–2, 190]

Dixon, W. J. and Massey, F. J., Jr. (1957). *Introduction to Statistical Analysis*, 2nd Ed. McGraw-Hill, New York. [226]

Dixon, W. J. and Tukey, J. W. (1968). Approximate behavior of the distribution of Winsorized *t* (trimming/Winsorization 2). *Technometrics* **10**, 83–98. [130]

Dodd, E. L. (1923). The greatest and the least variate under general laws of error. *Trans. Am. Math. Soc.* **25**, 525–39. [204]

Doksum, K. (1967). Asymptotically optimal statistics in some models with increasing failure rate average. *Ann. Math. Statist.* **38**, 1731–9. [124]

Doornbos, R. (1956). Significance of the smallest of a set of estimated normal variances. *Statist. Neerlandica* **10**, 117–26. [183]

Doornbos, R. (1966). *Slippage Tests*. Mathematical Centre Tracts **15**, Mathematisch Centrum, Amsterdam. [74–5, 170]

Doornbos, R. and Prins, H. J. (1956). Slippage tests for a set of gamma-variates. *Indag. Math.* **18**, 329–37. [74, 183]

Doornbos, R. and Prins, H. J. (1958). On slippage tests. *Indag. Math.* **20**, I. A general type of slippage test and a slippage test for normal variates. 38–46. II. Slippage tests for discrete variates. 47–55. III. Two distribution-free slippage tests and two tables. 438–47. [178, 183]

Doss, S. A. D. C. (1963). On the efficiency of best asymptotically normal estimates of the Poisson parameter based on singly and doubly truncated or censored samples. *Biometrics* **19**, 588–94. [120]

Downton, F. (1953). A note on ordered least-squares estimation. *Biometrika* **40**, 457–8. [105]

Downton, F. (1954). Least-squares estimates using ordered observations. *Ann. Math. Statist.* **25**, 303–16. [131]

Downton, F. (1966a). Linear estimates of parameters in the extreme value distribution. *Technometrics* **8**, 3–17. [109]

Downton, F. (1966b). Linear estimates with polynomial coefficients. *Biometrika* **53**, 129–41. [43, 146]

Dronkers, J. J. (1958). Approximate formulae for the statistical distributions of extreme values. *Biometrika* **45**, 447–70. [209]

Dubey, S. D. (1967). Some percentile estimators for Weibull parameters. *Technometrics* **9**, 119–29. [147]

Duncan, D. B. (1965). A Bayesian approach to multiple comparisons. *Technometrics* **7**, 171–222. [163]

Dunn, O. J. (1958). Estimation of the means of dependent variables. *Ann. Math. Statist.* **29**, 1095–1111. [79]

Dunnett, C. W. and Sobel, M. (1955). Approximations to the probability integral and certain percentage points of a multivariate analogue of Student's *t*-distribution. *Biometrika* **42**, 258–60. [79, 91]

Dunn-Rankin, P. and Wilcoxon, F. (1966). The true distributions of the range of rank totals in the two-way classification. *Psychometrika* **31**, 573–80. [230]

Durbin, J. (1961). Some methods of constructing exact tests. *Biometrika* **48**, 41–55. [89]

Dwass, M. (1964). Extremal processes. *Ann. Math. Statist.* **35**, 1718–25. [204]

Eilbott, Joan and Nadler, J. (1965). On precedence life testing. *Technometrics* **7**, 359–77. [124]

Eisenberger, I. (1968). Testing the mean and standard deviation of a normal distribution using quantiles. *Technometrics* **10**, 781–92. [155]

Eisenberger, I. and Posner, E. C. (1965). Systematic statistics used for data compression in space telemetry. *J. Amer. Statist. Ass.* **60**, 97–133. [2, 150, 154, 168, 231]

Eisenhart, C., Deming, Lola S., and Martin, C. S. (1963). Tables describing small-sample properties of the mean, median, standard deviation, and other statistics in sampling from various distributions. *Nat. Bur. Stand. Tech. Note* 191. [225]

Eisenhart, C., Hastay, M. W., and Wallis, W. A. (Eds.) (1947). *Selected Techniques of Statistical Analysis.* McGraw-Hill, New York. Ch. 5: Acceptance inspection when lot quality is measured by the range. [137]

Eisenhart, C. and Solomon, H. (1947). Significance of the largest of a set of sample estimates for variance. In: Eisenhart, C., Hastay, M. W., and Wallis, W. A.

(Eds.), *Selected Techniques of Statistical Analysis*. McGraw-Hill, New York. [81]

Elfving, G. (1947). The asymptotical distribution of range in samples from a normal population. *Biometrika* **34**, 111–9. [211]

Epstein, B. (1949a). A modified extreme value problem. *Ann. Math. Statist.* **20**, 99–103. [20]

Epstein, B. (1949b). The distribution of extreme values in samples whose members are subject to a Markov chain condition. *Ann. Math. Statist.* **20**, 590–4. Correction **22**, 133–4. [92]

Epstein, B. (1954). Truncated life tests in the exponential case. *Ann. Math. Statist.* **25**, 555–64. [122, 134]

Epstein, B. (1956). Simple estimators of the parameters of exponential distributions when samples are censored. *Ann. Inst. Statist. Math.* **8**, 15–26. [123]

Epstein, B. (1960a). Statistical life test acceptance procedures. *Technometrics* **2**, 435–46. [123]

Epstein, B. (1960b). Estimation from life test data. *Technometrics* **2**, 447–54. [123]

Epstein, B. (1967). Bacterial extinction time as an extreme value phenomenon. *Biometrics* **23**, 835–9. [200]

Epstein, B. and Sobel, M. (1953). Life testing. *J. Amer. Statist. Ass.* **48**, 486–502. [122]

Epstein, B. and Sobel, M. (1954). Some theorems relevant to life testing from an exponential distribution. *Ann. Math. Statist.* **25**, 373–81. [121–2]

Epstein, B. and Tsao, C. K. (1953). Some tests based on ordered observations from two exponential populations. *Ann. Math. Statist.* **24**, 458–66. [122, 133]

Esary, J. D. and Proschan, F. (1963). Relationship between system failure rate and component failure rates. *Technometrics* **5**, 183–9. [124]

Esary, J. D., Proschan, F., and Walkup, D. W. (1967). Association of random variables, with applications. *Ann. Math. Statist.* **38**, 1466–74. [78, 87]

Farrell, R. H. (1966). Bounded length confidence intervals for the p-point of a distribution function, III. *Ann. Math. Statist.* **37**, 586–92. [15]

Faulkenberry, G. D. and Weeks, D. L. (1968). Sample size determination for tolerance limits. *Technometrics* **10**, 343–8. [135]

Federer, W. T. (1963). Procedures and designs useful for screening material in selection and allocation, with a bibliography. *Biometrics* **19**, 553–87. [119]

Feldman, D. and Tucker, H. G. (1966). Estimation of non-unique quantiles. *Ann. Math. Statist.* **37**, 451–7. [203, 223]

Feller, W. (1951). The asymptotic distribution of the range of sums of independent random variables. *Ann. Math. Statist.* **22**, 427–32. [85, 215]

Feller, W. (1957). *An Introduction to Probability Theory and Its Applications*, Vol. I, 2nd Ed. Wiley, New York. [35]

Feller, W. (1966). *An Introduction to Probability Theory and Its Applications*, Vol. II. Wiley, New York. [82]

Ferguson, T. S. (1961a). Rules for rejection of outliers. *Revue Inst. Int. de Stat.* **29**, 29–43. [170, 172]

Ferguson, T. S. (1961b). On the rejection of outliers. *Proc. 4th Berkeley Symp.* **I**, 253–87. [175, 184–5, 187, 190]

Ferguson, T. S. (1967). On characterizing distributions by properties of order statistics. *Sankhyā A* **29**, 265–78. [19]

Filliben, J. J. (1969). Simple and robust linear estimation of the location parameter of a symmetric distribution. Ph.D. Thesis, Princeton University. [127]

Finney, D. J. (1941). The joint distribution of variance ratios based on a common error mean square. *Ann. Eugen.* **11**, 136–40. [78]

Fisher, R. A. (1929). Tests of significance in harmonic analysis. *Proc. Roy. Soc. A*, **125**, 54–9. [81, 177]

Fisher, R. A. (1940). On the similarity of the distributions found for the test of significance in harmonic analysis, and in Steven's problem in geometrical probability. *Ann. Eugen.* **10**, 14–17. [81, 84, 89]

Fisher, R. A. (1950). *Contributions to Mathematical Statistics*. Wiley, New York. [81, 229]

Fisher, R. A. and Tippett, L. H. C. (1928). Limiting forms of the frequency distribution of the largest or smallest member of a sample. *Proc. Camb. Phil. Soc.* **24**, 180–90. [204, 209]

Fraser, D. A. S. (1957). *Nonparametric Methods in Statistics.* Wiley, New York. [17]

Fréchet, M. (1927) Sur la loi de probabilité de l'écart maximum. *Ann. Société Polonaise de Mathématique* **6**, 92–116. [204]

Gallot, S. (1966). A bound for the maximum of a number of random variables. *J. Appl. Prob.* **3**, 556–8. [88]

Galton, F. (1902). The most suitable proportion between the values of first and second prizes. *Biometrika* **1**, 385–90. [39]

Gani, J. and Yeo, G. F. (1962). On the age distribution of *n* ranked elements after several replacements. *Aust. J. Statist.* **4**, 55–60. [124]

Garner, N. R. (1958). Curtailed sampling for variables. *J. Amer. Statist. Ass.* **53**, 862–7. [124]

Gastwirth, J. L. (1966). On robust procedures. *J. Amer. Statist. Ass.* **61**, 929–48. [129]

Gastwirth, J. L. and Cohen, M. L. (1968). The small sample behavior of some robust linear estimators of location. *Tech. Rep.* 91, Dept. Statistics, The John Hopkins University. [127, 227]

Gastwirth, J. L. and Rubin, H. (1969). On robust linear estimators. *Ann. Math. Statist.* **40**, 24–39. [129]

Gebhardt, F. (1964). On the risk of some strategies for outlying observations. *Ann. Math. Statist.* **35**, 1524–36. [195]

Gebhardt, F. (1966). On the effect of stragglers on the risk of some mean estimators in small samples. *Ann. Math. Statist.* **37**, 441–50. [195]

Geffroy, J. (1958). Contribution à la théorie des valeurs extrêmes. Ph.D. Thesis, University of Paris. [210]

Ghosh, B. K. (1963). On sequential tests of ratio of variances based on range. *Biometrika* **50**, 419–30. [142, 156]

Ghosh, B. K. (1965). Sequential range tests for components of variance. *J. Amer. Statist. Ass.* **60**, 826–36. [159]

Gilchrist, W. G. (1961). Some sequential tests using range. *J. R. Statist. Soc. B* **23**, 335–42. [158]

Gini, C. (1912). Variabilità é Mutabilità, contributo allo studio delle distribuzioni e relazioni statistiche. Studi Economico-Giuridici della R. Università di Cagliari. [147]

Gnanadesikan, R., Pinkham, R. S., and Hughes, L. P. (1967). Maximum likelihood estimation of the parameters of the beta distribution from smallest order statistics. *Technometrics* **9**, 607–20. [120]

Gnedenko, B. (1943). Sur la distribution limite du terme maximum d'une série aléatoire. *Ann. Math.* **44**, 423–53. [204, 206, 215, 223]

Godwin, H. J. (1949). Some low moments of order statistics. *Ann. Math. Statist.* **20**, 279–85. [31, 33]

Goodman, L. A. and Madansky, A. (1962). Parameter-free and nonparametric tolerance limits: the exponential case. *Technometrics* **4**, 75–95. [124]

Govindarajulu, Z. (1962). Exact lower moments of order statistics in samples from the chi-distribution (1 d.f.). *Ann. Math. Statist.* **33**, 1292–305. [227]

Govindarajulu, Z. (1963a). On moments of order statistics and quasi-ranges from normal populations. *Ann. Math. Statist.* **34**, 633–51. [39, 39, 40]

Govindarajulu, Z. (1963b). Relationships among moments of order statistics in samples from two related populations. *Technometrics* **5**, 514–8. [40]

Govindarajulu, Z. (1964). A supplement to Mendenhall's bibliography on life testing and related topics. *J. Amer. Statist. Ass.* **59**, 1231–91. [124]

Govindarajulu, Z. (1966). Best linear estimates under symmetric censoring of the parameters of a double exponential population. *J. Amer. Statist. Ass.* **61**, 248–58. [120, 227, 230]

Govindarajulu, Z. (1967). Characterization of the exponential and power distributions. *Skand. Aktuarietidskr.* 1966, 132–6. [19]

Govindarajulu, Z. (1968a). Certain general properties of unbiased estimates of location and scale parameters based on ordered observations. *SIAM J. Appl. Math.* **16**, 533–51. [40, 105]

Govindarajulu, Z. (1968b). Asymptotic normality of linear combinations of functions of order statistics, II. *Proc. Nat. Acad. Sci.* **59**, 713–9. [217]

Govindarajulu, Z. and Eisenstat, S. (1965). Best estimates of location and scale parameters of a chi (1 d.f.) distribution, using ordered observations. *Rep. Stat. Appl. Res., JUSE* **12**, 149–64. [120, 227, 230]

Govindarajulu, Z. and Hubacker, N. W. (1964). Percentiles of order statistics in samples from uniform, normal, chi (1 d.f.) and Weibull populations. *Rep. Stat. Appl. Res., JUSE* **11**, 64–90. [225]

Greig, Margaret (1967). Extremes in a random assembly. *Biometrika* **54**, 273–82. [85]

Grenander, U. (1965). A limit theorem for sums of minima of stochastic variables. *Ann. Math. Statist.* **36**, 1041–2. [224]

Grubbs, F. E. (1950). Sample criteria for testing outlying observations. *Ann. Math. Statist.* **21**, 27–58. [72, 173, 184, 228, 232]

Grubbs, F. E. (1964). *Statistical Measures of Accuracy for Rifleman and Missile Engineers*. Edwards Brothers, Ann Arbor, Mich. [148]

Grubbs, F. E. (1969). Procedures for detecting outlying observations in samples. *Technometrics* **11**, 1–21. [172]

Grubbs, F. E., Coon, Helen J., and Pearson, E. S. (1966). On the use of Patnaik type chi approximations to the range in significance tests. *Biometrika* **53**, 248–52. [144]

Grubbs, F. E. and Weaver, C. L. (1947). The best unbiased estimate of population standard deviation based on group ranges. *J. Amer. Statist. Ass.* **42**, 224–41. [141]

Grundy, P. M. (1952). The fitting of grouped truncated and grouped censored normal distributions. *Biometrika* **39**, 252–9. [119]

Gumbel, E. J. (1935). Les valeurs extrêmes des distributions statistiques. *Ann. Inst. Henri Poincaré* **5**, 115–58. [204, 210]

Gumbel, E. J. (1947). The distribution of the range. *Ann. Math. Statist.* **18**, 384–412. [211]

Gumbel, E. J. (1949). Probability tables for the range. *Biometrika* **36**, 14 2–8. [211]

Gumbel, E. J. (1954). The maxima of the mean largest value and of the range. *Ann. Math. Statist.* **25**, 76–84. [46, 48]

Gumbel, E. J. (1958). *Statistics of Extremes.* Columbia University Press, New York. [3, 21, 120, 200, 204, 207]

Gumbel, E. J. (1961). Statistical theory of breaking strength and fatigue failure. *Bull. Int. Statist. Inst.* **38** (3), 375–93. [200]

Gumbel, E. J. (1963). Statistical forecast of droughts. *Bull. I.A.S.H.* **8**, 5–23. [200]

Gumbel, E. J., Carlson, P. G., and Mustafi, C. K. (1965). A note on midrange. *Ann. Math. Statist.* **36**, 1052–4. [21]

Gumbel, E. J. and Goldstein, N. (1964). Analysis of empirical bivariate extremal distributions. *J. Amer. Statist. Ass.* **59**, 794–816. [214]

Gumbel, E. J. and Herbach, L. H. (1951). The exact distribution of the extremal quotient. *Ann. Math. Statist.* **22**, 418–26. [156]

Gumbel, E. J. and Keeney, R. D. (1950a). The geometric range for distributions of Cauchy's type. *Ann. Math. Statist.* **21**, 133–7. [212]

Gumbel, E. J. and Keeney, R. D. (1950b). The extremal quotient. *Ann. Math. Statist.* **21**, 523–8. [212]

Gumbel, E. J. and Mustafi, C. K. (1967). Some analytical properties of bivariate extremal distributions. *J. Amer. Statist. Ass.* **62**, 569–88. [214]

Gumbel, E. J. and Pickands, J., III (1967). Probability tables for the extremal quotient. *Ann. Math. Statist.* **38**, 1541–51. [212]

Gupta, A. K. (1952). Estimation of the mean and standard deviation of a normal population from a censored sample. *Biometrika* **39**, 260–73. [105, 109, 115–6]

Gupta, S. S. (1960). Order statistics from the gamma distribution. *Technometrics* **2**, 243–62. [225–6]

Gupta, S. S. (1961). Percentage points and modes of order statistics from the normal distribution. *Ann. Math. Statist.* **32**, 888–93. [225]

Gupta, S. S. (1962). Life test sampling plans for normal and lognormal distributions. *Technometrics* **4**, 151–75. [124]

Gupta, S. S. (1963a). Probability integrals of multivariate normal and multivariate *t*. *Ann. Math. Statist.* **34**, 792–828. [83, 229]

Gupta, S. S. (1963b). Bibliography on the multivariate normal integrals and related topics. *Ann. Math. Statist.* **34**, 829–38. [229]

Gupta, S. S. and Gnanadesikan, M. (1966). Estimation of the parameters of the logistic distribution. *Biometrika* **53**, 565–70. [154, 224]

Gupta, S. S. and Groll, Phyllis A. (1961). Gamma distribution in acceptance sampling based on life tests. *J. Amer. Statist. Ass.* **56**, 942–70. [124]

Gupta, S. S. and Pillai, K. C. S. (1965). On linear functions of ordered correlated normal random variables. *Biometrika* **52**, 367–79. [91]

Gupta, S. S., Pillai, K. C. S., and Steck, G. P. (1964). On the distribution of linear functions and ratios of linear functions of ordered correlated normal random variables with emphasis on range. *Biometrika* **51**, 143–51. [84]

Gupta, S. S., Qureishi, A. S., and Shah, B. K. (1967). Best linear unbiased estimators of the parameters of the logistic distribution using order statistics. *Technometrics* **9**, 43–56. [120, 227, 230]

Gupta, S. S. and Shah, B. K. (1965). Exact moments and percentage points of the order statistics and the distribution of the range from the logistic distribution. *Ann. Math. Statist.* **36**, 907–20. [227]

Gupta, S. S. and Sobel, M. (1958). On the distribution of a statistic based on ordered uniform chance variables. *Ann. Math. Statist.* **29**, 274–81. [124, 134]

Gupta, S. S. and Sobel, M. (1962). On the smallest of several correlated F statistics. *Biometrika* **49**, 509–23. [229]

Guttman, I. and Smith, D. (1966). Investigation of rejection rules for outliers in small samples from the normal distribution. *Tech. Reps.* 90–93, University of Wisconsin. [191–2, 195, 198]

Guttman, I. and Smith, D. E. (1969). Investigation of rules for dealing with outliers in small samples from the normal distribution, I: Estimation of the mean. *Technometrics* **11**, 527–50. [191, 193–5, 197]

Hájek, J. (1968). Asymptotic normality of simple linear rank statistics under alternatives. *Ann. Math. Statist.* **39**, 325–46. [217]

Haldane, J. B. S. and Jayakar, S. D. (1963). The distribution of extremal and nearly extremal values in samples from a normal distribution. *Biometrika* **50**, 89–94. [209]

Halperin, M. (1952). Maximum likelihood estimation in truncated samples. *Ann. Math. Statist.* **23**, 226–38. [115]

Halperin, M. (1960). Some asymptotic results for a coverage problem. *Ann. Math. Statist.* **31**, 1063–76. [22]

Halperin, M. (1967). An inequality on a bivariate Student's t distribution. *J. Amer. Statist. Ass.* **62**, 603–6. [87]

Halperin, M., Greenhouse, S. W., Cornfield, J., and Zalokar, J. (1955). Tables of percentage points for the studentized maximum absolute deviate in normal samples. *J. Amer. Statist. Ass.* **50**, 185–95. [173, 228]

Hammersley, J. M. and Morton, K. W. (1954). The estimation of location and scale parameters from grouped data. *Biometrika* **41**, 296–301. [119]

Han, C. P. (1968). Testing the homogeneity of a set of correlated variances. *Biometrika* **55**, 317–26. [156]

Han, C. P. (1969). Testing the homogeneity of variances in a two-way classification. *Biometrics* **25**, 153–8. [156]

Harter, H. L. (1959). The use of sample quasi-ranges in estimating population standard deviation. *Ann. Math. Statist.* **30**, 980–99. Correction **31**, 228. [71, 146, 228, 230]

Harter, H. L. (1960). Tables of range and studentized range. *Ann. Math. Statist.* **31**, 1122–47. [225, 228]

Harter, H. L. (1961a). Expected values of normal order statistics. *Biometrika* **48**, 151–65. Correction **48**, 476. [28, 67, 228]

Harter, H. L. (1961b). Estimating the parameters of negative exponential populations from one or two order statistics. *Ann. Math. Statist.* **32**, 1078–90. [140]

Harter, H. L. (1961c). The use of sample ranges in setting exact confidence bounds for the standard deviation of a rectangular population. *J. Amer. Statist. Ass.* **56**, 601–9. [225, 230]

Harter, H. L. (1963). Percentage points of the ratio of two ranges and power of the associated test. *Biometrika* **50**, 187–94. [155, 231]

Harter, H. L. (1964a). Exact confidence bounds, based on one order statistic, for the parameter of an exponential population. *Technometrics* **6**, 301–17. [140]

Harter, H. L. (1964b). Criteria for best substitute interval estimators, with an application to the normal distribution. *J. Amer. Statist. Ass.* **59**, 1133–40. [155, 231]

Harter, H. L. (1967). Maximum-likelihood estimation of the parameters of a four-parameter generalized gamma population from complete and censored samples. *Technometrics* **9**, 159–65. [119]

Harter, H. L. and Clemm, D. S. (1959). The probability integrals of the range and of the Studentized range—probability integrals, percentage points, and moments of the range. *Wright Air Development Center Tech. Rep.* 58–484, Vol. I. [225]

Harter, H. L., Clemm, D. S., and Guthrie, E. H. (1959). The probability integrals of the range and of the Studentized range—probability integral and percentage points of the Studentized range; critical values for Duncan's new multiple range test. *Wright Air Development Center Tech. Rep.* 58–484, Vol. II. [228]

Harter, H. L. and Moore, A. H. (1965). Point and interval estimators, based on m order statistics, for the scale parameter of a Weibull population with known shape parameter. *Technometrics* **7**, 405–22. [120]

Harter, H. L. and Moore, A. H. (1966). Local-maximum likelihood estimation of the parameters of three-parameter lognormal populations from complete and censored samples. *J. Amer. Statist. Ass.* **61**, 842–55. [119]

Harter, H. L. and Moore, A. H. (1967a). A note on estimation from a Type I extreme-value distribution. *Technometrics* **9**, 325–31. [120]

Harter, H. L. and Moore, A. H. (1967b). Asymptotic variances and covariances of maximum-likelihood estimators, from censored samples, of the parameters of Weibull and gamma populations. *Ann. Math. Statist.* **38**, 557–70. [119–20]

Harter, H. L. and Moore, A. H. (1967c). Maximum-likelihood estimation, from censored samples, of the parameters of a logistic distribution. *J. Amer. Statist. Ass.* **62**, 675–84. [120]

Harter, H. L. and Moore, A. H. (1968a). Conditional maximum-likelihood estimators, from singly censored samples, of the scale parameters of type II extreme-value distributions. *Technometrics* **10**, 349–59. [120]

Harter, H. L. and Moore, A. H. (1968b). Maximum-likelihood estimation, from doubly censored samples, of the parameters of the first asymptotic distribution of extreme values. *J. Amer. Statist. Ass.* **63**, 889–901. [120]

Hartley, H. O. (1938). Studentization and large-sample theory. *Suppl. J. R. Statist. Soc.* **5**, 80–8. [78]

Hartley, H. O. (1942). The range in random samples, *Biometrika* **32**, 334–48. [21, 117]

Hartley, H. O. (1944). Studentization or the elimination of the standard deviation of the parent population from the random sample-distribution of statistics. *Biometrika* **33**, 173–80. [71]

Hartley, H. O. (1949). Tests of significance in harmonic analysis. *Biometrika* **36**, 194–201. [81]

Hartley, H. O. (1950a). The use of range in analysis of variance. *Biometrika* **37**, 271–80. [85, 159, 231]

Hartley, H. O. (1950b). The maximum *F* ratio as a short-cut test for heterogeneity of variance. *Biometrika* **37**, 308–12. [155]

Hartley, H. O. (1955). Some recent developments in analysis of variance. *Comm. Pure and Appl. Math.* **8**, 47–72. [78, 166]

Hartley, H. O. and David, H. A. (1954). Universal bounds for mean range and extreme observation. *Ann. Math. Statist.* **25**, 85–99. [46, 50, 54]

Harvard Computation Laboratory. (1955). *Tables of the Cumulative Binomial Probability Distribution.* Harvard University Press, Cambridge, Mass. [7]

Hassanein, K. M. (1968). Analysis of extreme-value data by sample quantiles for very large samples. *J. Amer. Statist. Ass.* **63**, 877–88. [154]

Hastings, C., Jr., Mosteller, F., Tukey, J. W., and Winsor, C. P. (1947). Low moments for small samples: a comparative study of order statistics. *Ann. Math. Statist.* **18**, 413–26. [226]

Healy, M. J. R. (1968). Multivariate normal plotting. *Appl. Statist.* **17**, 157–61. [163]

Helmert, F. R. (1876). Die Berechnung des wahrscheinlichen Beobachtungsfehlers aus den ersten Potenzen der Differenzen gleichgenauer directer Beobachtungen. *Astron. Nach.* **88**, 127–32. [147]

Hill, B. M. (1963). The three-parameter lognormal distribution and Bayesian analysis of a point-source epidemic. *J. Amer. Statist. Ass.* **58**, 72–84. [101]

Hillier, F. S. (1964). $\bar{X}$ chart control limits based on a small number of subgroups. *Industr. Qual. Contr.* **20**, No. 8, 24–9. [164]

Hillier, F. S. (1967). Small sample probability limits for the range chart. *J. Amer. Statist. Ass.* **63**, 1488–93. Correction **63**, 1549–50. [164–5]

Hodges, J. L., Jr. (1967). Efficiency in normal samples and tolerance of extreme values for some estimates of location. *Proc. 5th Berkeley Symp.* **I**, 163–86. [128]

Hodges, J. L., Jr. and Lehmann, E. L. (1963). Estimates of location based on rank tests. *Ann. Math. Statist.* **34**, 598–611. [127]

Hodges, J. L., Jr. and Lehmann, E. L. (1967). On medians and quasi-medians. *J. Amer. Statist. Ass.* **62**, 926–31. [69, 139]

Hoeffding, W. (1948). A class of statistics with asymptotically normal distribution. *Ann. Math. Statist.* **19**, 293–325. [216]

Hoel, P. G. and Scheuer, E. M. (1961). Confidence sets for multivariate medians. *Ann. Math. Statist.* **32**, 477–84. [16]

Hogg, R. V. (1956). On the distribution of the likelihood ratio. *Ann. Math. Statist.* **27**, 529–32. [100, 130–1]

Hogg, R. V. (1960). Certain uncorrelated statistics. *J. Amer. Statist. Ass.* **55**, 265–7. [41]

Hogg, R. V. (1967). Some observations on robust estimation. *J. Amer. Statist. Ass.* **62**, 1179–86. [129]

Hogg, R. V. and Craig, A. T. (1956). Sufficient statistics in elementary distribution theory. *Sankhyā* **17**, 209–16. [96]

Hogg, R. V. and Craig, A. T. (1959). *Introduction to Mathematical Statistics.* MacMillan, New York. [95]

Hogg, R. V. and Tanis, E. A. (1963). An iterated procedure for testing the equality of several exponential distributions. *J. Amer. Statist. Ass.* **58**, 435–43. [124]

Hojo, T. (1931). Distribution of the median, quartiles and interquartile distance in samples from a normal population. *Biometrika* **23**, 315–60. [139]

Hojo, T. (1933). A further note on the relation between the median and the quartiles in small samples from a normal population. *Biometrika* **25**, 79–90. [139]

Howell, J. M. (1949). Control chart for largest and smallest values. *Ann. Math. Statist.* **21**, 615–6. [165]

Huber, P. J. (1964). Robust estimation of a location parameter. *Ann. Math. Statist.* **35**, 73–101. [129]

Huber, P. J. (1968). Robust estimation. In: *Selected Statistical Papers* 2, Mathematical Centre Tracts 27, Mathematisch Centrum, Amsterdam. [130]

Hudson, D. J. (1968). A short-cut method for estimating only one of two parameters from a set of order statistics. *Amer. Statist.* **22**, 23–5. [132]

Hume, M. W. (1965). The distribution of statistics expressible as maxima. *Virginia J. Sci.* **16**, 120–7. [74, 187]

Huzurbazar, V. S. (1955). Confidence intervals for the parameter of a distribution admitting a sufficient statistic when the range depends on the parameter. *J. R. Statist. Soc. B* **17**, 86–90. [97, 130]

Hyrenius, H. (1953). On the use of ranges, cross-ranges and extremes in comparing small samples. *J. Amer. Statist. Ass.* **48**, 534–45. Correction **48**, 907. [24, 100, 229]

Irwin, J. O. (1925). On a criterion for the rejection of outlying observations. *Biometrika* **17**, 238–50. [173]

Ishii, G. and Yamasaki, M. (1961). A note on the testing of homogeneity of k binomial experiments based on the range. *Ann. Inst. Stat. Math., Tokyo* **12**, 273–8. [160, 226]

Jackson, J. E. and Ross, Eleanor L. (1955). Extended tables for use with the "*G*" test for means. *J. Amer. Statist. Ass.* **50**, 416–33. [157, 231]

Jackson, O. A. Y. (1967). An analysis of departures from the exponential distribution. *J. R. Statist. Soc. B* **29**, 540–9. [162]

Jacobson, P. H. (1947). The relative power of three statistics for small sample destructive tests. *J. Amer. Statist. Ass.* **42**, 575–84. [124]

Jeffreys, H. and Jeffreys, B. S. (1946). *Methods of Mathematical Physics.* Cambridge University Press. [212]

Johns, M. V., Jr. and Lieberman, G. J. (1966). An exact asymptotically efficient confidence bound for reliability in the case of the Weibull distribution. *Technometrics* **8**, 135–75. [124]

Johnson, N. L. and Young, D. H. (1960). Some applications of two approximations to the multinomial distribution. *Biometrika* **47**, 463–9. [160]

Jones, A. E. (1946). A useful method for the routine estimation of dispersion from large samples. *Biometrika* **33**, 274–82. [146]

Jones, H. L. (1948). Exact lower moments of order statistics in small samples from a normal distribution. *Ann. Math. Statist.* **19**, 270–3. [31]

Joshi, P. C. (1969). Bounds and approximations for the moments of order statistics. *J. Amer. Statist. Ass.* **64**, 1617–24. [57]

Jung, J. (1955). On linear estimates defined by a continuous weight function. *Ark. Mat.* **3**, 199–209. [108, 216, 222]

Jung, J. (1962). Approximation of least-squares estimates of location and scale parameters. *SG*, 28–33. [108]

Kabe, D. G. (1968). Some distribution problems of order statistics from exponential and power function distributions. *Canad. Math. Bull.* **11**, 263–74. [196]

Kapur, M. N. (1957). A property of the optimum solution suggested by Paulson for the *k*-sample slippage problem for the normal distribution. *Ind. Soc. Agric. Statist.* **9**, 179–90. [181]

Karlin, S. and Studden, W. J. (1966). *Tchebycheff Systems: with applications in Analysis and Statistics*. Wiley, New York. [54]

Karlin, S. and Truax, D. R. (1960). Slippage problems. *Ann. Math. Statist.* **31**, 296–324. [183]

Kemperman, J. H. B. (1959). Asymptotic expansions for the Smirnov test and for the range of cumulative sums. *Ann. Math. Statist.* **30**, 448–62. [215]

Kendall, M. G. (1954). Two problems in sets of measurements. *Biometrika* **41**, 560–4. [42, 84]

Kendall, M. G. and Stuart, A. (1961). *The Advanced Theory of Statistics*. Vol. 2. Griffin, London; Hafner, New York. [95, 130]

Khatri, C. G. (1960). On testing the equality of parameters in *k* rectangular populations. *J. Amer. Statist. Ass.* **55**, 144–7. [100, 229]

Khatri, C. G. (1962). Distributions of order statistics for discrete case. *Ann. Inst. Statist. Math., Tokyo* **14**, 167–71. [12]

Khatri, C. G. (1965). On the distributions of certain statistics derived by the union-intersection principle for the parameters of *k* rectangular populations. *J. Ind. Statist. Ass.* **3**, 158–64. [100]

Kiefer, J. (1967). On Bahadur's representation of sample quantiles. *Ann. Math. Statist.* **38**, 1323–42. [203]

Kimball, A. W. (1951). On dependent tests of significance in the analysis of variance. *Ann. Math. Statist.* **22**, 600–2. [78, 87]

Kimball, B. F. (1960). On the choice of plotting positions on probability paper. *J. Amer. Statist. Ass.* **55**, 546–60. [162]

King, E. P. (1952). The operating characteristic of the control chart for sample means. *Ann. Math. Statist.* **23**, 384–95. [165]

King, E. P. (1953). On some procedures for the rejection of suspected data. *J. Amer. Statist. Ass.* **48**, 531–3. [175]

Knight, W. (1963). The use of the range in place of the standard deviation in Stein's test. *Ann. Math. Statist.* **34**, 346–7. [158]

Kounias, E. G. (1968). Bounds for the probability of a union of events, with applications. *Ann. Math. Statist.* **39**, 2154–8. [75, 88]

Kozelka, R. M. (1956). Approximate upper percentage points for extreme values in multinomial sampling. *Ann. Math. Statist.* **27**, 507–12. [88]

Krem, A. (1963). On the independence in the limit of extreme and central order statistics. *Publ. Math. Inst. Acad. Sci.* **8**, 469–74. [213]

Krishnaiah, P. R. and Armitage, J. V. (1964). Distribution of the studentized smallest chi-square, with tables and applications. Aerospace Research Laboratories 64–218. [229]

Krishnaiah, P. R. and Armitage, J. V. (1965a). Tables for the distribution of the maximum of correlated chi-square variates with one degree of freedom. *Aerospace Research Laboratories* 65–136. [84, 229, 232]

Krishnaiah, P. R. and Armitage, J. V. (1965b). Percentage points of the multivariate *t* distribution. *Aerospace Research Laboratories* 65–199. [229]

Krishnaiah, P. R. and Rizvi, M. H. (1966). A note on recurrence relations between expected values of functions of order statistics. *Ann. Math. Statist.* **37**, 733–4. [39]

Kruskal, W. H. (1960). Some remarks on wild observations. *Technometrics* **2**, 1–3. [172]

Kudô, A. (1956a). On the testing of outlying observations. *Sankhyā* **17**, 67–76. [182]

Kudô, A. (1956b). On the invariant multiple decision procedures. *Bull. Math. Statist.* **6**, 57–68. [18]

Kudô, A. (1956c). Tables for studentization. *Sankhyā* **18**, 163–6. [71]

Kudô, A. (1957). The extreme value in a multivariate normal sample. *Mem. Fac. Sci. Kyushu Univ.* (*A*) **11**, 143–56. [184]

Kurtz, T. E., Link, R. F., Tukey, J. W., and Wallace, D. L. (1965a). Short-cut multiple comparisons for balanced single and double classifications, Part 1: Results. *Technometrics* **7**, 95–161. [159]

Kurtz, T. E., Link, R. F., Tukey, J. W., and Wallace, D. L. (1965b). Short-cut multiple comparisons for balanced single and double classifications, Part 2: Derivations and approximations. *Biometrika* **52**, 485–98. [159]

Kurtz, T. E., Link, R. F., Tukey, J. W., and Wallace, D. L. (1966). Correlation or ranges of correlated deviates. *Biometrika* **53**, 191–7. [166]

Lachenbruch, P. A. and David, H. A. (1968). The non-central distribution of range and studentized range in normal samples. (Abstract) *Ann. Math. Statist.* **39**, 1092. [160]

Lambert, J. A. (1964). Estimation of parameters in the three-parameter log-normal distribution. *Aust. J. Statist.* **6**, 29–32. [101]

Lamperti, J. (1964). On extreme order statistics. *Ann. Math. Statist.* **35**, 1726–37. [204]

Laurent, A. G. (1963). Conditional distribution of order statistics and distribution of the reduced *i*th order statistic of the exponential model. *Ann. Math. Statist.* **34**, 652–7. [177, 196, 232]

Lehmann, E. L. (1959). *Testing Statistical Hypotheses.* Wiley, New York. [95, 99, 130]

Lehmann, E. L. (1966). Some concepts of dependence. *Ann. Math. Statist.* **37**, 1137–53. [74]

Lentner, M. M. and Buehler, R. J. (1963). Some inferences about gamma parameters with an application to a reliability problem. *J. Amer. Statist. Ass.* **58**, 670–7. [124]

Leone, F. C., Jayachandran, T., and Eisenstat, S. (1967). A study of robust estimators. *Technometrics* **9**, 652–60. [129]

Leslie, R. T. and Brown, B. M. (1966). Use of range in testing heterogeneity of variance. *Biometrika* **53**, 221–7. [156, 231]

Lieblein, J. (1952). Properties of certain statistics involving the closest pair in a sample of three observations. *J. Res., Nat. Bur. Stand.* **48**, 255–68. [23, 195]

Lieblein, J. (1954a). A new method of analyzing extreme-value data. *Nat. Advisory Comm. Aeronaut. Tech. Note* 3053. [120]

Lieblein, J. (1954b). Two early papers on the relation between extreme values and tensile strength. *Biometrika* **41**, 559–60. [200]

Lieblein, J. (1955). On moments of order statistics from the Weibull distribution. *Ann. Math. Statist.* **26**, 330–3. [226]

Lieblein, J. (1962). The closest two out of three observations. *SG*, 129–35. [195]

Lieblein, J. and Salzer, H. E. (1957). Table of the first moment of ranked extremes. *J. Res., Nat. Bur. Stand.* **59**, 203–6. [226]

Lieblein, J. and Zelen, M. (1956). Statistical investigation of the fatigue life of deep-groove ball bearings. *J. Res., Nat. Bur. Stand.* **57**, 273–316. [120, 226]

Likeš, J. (1962). On the distribution of certain linear functions of ordered sample from exponential population. *Ann. Inst. Statist. Math., Tokyo* **13**, 225–30. [124]

Likeš, J. (1967). Distributions of some statistics in samples from exponential and power-function populations. *J. Am. Statist. Ass.* **62**, 259–71. [120, 122, 124]

Link, R. F. (1950). The sampling distribution of the ratio of two ranges from independent samples. *Ann. Math. Statist.* **21**, 112–6. Correction **23**, 298–9. [22, 155]

Lloyd, E. H. (1952). Least-squares estimation of location and scale parameters using order statistics. *Biometrika* **39**, 88–95. [93, 105, 131, 218]

Lord, E. (1947). The use of range in place of standard deviation in the *t* test. *Biometrika* **34**, 41–67. Correction **39**, 442. [157]

Lord, E. (1950). Power of the modified *t* test (*u* test) based on range. *Biometrika* **37**, 64–77. [157]

Loynes, R. M. (1965). Extreme values in uniformly mixing stationary stochastic processes. *Ann. Math. Statist.* **36**, 993–9. [215]

Loynes, R. M. (1966). Some aspects of the estimation of quantiles. *J. R. Statist. Soc.* B **28**, 497–512. [16]

Ludwig, O. (1959). Ungleichungen für Extremwerte und andere Ranggrößen in Anwendung auf biometrische Probleme. *Biom. Zeit.* **1**, 203–9. [51]

Ludwig, O. (1960). Über Erwartungswerte und Varianzen von Ranggrößen in kleinen Stichproben. *Metrika* **3**, 218–33. [51]

McCarthy, P. J. (1965). Stratified sampling and distribution-free confidence intervals for a median. *J. Amer. Statist. Ass.* **60**, 772–83. [16]

McCool, J. I. (1965). The construction of good linear unbiased estimates from the best linear estimates for a smaller sample size. *Technometrics* 7, 543–52. [109]

McCord, J. R. (1964). On asymptotic moments of extreme statistics. *Ann. Math. Statist.* 35, 1738–45. [210]

McKay, A. T. (1935). The distribution of the difference between the extreme observation and the sample mean in samples of *n* from a normal universe. *Biometrika* 27, 466–71. [86]

McKay, A. T. and Pearson, E. S. (1933). A note on the distribution of range in samples of *n*. *Biometrika* 25, 415–20. [21]

MacKinnon, W. J. (1964). Table for both the sign test and distribution-free confidence intervals of the median for sample sizes to 1000. *J. Amer. Statist. Ass.* 59, 935–56. [14, 226]

McMillan, R. G. (1968). Tests for one or two outliers. Ph.D. Thesis, North Carolina State University. [191]

Madansky, A. (1962). More on length of confidence intervals. *J. Amer. Statist. Ass.* 57, 586–9. [124]

Maguire, B. A., Pearson, E. S., and Wynn, A. H. A. (1952). The time intervals between industrial accidents. *Biometrika* 39, 168–80. [124]

Maguire, B. A., Pearson, E. S., and Wynn, A. H. A. (1953). Further notes on the analysis of accident data. *Biometrika* 40, 213–6. [124]

Malik, H. J. (1966). Exact moments of order statistics from the Pareto distribution. *Skand. Aktuarietidskr.* 1966, 144–57. [226]

Malik, H. J. (1967). Exact moments of order statistics from a power-function distribution. *Skand. Aktuarietidskr.* 1967, 64–9. [39]

Mallows, C. L. (1968). An inequality involving multinomial probabilities. *Biometrika* 55, 422–4. [75]

Malmquist, S. (1950). On a property of order statistics from a rectangular distribution. *Skand. Aktuarietidskr.* 33, 214–22. [19]

Mann, Nancy R. (1967a). Tables for obtaining the best linear invariant estimates of parameters of the Weibull distribution. *Technometrics* 9, 629–45. [120]

Mann, Nancy R. (1967b). Results on location and scale parameter estimation with application to the extreme-value distribution. Aerospace Research Laboratories 67-0023. [120]

Mann, Nancy R. (1968). Point and interval estimation procedures for the two-parameter Weibull and extreme-value distributions. *Technometrics* 10, 231–56. [200]

Mantel, N. (1951). Rapid estimation of standard errors of means for small samples. *Amer. Statist.* 5, No. 14, 26–7. [145]

Mantel, N. and Pasternak, B. S. (1966). Light bulb statistics. *J. Amer. Statist. Ass.* 61, 633–9. [124]

Mardia, K. V. (1964a). Asymptotic independence of bivariate extremes. *Calcutta Statist. Ass. Bull.* 13, 172–8. [214]

Mardia, K. V. (1964b). Some results on the order statistics of the multivariate normal and Pareto type 1 populations. *Ann. Math. Statist.* 35, 1815–8. [91]

Mardia, K. V. (1967). Correlation of the ranges of correlated samples. *Biometrika* 54, 529–39. [159, 167, 231]

Margolin, B. H. and Winokur, H. S., Jr. (1967). Exact moments of the order statistics of the geometric distribution and their relation to inverse sampling and reliability of redundant systems. *J. Amer. Statist. Ass.* **62**, 915–25. [19]

Maritz, J. S. and Munro, A. H. (1967). On the use of the generalized extreme-value distribution in estimating extreme percentiles. *Biometrics* **23**, 79–103. [200, 226]

Mead, R. (1966). A quick method of estimating the standard deviation. *Biometrika* **53**, 559–64. [147]

Mejzler, D. and Weissman, I. (1969). On some results of N. V. Smirnov concerning limit distributions for variational series. *Ann. Math. Statist.* **40**, 480–91. [210]

Melnick, E. L. (1964). Moments of ranked Poisson variates. M.S. Thesis, Virginia Polytechnic Institute. [37]

Mendenhall, W. (1958). A bibliography on life testing and related topics. *Biometrika* **45**, 521–43. [124]

Menon, M. V. (1963). Estimation of the shape and scale parameters of the Weibull distribution. *Technometrics* **5**, 175–82. [120]

Miller, R. G., Jr. (1960). Early failures in life testing. *J. Amer. Statist. Ass.* **55**, 491–502. [124]

Miller, R. G., Jr. (1966). *Simultaneous Statistical Inference.* McGraw-Hill, New York. [3, 158]

Miller, R. G., Jr. (1968). Jacknifing variances. *Ann. Math. Statist.* **39**, 567–82. [156]

Mitra, S. K. (1957). Tables for tolerance limits for a normal population based on sample mean and range or mean range. *J. Amer. Statist. Ass.* **52**, 88–94. [165, 231]

Moore, D. S. (1968). An elementary proof of asymptotic normality of linear functions of order statistics. *Ann. Math. Statist.* **39**, 263–5. [217]

Moore, P. G. (1957). The two-sample *t*-test based on range. *Biometrika* **44**, 482–5. [157, 231]

Moran, P. A. P. (1964). On the range of cumulative sums. *Ann. Inst. Statist. Math., Tokyo* **16**, 109–12. [85]

Moranda, P. B. (1959). Comparison of estimates of circular probable error. *J. Amer. Statist. Ass.* **54**, 794–800. [148]

Moriguti, S. (1951). Extremal properties of extreme value distributions. *Ann. Math. Statist.* **22**, 523–36. [49, 50, 67–8]

Moriguti, S. (1953a). A modification of Schwarz's inequality with applications to distributions. *Ann. Math. Statist.* **24**, 107–13. [51, 53, 68]

Moriguti, S. (1953b). A note on Hartley's formula of studentization. *Rep. Stat. Appl. Res., JUSE* **2**, 99–103. [71]

Moriguti, S. (1954). Bounds for second moments of the sample range. *Rep. Stat. Appl. Res., JUSE* **3**, 57–64. [50]

Morrison, D. F. and David, H. A. (1960). The life distribution and reliability of a system with spare components. *Ann. Math. Statist.* **31**, 1084–94. [124]

Morrison, M. and Tobias, F. (1965). Some statistical characteristics of a peak to average ratio. *Technometrics* **7**, 379–85. [22]

Mosteller, F. (1946). On some useful "inefficient" statistics. *Ann. Math. Statist.* **17**, 377–408. [199, 201]

Mosteller, F. (1948). A *k*-sample slippage test for an extreme population. *Ann. Math. Statist.* **19**, 58–65. [178, 233]

Mosteller, F. and Tukey, J. W. (1950). Significance levels for a k-sample slippage test. *Ann. Math. Statist.* **21**, 120–3. [178, 196]

Murphy, R. B. (1948). Non-parametric tolerance limits. *Ann. Math. Statist.* **19**, 581–9. [17, 226]

Murphy, R. B. (1951). On tests for outlying observations. Ph.D. Thesis, Princeton University. [185, 190]

Murty, V. N. (1955). The distribution of the quotient of maximum values in samples from a rectangular distribution. *J. Amer. Statist. Ass.* **50**, 1136–41. [22, 100]

Nair, K. R. (1948). The distribution of the extreme deviate from the sample mean and its studentized form. *Biometrika* **35**, 118–44. [72, 182]

Nair, K. R. (1950). Efficiencies of certain linear systematic statistics for estimating dispersion from normal samples. *Biometrika* **37**, 182–3. [147]

Nair, U. S. (1936). The standard error of Gini's mean difference. *Biometrika* **28**, 428–36. [167]

Naus, J. I. (1966). Some probabilities, expectations and variances for the size of largest clusters and smallest intervals. *J. Amer. Statist. Ass.* **61**, 1191–9. [81]

Newell, G. F. (1964). Asymptotic extremes for m-dependent random variables. *Ann. Math. Statist.* **35**, 1322–5. [215]

Neyman, J. and Pearson, E. S. (1928). On the use and interpretation of certain test criteria for purposes of statistical inference, I. *Biometrika* **20A**, 175–240. [21]

Noether, G. E. (1955). Use of the range instead of the standard deviation. *J. Amer. Statist. Ass.* **50**, 1040–55. [158]

Odeh, R. E. (1967). The distribution of the maximum sum of ranks. *Technometrics* **9**, 271–8. [178, 196, 233]

Ogawa, J. (1951). Contributions to the theory of systematic statistics, I. *Osaka Math. J.* **3**, 175–213. [151]

Ogawa, J. (1962). In *SG*: (i) Estimation of the location and scale parameters by sample quantiles (for large samples). 47–55. (ii) Determinations of optimum spacings in the case of normal distribution. 272–83. (iii) Tests of significance using sample quantiles. 291–9. (iv) Optimum spacing and grouping for the exponential distribution. 371–80. (v) Tests of significance and confidence intervals. 380–2. [151, 154–5, 167, 231]

Owen, D. B. (1962). *Handbook of Statistical Tables.* Addison-Wesley, Reading, Mass. [226]

Owen, D. B. and Steck, G. P. (1962). Moments of order statistics from the equicor-related multivariate normal distribution. *Ann. Math. Statist.* **33**, 1286–91. [91]

Patnaik, P. B. (1950). The use of mean range as an estimator of variance in statistical tests. *Biometrika* **37**, 78–87. [142, 159]

Paulson, E. (1952). An optimum solution to the k-sample slippage problem for the normal distribution. *Ann. Math. Statist.* **23**, 610–6. [178]

Paulson, E. (1961). A non-parametric solution for the k-sample slippage problem. In: Solomon, H. (Ed.), *Studies in Item Analysis and Prediction*, 233–8. Stanford University Press. [178]

Pearson, E. S. (1929). The distribution of frequency constants in small samples from non-normal symmetrical and skew populations. *Biometrika* **21**, 280–6. [158]

Pearson, E. S. (1950). Some notes on the use of range. *Biometrika* **37**, 88–92. [144]

Pearson, E. S. (1952). Comparison of two approximations to the distribution of the range in small samples from normal populations. *Biometrika* **39**, 130–6. [141]

Pearson, E. S. (1966). Alternative tests for heterogeneity of variance; some Monte Carlo results. *Biometrika* **53**, 229–34. [156]

Pearson, E. S. and Adyanthāya, N. K. (1928). The distribution of frequency constants in small samples from symmetrical populations. *Biometrika* **20A**, 356–60. [144]

Pearson, E. S. and Chandra Sekar, C. (1936). The efficiency of statistical tools and a criterion for the rejection of outlying observations. *Biometrika* **28**, 308–20. [87, 182, 185]

Pearson, E. S. and Haines, Joan. (1935). The use of range in place of standard deviation in small samples. *Suppl. J. R. Statist. Soc.* **2**, 83–98. [138]

Pearson, E. S. and Hartley, H. O. (1942). The probability integral of the range in samples of *n* observations from a normal population. *Biometrika* **32**, 301–10. [225]

Pearson, E. S. and Hartley, H. O. (1943). Tables of the probability integral of the studentized range. *Biometrika* **33**, 89–99. [71]

Pearson, E. S. and Hartley, H. O. (1966). *Biometrika Tables for Statisticians.* Vol. I, 3rd Ed. Cambridge University Press. [3, 6, 8, 117, 145, 155, 172–3, 225, 232]

Pearson, K. (1902). Note on Francis Galton's difference problem. *Biometrika* **1**, 390–9. [39]

Pearson, K. (1920). On the probable errors of frequency constants, III. *Biometrika* **13**, 113–32. [154]

Pearson, K. (1934). *Tables of the Incomplete B-function.* Cambridge University Press. [8]

Pearson, K. and Pearson, M. V. (1931). On the mean character and variance of a ranked individual, and on the mean and variance of the intervals between ranked individuals. I (1931): Symmetrical distributions (normal and rectangular), *Biometrika* **23**, 364–97. II (1932): Case of certain skew curves, *Biometrika* **24**, 203–79. [66, 139]

Pettigrew, H. M. and Mohler, W. C. (1967). A rapid test for the Poisson distribution using the range. *Biometrics* **23**, 685–92. [160]

Pfanzagl, J. (1959). Ein kombiniertes Test & Klassifikations-Problem. *Metrika* **2**, 11–45. [182–3]

Pickands, J., III (1967a). Maxima of stationary Gaussian processes. *Z. Wahrscheinlichkeitstheorie verw. Geb.* **7**, 190–223. [215]

Pickands, J., III (1967b). Sample sequences of maxima. *Ann. Math. Statist.* **38**, 1570–4. [210]

Pickands, J., III (1968). Moment convergence of sample extremes. *Ann. Math. Statist.* **39**, 881–9. [210]

Pike, M. C. (1966). A method of analysis of a certain class of experiments in carcinogenesis. *Biometrics* **22**, 142–61. [200]

Pillai, K. C. S. (1950). On the distributions of midrange and semi-range in samples from a normal population. *Ann. Math. Statist.* **21**, 100–5. [142]

Pillai, K. C. S. and Ramachandran, K. V. (1954). On the distribution of the ratio of the *i*th observation in an ordered sample from a normal population to an independent estimate of the standard deviation. *Ann. Math. Statist.* **25**, 565–72. [229]

Pitman, E. J. G. (1936). Sufficient statistics and intrinsic accuracy. *Proc. Camb. Phil. Soc.* **32**, 567–79. [95]

Plackett, R. L. (1947). Limits of the ratio of mean range to standard deviation. *Biometrika* **34**, 120–2. [50]

Plackett, R. L. (1958). Linear estimation from censored data. *Ann. Math. Statist.* **29**, 131–42. [66, 112]

Proschan, F. and Pyke, R. (1967). Tests for monotone failure rate. *Proc. 5th Berkeley Symp.* **III**, 293–312. [124]

Pyke, R. (1965). Spacings. *J. R. Statist. Soc. B* **27**, 395–436. Discussion: 437–49. [81, 90, 213]

Quenouille, M. H. (1956). Notes on bias in estimation. *Biometrika* **43**, 353–60. [101]

Quesenberry, C. P. and David, H. A. (1961). Some tests for outliers. *Biometrika* **48**, 379–90. [175, 182, 197]

Rahman, N. A. (1964). Some generalisations of the distributions of product statistics arising from rectangular populations. *J. Amer. Statist. Ass.* **59**, 557–63. [22]

Ramachandran, K. V. and Khatri, C. G. (1957). On a decision procedure based on the Tukey statistic. *Ann. Math. Statist.* **28**, 802–6. [197]

Rao, M. M. (1962). Theory of order statistics. *Math. Annalen* **147**, 298–312. [124]

Rényi, A. (1953). On the theory of order statistics. *Acta Math. Acad. Sci. Hung.* **4**, 191–231. [18]

Richter, W. (1964). Ein zentraler Grenzwertsatz für das Maximum einer zufälligen Anzahl unabhängiger Zufallsgrößen. *Wiss. Zeit. Tech. Univ. Dresden* **13**, 1343–6. [214]

Rider, P. R. (1951). The distribution of the quotient of ranges in samples from a rectangular population. *J. Amer. Statist. Ass.* **46**, 502–7. [22, 100, 229]

Rider, P. R. (1955). The distribution of the product of maximum values in samples from a rectangular distribution. *J. Amer. Statist. Ass.* **51**, 1142–3. [22]

Rider, P. R. (1957). The midrange of a sample as an estimator of the population midrange. *J. Amer. Statist. Ass.* **52**, 537–42. [140]

Rider, P. R. (1960). Variance of the median of samples from a Cauchy distribution. *J. Amer. Statist. Ass.* **55**, 322–3. [40]

Robbins, H. (1944). On distribution-free tolerance limits in random sampling. *Ann. Math. Statist.* **15**, 214–6. [16]

Robson, D. S. and Whitlock J. H. (1964). Estimation of a truncation point. *Biometrika* **51**, 33–9. [101]

Romanovsky, V. (1933). On a property of the mean ranges in samples from a normal population and on some integrals of Prof. T. Hojo. *Biometrika* **25**, 195–7. [44]

Rosengard, A. (1964a). Indépendance limite uniforme de la moyenne et des valeurs extrêmes d'un échantillon. *C. R. Acad. Sci. Paris* **258**, 5786–8. [213]

Rosengard, A. (1964b). Indépendance limite uniforme d'un quantile et des valeurs extrêmes d'un échantillon. *C. R. Acad. Sci. Paris* **259**, 2955–6. [213]

Rossberg, H. J. (1963). Über das asymptotische Verhalten der Rand- und Zentral-glieder einer Variationsreihe. *Publ. Math. Inst. Hung. Acad. Sci.* **8**, 463–8. [213]

Rossberg, H. J. (1965a). Über die stochastische Unabhängigkeit gewisser Funktionen von Ranggrößen. *Math. Nachr.* **28**, 157–67. [19]

Rossberg, H. J. (1965b). Die asymptotische Unabhängigkeit der kleinsten und größten Werte einer Stichprobe vom Stichprobenmittel. *Math. Nachr.* **28**, 305–18. [213]

Rotherberg, T. J., Fisher, F. M., and Tilanus, C. B. (1964). A note on estimation from a Cauchy sample. *J. Amer. Statist. Ass.* **59**, 460–3. [136]

Ruben, H. (1954). On the moments of order statistics in samples from normal populations. *Biometrika* **41**, 200–27. [34, 86, 228]

Ruben, H. (1956a). On the sum of squares of normal scores. *Biometrika* **43**, 456–8. Correction **52**, 669. [42]

Ruben, H. (1956b). On the moments of the range and product moments of extreme order statistics in normal samples. *Biometrika* **43**, 458–60. [34]

Rushton, S. (1952). On sequential tests of the equality of variances of two normal populations with known means. *Sankhyā* **12**, 63–78. [156]

Rustagi, J. S. (1957). On minimizing and maximizing a certain integral with statistical applications. *Ann. Math. Statist.* **28**, 309–28. [54]

Rutemiller, H. C. (1966). Point estimation of reliability of a system comprised of k elements from the same exponential distribution. *J. Amer. Statist. Ass.* **61**, 1029–32. [124]

Saleh, A. K. M. E. (1967). Determination of the exact optimum order statistics for estimating the parameters of the exponential distribution from censored samples. *Technometrics* **9**, 279–92. [154]

Saleh, A. K. M. E. and Ali, M. M. (1966). Asymptotic optimum quantiles for the estimation of the parameters of the negative exponential distribution. *Ann. Math. Statist.* **37**, 143–51. [154]

Sarhan, A. E. (1954). Estimation of the mean and standard deviation by order statistics. *Ann. Math. Statist.* **25**, 317–28. [227]

Sarhan, A. E. (1955). Estimation of the mean and standard deviation by order statistics, Part III. *Ann. Math. Statist.* **26**, 576–92. [110, 123, 132, 135]

Sarhan, A. E. and Greenberg, B. G. (1956). Estimation of location and scale parameters by order statistics from singly and doubly censored samples. Part I: The normal distribution up to samples of size 10. *Ann. Math. Statist.* **27**, 427–51. Correction **40**, 325. [28, 106, 228]

Sarhan, A. E. and Greenberg, B. G. (1957). Tables for best linear estimates by order statistics of the parameters of single exponential distributions from singly and doubly censored samples. *J. Amer. Statist. Ass.* **52**, 58–87. [124]

Sarhan, A. E. and Greenberg, B. G. (1959). Estimation of location and scale parameters for the rectangular population from censored samples. *J. R. Statist. Soc. B* **21**, 356–63. [132]

Sarhan, A. E. and Greenberg, B. G. (Eds.). (1962). *Contributions to Order Statistics*. Wiley, New York. [3, 6, 93, 109–10, 115, 118–9, 123, 159, 176, 200, 222, 225, 230, 232]

Sarkadi K., Schnell, E., and Vincze, I. (1962). On the position of the sample mean among the ordered sample elements. *Publ. Math. Inst. Hung. Acad. Sci.* **7A**, 239–54. [84]

Särndal, C. E. (1962). *Information from Censored Samples*. Almqvist and Wiksell, Stockholm. [154–5]

Särndal, C. E. (1964). Estimation of the parameters of the gamma distribution by sample quantiles. *Technometrics* 6, 405–14. [154]

Sathe, Y. S. and Varde, S. D. (1969). Minimum variance unbiased estimation of reliability for the truncated exponential distribution. *Technometrics* 11, 609–12. [123]

Saunders, S. C. (1963). On the sample size and coverage for the Jirina sequential procedure. *Ann. Math. Statist.* 34, 847–56. [17]

Saunders, S. C. (1968). On the determination of a safe life for distributions classified by failure rate. *Technometrics* 10, 361–77. [124]

Saw, J. G. (1959). Estimation of the normal population parameters given a singly censored sample. *Biometrika* 46, 150–9. [133]

Saw, J. G. (1960). A note on the error after a number of terms of the David-Johnson series for the expected values of normal order statistics. *Biometrika* 47, 79–86. [66]

Saw, J. G. (1961). The bias of the maximum likelihood estimates of the location and scale parameters given a type II censored normal sample. *Biometrika* 48, 448–51. [114]

Saw, J. G. and Chow, B. (1966). The curve through the expected values of ordered variates and the sum of squares of normal scores. *Biometrika* 53, 252–5. [42]

Scheffé, H. and Tukey, J. W. (1945). Non-parametric estimation. I. Validation of order statistics. *Ann. Math. Statist.* 16, 187–92. [14, 17]

Sen, P. K. (1959). On the moments of the sample quantiles. *Calcutta Statist. Ass. Bull.* 9, 1–19. [26, 210]

Sen, P. K. (1961). A note on the large-sample behavior of extreme sample values from distribution with finite end-points. *Calcutta Statist. Ass. Bull.* 10, 106–15. [210]

Sen, P. K. (1964). On stochastic convergence of the sample extreme values from distributions with infinite extremities. *J. Ind. Soc. Agric. Statist.* 16, 189–201. [210]

Sen, P. K. (1968). Asymptotic normality of sample quantiles for m-dependent processes. *Ann. Math. Statist.* 39, 1724–30. [203]

Seth, G. R. (1950). On the distribution of the two closest among a set of three observations. *Ann. Math. Statist.* 21, 298–301. [23, 195]

Shah, B. K. (1965). Distribution of midrange and semirange from logistic population. *J. Ind. Statist. Ass.* 3, 185–8. [225]

Shah, B. K. (1966a). A note on Craig's paper on the minimum of binomial variates. *Biometrika* 53, 614–5. [36]

Shah, B. K. (1966b). On the bivariate moments of order statistics from a logistic distribution. *Ann. Math. Statist.* 37, 1002–10. [227]

Shapiro, S. S. and Wilk, M. B. (1965). An analysis of variance test for normality (complete samples). *Biometrika* 52, 591–611. [169, 231]

Shapiro, S. S. and Wilk, M. B. (1968). Approximations for the null distribution of the W statistic. *Technometrics* 10, 861–6. [168]

Shapiro, S. S., Wilk, M. B., and Chen, H. J. (1968). A comparative study of various tests for normality. *J. Amer. Statist. Ass.* 63, 1343–72. [169]

Shimada, S. (1957). Bias included in the estimator of standard deviation using range. *Rep. Statist. Appl. Res., JUSE* **5**, 21–6. [165]

Shorack, R. A. (1967). On the power of precedence life tests. *Technometrics* **9**, 154–8. [124]

Šidák, Z. (1968). On multivariate normal probabilities of rectangles. *Ann. Math. Statist.* **39**, 1425–34. [79]

Siddiqui, M. M. (1960). Distribution of quantiles in samples from a bivariate population. *J. Res. Nat. Bur. Stand.* **64B**, 145–50. [204]

Siddiqui, M. M. (1962). Approximations to the moments of the sample median. *Ann. Math. Statist.* **33**, 157–68. [139]

Siddiqui, M. M. and Raghunandanan, K. (1967). Asymptotically robust estimators of location. *J. Amer. Statist. Ass.* **62**, 950–3. [126]

Sillitto, G. P. (1951). Interrelations between certain linear systematic statistics of samples from any continuous population. *Biometrika* **38**, 377–82. [44]

Sillitto, G. P. (1964). Some relations between expectations of order statistics in samples of different sizes. *Biometrika* **51**, 259–62. [43]

Singh, C. (1967). On the extreme values and range of samples from non-normal populations. *Biometrika* **54**, 541–50. [145]

Singh, N. (1960). Estimation of parameters of a multivariate normal population from truncated and censored samples. *J. R. Statist. Soc. B* **22**, 307–11. [119]

Siotani, M. (1957). Order statistics for discrete case with a numerical application to the binomial distribution. *Ann. Inst. Statist. Math., Tokyo* **8**, 95–104. [23, 43, 160]

Siotani, M. (1959). The extreme value of the generalized distances of the individual points in the multivariate normal sample. *Ann. Inst. Statist. Math., Tokyo* **10**, 183–208. [77, 184, 228]

Siotani, M. and Ozawa, M. (1958). Tables for testing the homogeneity of *k* independent binomial experiments on a certain event based on the range. *Ann. Inst. Statist. Math., Tokyo* **10**, 47–63. [160, 226]

Slepian, D. (1962). The one-sided barrier problem for Gaussian noise. *Bell System Tech. J.* **41**, 463–501. [79]

Smirnov, N. V. (1952). Limit distributions for the terms of a variational series. *Amer. Math. Soc. Transl. Ser.* **1**, No. 67. Original published in 1949. [210]

Smirnov, N. V. (1966). Convergence of distributions of order statistics to the normal distribution. *Izv. Akad. Nauk Uz SSR Ser. Fiz.-Mat. Nauk* **10** (3), 24–32. [203]

Smirnov, N. V. (1967). Some remarks on limit laws for order statistics. *Theory Prob. and Its Applications* **12**, 337–9. [203]

Smith, W. B. and Hartley, H. O. (1968). A note on the correlation of ranges in correlated normal samples. *Biometrika* **55**, 595–7. [159]

Solari, M. E. and Anis, A. A. (1957). The mean and variance of the maximum of the adjusted partial sums of a finite number of independent normal variates. *Ann. Math. Statist.* **28**, 706–16. [85]

Somerville, P. N. (1958). Tables for obtaining non-parametric tolerance limits, *Ann. Math. Statist.* **29**, 599–601. [17, 226]

Srikantan, K. S. (1961). Testing for the single outlier in a regression model, *Sankhyā A* **23**, 251–60. [171]

Srikantan, K. S. (1962). Recurrence relations between the PDF's of order statistics, and some applications. *Ann. Math. Statist.* **33**, 169–77. [38]

Srivastava, O. P. (1967). Asymptotic independence of certain statistics connected with the extreme order statistics in a bivariate distribution. *Sankhyā* **29**, 175–82. [214]

Srivastava, O. P., Harkness, W. L., and Bartoo, J. B. (1964). Asymptotic distribution of distances between order statistics from bivariate populations. *Ann. Math. Statist.* **35**, 748–54. [214]

Staude, H. (1959). Abkürzung des Range-Verfahrens von H. O. Hartley zur Auswertung von Blockversuchen. *Biom. Zeit.* **1**, 261–75. [159]

Stevens, W. L. (1939). Solution to a geometrical problem in probability. *Ann. Eugen.* **9**, 315–20. [81, 89]

Stigler, S. M. (1969). Linear functions of order statistics. *Ann. Math. Statist.* **40**, 770–88. [217]

Stuart, A. (1958). Equally correlated variates and the multinormal integral. *J. R. Statist. Soc. B* **20**, 373–8. [91]

Sugiura, N. (1962). On the orthogonal inverse expansion with an application to the moments of order statistics. *Osaka Math. J.* **14**, 253–63. [54, 58, 68]

Sugiura, N. (1964). The bivariate orthogonal inverse expansion and the moments of order statistics. *Osaka J. Math.* **1**, 45–59. [54]

Sukhatme, P. V. (1937). Tests of significance for samples of the χ^2 population with two degrees of freedom. *Ann. Eugen.* **8**, 52–6. [17, 121, 133]

Swamy, P. S. (1962). On the amount of information supplied by censored samples of grouped observations in the estimation of statistical parameters. *Biometrika* **49**, 245–9. [119]

Takács, L. (1967). On the method of inclusion and exclusion. *J. Amer. Statist. Ass.* **62**, 102–13. [77]

Tanis, E. A. (1964). Linear forms in the order statistics from an exponential distribution. *Ann. Math. Statist.* **35**, 270–6. [124]

Tarter, M. E. and Clark, V. A. (1965). Properties of the median and other order statistics of logistic variates. *Ann. Math. Statist.* **36**, 1779–86. [227]

Teichroew, D. (1955). Probabilities associated with order statistics in samples from two normal populations with equal variance. Army Chemical Center, Maryland, Chemical Corps Engineering Agency. [186]

Teichroew, D. (1956). Tables of expected values of order statistics and products of order statistics for samples of size twenty and less from the normal distribution. *Ann. Math. Statist.* **27**, 410–26. [28, 30, 228]

Thigpen, C. C. (1961). Distribution of the largest observation in normal samples under non-standard conditions. Ph.D. Thesis, Virginia Polytechnic Institute. [229]

Thompson, W. A., Jr. and Willke, T. A. (1963). On an extreme rank sum test for outliers. *Biometrika* **50**, 375–83. [178, 233]

Thompson, W. R. (1936). On confidence ranges for the median and other expectation distributions for populations of unknown distribution form. *Ann. Math. Statist.* **7**, 122–8. [14]

Thomson, G. W. (1955). Bounds for the ratio of range to standard deviation. *Biometrika* **42**, 268–9. [145]

Tiago de Oliveira, J. (1963). Structure theory of bivariate extremes; extensions. *Estudos de Mathematica, Estatistica E Econometria* **7**, 165–95. [214]

Tiku, M. L. (1967a). Estimating the mean and standard deviation from a censored normal sample. *Biometrika* **54**, 155–65. [112]

Tiku, M. L. (1967b). A note on estimating the location and scale parameters of the exponential distribution from a censored sample. *Aust. J. Statist.* **9**, 49–54. [123]

Tiku, M. L. (1968a). Estimating the parameters of log-normal distribution from censored samples. *J. Amer. Statist. Ass.* **63**, 134–40. [119]

Tiku, M. L. (1968b). Estimating the parameters of normal and logistic distributions from censored samples. *Aust. J. Statist.* **10**, 64–74. [120]

Tiku, M. L. (1968c). Estimating the mean and standard deviation from progressively censored normal samples. *J. Ind. Soc. Agric. Statist.* **20**, 20–5. [124]

Tippett, L. H. C. (1925). On the extreme individuals and the range of samples taken from a normal population. *Biometrika* **17**, 364–87. [29, 43, 50, 225, 228]

Truax, D. R. (1953). An optimum slippage test for the variances of k normal distributions. *Ann. Math. Statist.* **24**, 669–74. [183]

Tsukibayashi, S. (1958). Estimation of variance and standard deviation based on range. *Rep. Statist. Appl. Res. JUSE* **5**, 59–67. [145]

Tsukibayashi, S. (1962). Estimation of bivariate parameters based on range. *Rep. Statist. Appl. Res. JUSE* **9**, 10–23. [149, 167]

Tukey, J. W. (1947). Non-parametric estimation, II: Statistically equivalent blocks and tolerance regions—the continuous case. *Ann. Math. Statist.* **18**, 529–39. [18]

Tukey, J. W. (1949). The simplest signed-rank tests. *Memo Rep.* 17, Statist. Res. Group, Princeton University (duplicated). [15]

Tukey, J. W. (1955). Interpolations and approximations related to the normal range. *Biometrika* **42**, 480–5. [144]

Tukey, J. W. (1958). A problem of Berkson, and minimum variance orderly estimators. *Ann. Math. Statist.* **29**, 588–92. [41]

Tukey, J. W. (1960). A survey of sampling from contaminated distributions. In: *Contributions to Probability and Statistics*, Olkin *et al.* (Eds.), 448–85. Stanford University Press. [124]

Tukey, J. W. (1962). The future of data analysis. *Ann. Math. Statist.* **33**, 1–67. [2]

Tukey, J. W. and McLaughlin, D. H. (1963). Less vulnerable confidence and significance procedures for location based on a single sample: trimming/Winsorization, 1. *Sankhyā* A **25**, 331–52. [130]

Uzgören, Nakibe T. (1954). The asymptotic development of the distribution of the extreme values of a sample. In: *Studies in Mathematics and Mechanics Presented to Richard von Mises*. Academic Press, New York. [209]

van der Vaart, H. R. (1961a). A simple derivation of the limiting distribution function of a sample quantile with increasing sample size. *Statist. Neerlandica* **15**, 239–42. [203]

van der Vaart, H. R. (1961b). Some extensions of the idea of bias. *Ann. Math. Statist.* **32**, 436–47. [20]

van Zwet, W. R. (1964). *Convex Transformations of Random Variables*. Mathematical Center Tracts **7**, Mathematisch Centrum, Amsterdam. [58, 63, 66, 68–9] van Zwet, W. R. (1966). Bias in estimation from type I censored samples. *Statist. Neerlandica* **20**, 143–8. [110]

van Zwet, W. R. (1967). An inequality for expected values of sample quantiles. *Ann. Math. Statist.* **38**, 1817–21. [65]

Venter, J. H. (1967). On estimation of the mode. *Ann. Math. Statist.* **38**, 1446–55. [204]

von Andrae (1872). Ueber die Bestimmung des wahrscheinlichen Fehlers durch die gegebenen Differenzen von *m* gleich genauen Beobachtungen einer Unbekannten. *Astron. Nach.* **79**, 257–72. [147]

von Mises, R. (1923). Über die Variationsbreite einer Beobachtungsreihe. *Sitzungsberichte der Berliner Math. Gesellschaft* **22**, 3–8. [Reproduced in von Mises (1964), pp. 129–34.] [204]

von Mises, R. (1936). La distribution de la plus grande de *n* valeurs. *Rev. Math. Union Interbalkanique* **1**, 141–60. [Reproduced in von Mises (1964), pp. 271–94.] [204, 207]

von Mises, R. (1964). *Selected Papers of Richard von Mises*, Vol. 2. American Mathematical Society, Providence. [266]

Walker, A. M. (1968). A note on the asymptotic distribution of sample quantiles. *J. R. Statist. Soc.* **30**, 570–5. [203]

Walsh, J. E. (1949a). Some significance tests for the median which are valid under very general conditions. *Ann. Math. Statist.* **20**, 64–81. [15]

Walsh, J. E. (1949b). Applications of some significance tests for the median which are valid under very general conditions. *J. Amer. Statist. Ass.* **44**, 342–55. [15]

Walsh, J. E. (1949c). On the range-midrange test and some tests with bounded significance levels. *Ann. Math. Statist.* **20**, 257–67. [158]

Walsh, J. E. (1956). Asymptotic efficiencies of a nonparametric life test for smaller percentiles of a gamma distribution. *J. Amer. Statist. Ass.* **51**, 467–80. [124]

Walsh, J. E. (1958). Nonparametric estimation of sample percentage point standard deviation. *Ann. Math. Statist.* **29**, 601–4. [147]

Walsh, J. E. (1962). Distribution-free tolerance intervals for continuous symmetrical populations. *Ann. Math. Statist.* **33**, 1167–74. [24]

Watanabe, Y. *et al.* (1957). Some contributions to order statistics. *J. Gakugei, Tokushima Univ.* **8**, 41–90. [34]

Watanabe, Y. *et al.* (1958). Some contributions to order statistics (continued). *J. Gakugei, Tokushima Univ.* **9**, 31–86. [34]

Watson, G. S. (1954). Extreme values in samples from *m*-dependent stationary stochastic processes. *Ann. Math. Statist.* **25**, 798–800. [214]

Watterson, G. A. (1959). Linear estimation in censored samples from multivariate normal populations. *Ann. Math. Statist.* **30**, 814–24. [41, 119]

Weiler, H. (1954). A new type of control chart limits for means, ranges, and sequential runs. *J. Amer. Statist. Ass.* **49**, 298–314. [165]

Weiss, L. (1964). On the asymptotic joint normality of quantiles from a multivariate distribution. *J. Res., Nat. Bur. Stand.* **68B**, 65–6. [204]

Weiss, L. (1965). On the asymptotic distribution of the largest sample spacing. *J. Soc. Indust. Appl. Math.* **13**, 720–31. [213]

White, J. S. (1964). Least squares unbiased censored linear estimation for the log Weibull (extreme value) distribution. *Indust. Math.* **14,** 21–60. [120, 230]

White, J. S. (1969). The moments of log-Weibull order statistics. *Technometrics* **11,** 373–86. [226]

Wilk, M. B. and Gnanadesikan, R. (1968). Probability plotting methods for the analysis of data. *Biometrika* **55,** 1–17. [163]

Wilk, M. B., Gnanadesikan, R., and Freeny, Anne E. (1963a). Estimation of error variance from smallest ordered contrasts. *J. Amer. Statist. Ass.* **58,** 152–60. [163]

Wilk, M. B., Gnanadesikan, R., and Huyett, Marilyn J. (1963b). Separate maximum likelihood estimation of scale or shape parameters of the gamma distribution using order statistics. *Biometrika* **50,** 217–21. [119]

Wilk, M. B., Gnanadesikan, R., and Huyett, Marilyn J. (1962a). Probability plots for the gamma distribution. *Technometrics* **4,** 1–20. [162]

Wilk, M. B., Gnanadesikan, R., and Huyett, Marilyn J. (1962b). Estimation of parameters of the gamma distribution using order statistics. *Biometrika* **49,** 525–45. [119]

Wilk, M. B., Gnanadesikan, R., and Lauh, Elizabeth. (1966). Scale parameter estimation from the order statistics of unequal gamma components. *Ann. Math. Statist.* **37,** 152–76. [119]

Wilk, M. B. and Shapiro, S. C. (1968). The joint assessment of normality of several independent samples. *Technometrics* **10,** 825–39. [169]

Wilks, S. S. (1941). Determination of sample sizes for setting tolerance limits. *Ann. Math. Statist.* **12,** 91–6. [24]

Wilks, S. S. (1942). Statistical prediction with special reference to the problem of tolerance limits. *Ann. Math. Statist.* **13,** 400–9. [16]

Wilks, S. S. (1948). Order statistics. *Bull. Amer. Math. Soc.* **5,** 6–50. [2, 80]

Wilks, S. S. (1962). *Mathematical Statistics.* Wiley, New York. [17, 20, 23, 80]

Wilks, S. S. (1963). Multivariate statistical outliers. *Sankhyā A* **25,** 407–26. [177, 232]

Willke, T. A. (1966). A note on contaminated samples of size three. *J. Res., Nat. Bur. Standards* **70B,** 149–51. [195]

Winer, P. (1963). The estimation of the parameters of the iterated exponential distribution from singly censored samples. *Biometrics* **19,** 460–4. [120]

Youden, W. J. (1963). Ranking laboratories by round-robin tests. *Materials Res. and Stand.* **3,** 9–13. [178]

Young, D. H. (1967). Recurrence relations between the P.D.F's of order statistics of dependent variables, and some applications. *Biometrika* **54,** 283–92. [84]

Zacks, S. and Even, M. (1966). The efficiencies in small samples of the maximum likelihood and best unbiased estimators of reliability functions. *J. Amer. Statist. Ass.* **61,** 1033–51. [124]

Zelen, M. (1959). Factorial experiments in life testing. *Technometrics* **1,** 269–88. [124]

Zelen, M. and Dannemiller, Mary C. (1961). The robustness of life testing procedures derived from the exponential distribution. *Technometrics* **3,** 29–49. [121]

Subject Index

For authors *see* under References
For abbreviations *see* p. 6
OS = order statistics

Analysis of variance, by range methods, 158–160, 231
Antisymmetrical function, 58
Applications, general, 1–2
  of extremes, 199–200
  of range, 137
Approximations, for moments of OS, 45ff
  to upper percentage points of statistics expressible as maxima, 73 ff
Asymptotic distribution, for dependent variates, 214–216
  in bivariate case, 214
  of extremes, 204–211, 223
  of linear functions of OS, 216–217
  of quantiles, 201–204, 223
  of range, 211–213, 224
  of sample spacings, 213
  of studentized extreme deviate, 213
Asymptotic estimation, 218–222, 224
Asymptotic methods, 3, 199 ff

Bahadur representation of a quantile, 203, 223
Basu's theorem, 96
Bayesian methods, 172, 183
Best linear estimates, *see* Linear estimators
Beta distribution, 5, 69
  bounds for $\mathcal{E}X_{r:n}$ in, 69
  censoring in, 120
Binomial distribution, range in, 160, 226
  smaller of two from, 36
Bivariate extremal distributions, 214
Bivariate normal distribution, circular, 148
  estimation by range in, 167
  linear function of OS in, 91
  moments of OS in, 41
Blom's estimates, 106–108
BLUE, 230
  tables of coefficients of, 230
  see also Linear estimators

Bonferroni inequalities, 74, 79, 88
Boole formula, 73
Bounds, for moments of OS, 45 ff
  for upper percentage points of statistics expressible as maxima, 73 ff

c–comparison, 60–65
  ordering, 60
  precedence, 60
Cauchy distribution, 40
  moments of OS in, 227
  optimal spacing of OS in, 154
  trimmed mean in, 136
Censoring, 109 ff
  BLUE in presence of, 230
  double, 109
  estimation in presence of, 109–124, 133–135, 218–222
  in exponential distribution, 120–124, 132–134
  in non-normal distribution, 119–120
  of multivariate normal, 119
  single, 109
  Type I, 109, 112
  Type II, 109, 112, 114
Characterizations, 19
Chi (1 DF) distribution, 227
  censoring in, 120
  moments of OS in, 227
Circular normal distribution, 148
Closest two out of three observations, 23
Combinatorial extreme-value distributions, 76, 82
Completeness, 96
Concave-convex function, 59
Concave function, 58
Conditional distribution, of OS, 18
Confidence intervals, distribution-free, 13–16, 23
  for σ in normal samples, 155, 231

267